# Introduction to the Theory of Formal Groups

PURE AND APPLIED MATHEMATICS

A Series of Monographs and Textbooks



1. K. Yano. Integral Formulas in Riemannian Geometry (1970)
2. S. Kobayashi. Hyperbolic Manifolds and Holomorphic Mappings (1970)
3. V. S. Vladimirov. Equations of Mathematical Physics (A. Jeffrey, editor; A. Littlewood, translator) (1970)
4. B. N. Pshenichnyi. Necessary Conditions for an Extremum (L. Neustadt, translation editor; K. Makowski, translator) (1971)
5. L. Narici, E. Beckenstein, and G. Bachman. Functional Analysis and Valuation Theory (1971)
6. D. S. Passman. Infinite Group Rings (1971)
7. L. Dornhoff. Group Representation Theory (in two parts). Part A: Ordinary Representation Theory. Part B: Modular Representation Theory (1971, 1972)
8. W. Boothby and G. L. Weiss (eds.). Symmetric Spaces: Short Courses Presented at Washington University (1972)
9. Y. Matsushima. Differentiable Manifolds (E. T. Kobayashi, translator) (1972)
10. L. E. Ward, Jr. Topology: An Outline for a First Course (1972)
11. A. Babakhanian. Cohomological Methods in Group Theory (1972)
12. R. Gilmer. Multiplicative Ideal Theory (1972)
13. J. Yeh. Stochastic Processes and the Wiener Integral (1973)
14. J. Barros-Neto. Introduction to the Theory of Distributions (1973)
15. R. Larsen. Functional Analysis: An Introduction (1973)
16. K. Yano and S. Ishihara. Tangent and Cotangent Bundles: Differential Geometry (1973)
17. C. Procesi. Rings with Polynomial Identities (1973)
18. R. Hermann. Geometry, Physics, and Systems (1973)
19. N. R. Wallach. Harmonic Analysis on Homogeneous Spaces (1973)
20. J. Dieudonné. Introduction to the Theory of Formal Groups (1973)
21. I. Vaisman. Cohomology and Differential Forms (1973)
22. B.-Y. Chen. Geometry of Submanifolds (1973)
23. M. Marcus. Finite Dimensional Multilinear Algebra (in two parts) (1973)
24. R. Larsen. Banach Algebras: An Introduction (1973)

*In Preparation:*

K. B. Stolarsky. Algebraic Numbers and Diophantine Approximation

# Introduction to the Theory of Formal Groups

*J. Dieudonné*
*Nice, France*

MARCEL DEKKER, INC. New York 1973

MARCEL DEKKER, INC.
*95 Madison Avenue, New York, New York 10016*

LIBRARY OF CONGRESS CATALOG CARD NUMBER: 72-90372
ISBN: 0-8247-6011-5

*Printed in the United States of America*

# Contents

# Foreword

The concept of formal Lie group was derived in a natural way from classical Lie theory by S. Bochner in 1946 [3], for fields of characteristic 0. Its study over fields of characteristic $p > 0$ began in the early 1950's, when it was realized, through the work of Chevalley [14], that the familiar "dictionary" between Lie groups and Lie algebras completely broke down for Lie algebras of algebraic groups over such a field. In the search for a new "infinitesimal object" which would take the place of the failing Lie algebra, a helpful guide was found in a conception of the "enveloping algebra" of a Lie algebra (in the classical case), due to L. Schwartz, and slightly different from the usual one: instead of considering that enveloping algebra as consisting of the left invariant differential operators on the group, one may also consider it as consisting of *distributions* with support at the neutral element $e$, the operation defining the algebra being the *convolution* of distributions. The space of distributions with support at $e$ being the dual of the space of germs of $C^\infty$ functions at $e$, it was natural to replace the latter, in the case of formal groups, by the space of formal power series over a field of arbitrary characteristic. It was then immediately recognized that the "infinitesimal (associative) algebra" $\mathfrak{G}$ thus defined had a far more complex structure, for fields of characteristic $p > 0$, than in the classical case; its most remarkable feature was that the Lie algebra, instead of generating the whole algebra $\mathfrak{G}$, only generates a tiny subalgebra of $\mathfrak{G}$, and of course this explained the pathology discovered by Chevalley. A natural conjecture at that stage was that the "infinitesimal algebra" $\mathfrak{G}$ would completely determine the formal group up to isomorphism (just as the Lie algebra determines the group germ in the classical case). However, it was found that this was not so: complex as it was, the algebra $\mathfrak{G}$ needed an additional structure in order to reconstruct the group law, namely a "comultiplication" which, by duality, would give the product law of the ring of formal power series (*). Thus, gradually and somewhat experimentally, the concept of *bigebra* of a formal group emerged, and its usefulness soon became apparent, in particular in the study of commutative formal groups. But the true nature of its relationship to the idea of group was only understood a few years later, with the development of the theory of categories, and in particular of the idea of group (or more generally of algebraic structure) in a category (Eckmann-Hilton, Ehresmann, Grothendieck), which showed that the concept of (cocommutative) bigebra was equivalent to the notion of "group in the dual category of the category of commutative algebras."

In this volume, we therefore start with the concept of **C**-group for any category **C** (with products and final object), but we do not exploit it in its full generality (for a general point of view, see Gabriel [27]). The book is meant to be introductory to the theory, and therefore we have tried to keep the necessary background to its minimum possible level: no algebraic geometry and very little commutative algebra is required in chapters I to III, and the algebraic geometry used in chapter IV is limited to the Serre-Chevalley type (varieties over an algebraically closed field).

It was early realized by Cartier and Gabriel that the concept of **C**-group, even when restricted to the category **C** of cogebras over a field $k$, led to types of formal groups more general than the "naive" ones of Bochner's definition. Chapter I is devoted to the study of the properties of these general formal groups. Beginning in chapter II, the field $k$ is supposed to be *perfect*; a fundamental theorem due to Cartier and Gabriel then allows one to split any formal group into a semi-direct product of an "etale" and an "infinitesimal" group. The "etale" part essentially corresponds to a "set-theoretic" group, and apparently no new results may be expected from its study. The remainder of the book is therefore limited to the theory of infinitesimal formal groups over a perfect field (the latter being even supposed to be algebraically closed in the last part of chapter III and in chapter IV). Even so, this notion is still more general than Bochner's, in two directions: first it includes "non-reduced" groups, which are unavoidable in characteristic $p > 0$, and correspond to the "inseparability" phenomena. Second, it includes groups whose Lie algebra is infinite dimensional. This might seem to be pointless generality, were it not that, even in the commutative case, one cannot help using such groups if one

(*) The first example of "comultiplication" had been met earlier by H. Hopf in his pioneering work on the cohomology of Lie groups; hence the name "Hopf algebras" used by several mathematicians to designate bigebras.

wants to obtain "free" objects in the category of infinitesimal groups. This is again a surprising phenomenon linked to the characteristic (since in characteristic 0 the "free" indecomposable commutative groups are of course one-dimensional). But it is my opinion that it yields the only *natural* introduction of the "Witt vectors" (which usually seem to come out of nowhere); it also explains their fundamental (and unexpected) part in the study of the structure of commutative infinitesimal groups, which is described in detail in chapter III, and reveals features which have no counterpart in characteristic 0 (which in a sense appears as a "degenerate" case). By way of contrast, the theory of "linear" reduced infinitesimal groups, developed in chapter IV (over an algebraically closed field), turns out to be merely a small extension of the Borel-Chevalley theory of *algebraic* affine groups, to which they are even closer in characteristic $p > 0$ than in characteristic 0.

It should be emphasized that, although the motivation for the introduction of formal groups in characteristic $p > 0$ originally comes from the theory of algebraic groups, no applications to that theory are given in this book. This is due to the fact that the most interesting of these applications at present concern the theory of abelian varieties, and therefore are on a much higher level than this volume and than its author's knowledge (**). Furthermore, most of these applications use the theory of formal groups, not only over a field, but over a *local ring* (the theory over a field coming into the picture only by "reduction" modulo the maximal ideal). The readers who want to get acquainted with that more difficult theory and its consequences should first consult the book of Fröhlich [26] which is limited to the one-dimensional case, well developed through the work of Lubin, Tate and Serre; the more general theory and its applications will hopefully be treated in the long-awaited book by Cartier developing his recent notes ([10], [11], [12], [13]). I have tried to include in the Bibliography all the relevant material (***).

Nice, May 1972

(**) Other recent applications are to the cohomology of schemes ([42], [34]) and to homotopy theory [33].
(***) Numbers in square brackets refer to the Bibliography.

# Notations

In the following definitions, Roman numerals indicate the chapter and arabic numerals the section and subsection within the chapter.

| | |
|---|---|
| $\boldsymbol{C}$, $\boldsymbol{C}^0$ | category and dual category: I, 1, 1 and 2. |
| $\mathrm{Mor}_{\boldsymbol{C}}(X, G)$, $\mathrm{Mor}(X, G)$ | set of morphisms of $\boldsymbol{C}$ of X into G: I, 1, 2. |
| ***Gr*** | category of groups: I, 1, 2. |
| ***GC*** | category of $\boldsymbol{C}$-groups: I, 1, 4. |
| $\boldsymbol{Alg}_k$ | category of commutative $k$-algebras: I, 1, 6. |
| $\boldsymbol{CAlg}_k$ | category of $\boldsymbol{Alg}_k$-cogroups: I, 1, 6. |
| $\mathrm{Hom}_{k\text{-alg.}}(B, X)$ | set of $k$-algebra homomorphisms of B into X: I, 1, 6. |
| $\boldsymbol{GAff}_k$ | category of affine group schemes over $k$: I, 1, 6. |
| $\mathbf{G}_a$ | additive group scheme: I, 1, 6. |
| $\mathbf{E}^*$ | dual vector space of E: I, 2, 1. |
| $\langle x, x^* \rangle$, $\tilde{x}$ | $x$ vector of E, $x^*$ vector of E*: I, 2, 1. |
| $\sigma(E^*, E)$ | weak topology on E*: I, 2, 1. |
| $M^0$, $M'^0$ | orthogonal subspaces of $M \subset E$ and $M' \subset E^*$: I, 2, 2. |
| ${}^t u$ | transposed mapping: I, 2, 3 and 6. |
| $E^* \hat{\otimes}_k F^*$ | completed tensor product of linearly compact spaces: I, 2, 4. |
| $\boldsymbol{Alc}_k$ | category of commutative linearly compact $k$-algebras: I, 2, 10. |
| $\boldsymbol{CAlc}_k$ | category of $\boldsymbol{Alc}_k$-cogroups: I, 2, 11. |
| $B^+$ | augmentation ideal: I, 2, 11. |
| $\boldsymbol{Cog}_k$ | category of $k$-cogebras: I, 2, 13. |
| $\boldsymbol{GCog}_k$ | category of $\boldsymbol{Cog}_k$-groups: I, 2, 13. |
| **G** | formal group: I, 3, 1. |
| $\mathfrak{G}$, $\mathfrak{G}^*$ | covariant and contravariant bigebras of a formal group **G**: I, 3, 1. |

| | |
|---|---|
| $\mathbf{G}_A$ | group $\mathrm{Hom}_{\mathit{Alc}_k}(\mathfrak{G}^*, A)$ of points of $\mathbf{G}$ with values in A: I, 3, 1. |
| $\mathbf{u}_A(x)$ | image $x \circ {}^t u$ of a point of $\mathbf{G}_A$ by a formal homomorphism $\mathbf{u}$: I, 3, 1. |
| $\mathbf{G}/\mathbf{N}$ | quotient of a formal group by an invariant formal subgroup: I, 3, 5. |
| $\mathbf{H} \subset \mathbf{H}'$ | inclusion relation between formal subgroups: I, 3, 7. |
| $\inf(\mathbf{H}_\lambda)$, $\sup(\mathbf{H}_\lambda)$ | greatest lower bound and least upper bound of a family of formal subgroups: I, 3, 7. |
| $\mathscr{Z}(\mathbf{H})$ | centralizer of a formal subgroup: I, 3, 10. |
| $\mathbf{G}_1 \times \mathbf{G}_2$ | product of formal groups: I, 3, 12. |
| $(\mathbf{u}_1, \mathbf{u}_2)$ | formal homomorphism into $\mathbf{G}_1 \times \mathbf{G}_2$: I, 3, 13. |
| $[\mathbf{N}_1, \mathbf{N}_2]$ | commutator subgroup of two invariant formal subgroups: I, 3, 16. |
| $[\mathfrak{N}_1, \mathfrak{N}_2]$ | covariant bigebra of $[\mathbf{N}_1, \mathbf{N}_2]$: I, 3, 16. |
| $\mathscr{D}(\mathbf{G})$ | derived group of a formal group: I, 3, 18. |
| $\alpha \leqslant \beta$, $\|\alpha\|$, $\alpha + \beta$, $\alpha - \beta$ | $\alpha$, $\beta$ multiindices: II, 2, 1. |
| $\mathrm{ht}(\alpha)$ | height of a multiindex $\alpha$: II, 2, 1. |
| $\alpha!$ | factorial of a multiindex: II, 2, 1. |
| $\varepsilon_i$ | multiindices of degree 1: II, 2, 1. |
| $\Delta$, $\gamma$ | comultiplication and counit of the covariant bigebra $\mathfrak{G}$: II, 2, 2. |
| $c$, $\eta$ | comultiplication and counit of the contravariant bigebra $\mathfrak{G}^*$: II, 2, 2. |
| $\mathfrak{R}$ | radical of $\mathfrak{G}^*$: II, 2, 2. |
| $\mathfrak{R}^n$ | closure of the product of $n$ ideals equal to $\mathfrak{R}$: II, 2, 2. |
| $M_n$ | $(\mathfrak{R}^{n+1})^0$, cogebra of the smallest increasing filtration of $\mathfrak{G}$: II, 2, 2. |
| $\mathfrak{g}_0$ | Lie algebra of $\mathfrak{G}$: II, 2, 2. |
| $\mathfrak{G}^{(r)}$, $\mathfrak{G}^{*(r)}$ | bigebras deduced from $\mathfrak{G}$ and $\mathfrak{G}^*$ by the automorphism $\xi \mapsto \xi^{p^r}$ of $k$: II, 2, 3. |
| F | Frobenius homomorphism $\mathfrak{G}^* \to \mathfrak{G}^{*(1)}$: II, 2, 3. |
| $\mathfrak{B}_r$ | closed ideal generated by $F^r(\mathfrak{R})$: II, 2, 3. |
| $\mathfrak{s}_{r-1}$, $\mathfrak{s}_{r-1}(\mathfrak{G})$ | Frobenius subbigebra $\mathfrak{B}_r^0$ of $\mathfrak{G}$: II, 2, 3. |
| $x^\alpha$ | $\prod_{i \in I} x_i^{\alpha(i)}$ for a multiindex $\alpha$: II, 2, 6. |
| $N_r$ | kernel of the homomorphism $\overline{F}_r$: $\mathfrak{R}/\mathfrak{R}^2 \to \mathfrak{R}^{p^r}/\mathfrak{R}^{p^r+1}$: II, 2, 3. |

| | |
|---|---|
| $\mathfrak{g}'_r$ | orthogonal to $N_r$ in $\mathfrak{g}_0$: II, 2, 3. |
| $X_{si}$, $X_\alpha$ | II, 2, 6. |
| $\mathbf{G}_a$, $\mathbf{G}_a(k)$ | additive formal group: II, 2, 6. |
| V | shift in a covariant bigebra: II, 2, 7. |
| $\mathfrak{g}_k$, $\mathfrak{g}_k(\mathfrak{G})$ | higher Lie algebras of a covariant bigebra: II, 2, 7. |
| dim $\mathfrak{G}$, dim $\mathbf{G}$ | dimension of a covariant bigebra, of an infinitesimal group: II, 2, 9. |
| $\mathbf{u}^{-1}(\mathbf{H}')$ | inverse image of a formal subgroup by a formal homomorphism: II, 3, 4. |
| $\mathscr{N}(\mathbf{H})$ | normalizer of a reduced subgroup: II, 3, 5. |
| ${}^r\mathscr{Z}(\mathbf{H})$ | reduced centralizer of a reduced subgroup: II, 3, 5. |
| Int($z$), **Int**($z$) | II, 3, 8. |
| $\mathscr{D}^n(\mathbf{G})$ | $n$-th derived formal subgroup of a reduced group: II, 3, 9. |
| $\mathscr{C}^n(\mathbf{G})$ | $n$-th term of the descending central series: II, 3, 9. |
| End($\mathfrak{G}$) | ring of endomorphisms of a commutative bigebra: III, 1, 2. |
| $[r]$, $[r]_\mathfrak{G}$ | addition of $r$ equal terms: III, 1, 3. |
| $\mathfrak{N}_r$ | ideal in End($\mathfrak{G}$) of all endomorphisms vanishing in $\mathfrak{s}_r(\mathfrak{G}) \cap \mathfrak{G}^+$: III, 1, 5. |
| $\mathfrak{G}_r$ ($r$ finite or $\infty$) | free commutative bigebra: III, 2, 1. |
| $v(\alpha)$ | lateral shift of a multiindex: III, 2, 1. |
| $\pi$, $\pi(\alpha)$ | graduation on $\mathfrak{G}_r$, degree (for $\pi$) of $Z_\alpha$: III, 2, 1. |
| $\mathfrak{G}$, $\mathfrak{G}_\infty(k)$ | Witt bigebra: III, 2, 1. |
| W($k$), W(A) | Witt ring: III, 2, 3. |
| $\omega$ | uniformizing element of W($k$): III, 2, 3. |
| $w$ | valuation on W($k$): III, 2, 3. |
| $E_\alpha(T_0, T_1, ..)$ | privileged basis of the Witt bigebra: III, 2, 3. |
| $\boldsymbol{\pi}$, $\mathbf{t}$ | endomorphisms of $\mathfrak{G}_\infty(k)$: III, 2, 4. |
| $\sigma$ | canonical automorphism of W($k$): III, 2, 4. |
| **T**, **X**, **Y** | systems of indeterminates: III, 3, 1. |
| $E_m(T_0, T_1, ...)$ | hyperexponential polynomials: III, 3, 1. |
| $\mathscr{E}(\mathfrak{G})$ | module of hyperexponential vectors: III, 3, 2. |
| $\mathscr{A}$, $\mathscr{A}(k)$ | subring of End($\mathfrak{G}_\infty$): III, 3, 2. |
| $\mathscr{E}'(\mathfrak{G}_\infty)$ | submodule of $\mathscr{E}(\mathfrak{G}_\infty)$ generated by **T**: III, 3, 4. |
| $\mathscr{R}$, $\mathscr{R}(k)$ | localized ring of $\mathscr{A}$: III, 3, 5. |
| $\mathscr{E}_{(\mathscr{R})}(\mathfrak{G})$ | $\mathscr{R}$-module $\mathscr{E}(\mathfrak{G}) \otimes_\mathscr{A} \mathscr{R}$: III, 3, 5. |

| | |
|---|---|
| K | field of fractions of W($k$): III, 4, 1. |
| $K_e$ | completely ramified extension of K: III, 4, 1. |
| $W_e$ | integral closure of W($k$) in $K_e$: III, 4, 1. |
| $\omega_e$ | root of $T^e - \omega$ in $K_e$: III, 4, 1. |
| A | Hilbert-Witt ring: III, 4, 2. |
| R | localized Hilbert-Witt ring: III, 4, 2. |
| $v(a)$ | stathm of $a$ in R: III, 4, 2. |
| $c(a)$ | costathm of $a$ in R: III, 4, 1. |
| rk(N) | rank of a torsion R-module: III, 4, 2. |
| crk(N) | corank of a torsion R-module: III, 4, 2. |
| $\pi$ | element $t^{-1}\omega = \omega t^{-1}$ in R: III, 4, 3. |
| rk(M) | rank of a distinguished A-module: III, 4, 3. |
| crk(M) | corank of an equidimensional A-module: III, 4, 3. |
| $\bar{z}$ | class of $z \in W(k)$ in $k$: III, 5, 1. |
| $\gamma(x)$, $r(x)$, $s(x)$, $j(x)$ | numbers attached to $x \in A$: III, 5, 1. |
| $K_n$ | field of fractions of $W(\mathbf{F}_{p^n})$: III, 5, 3. |
| $K_{e,n}$ | completely ramified extension of $K_n$: III, 5, 3. |
| $W_{e,n}$ | ring of integers in $K_{e,n}$: III, 5, 3. |
| $V_e$ | completely ramified extension of $\mathbf{Q}_p$ of degree $e$: III, 5, 3. |
| $\tau$ | image of $t$ in the sfield of endomorphisms of $R/(\omega^m - t^n)R$: III, 5, 3. |
| $M_{m,n}(k)$ | distinguished A-module: III, 5, 4 and 5. |
| $\varpi$ | element $\omega\tau^{-1}$ in End($M_{m,n}(k)$): III, 5, 5. |
| $\mathfrak{G}_{m,n}(k)$, $\mathbf{G}_{m,n}(k)$ | bigebra and formal group corresponding to $M_{m,n}(k)$: III, 6, 1. |
| $\Phi$(G) | formalization of an affine group: IV, 2. |
| $\Phi(u)$ | formalization of a homomorphism of affine groups: IV, 2. |
| **GL**(N) | formalization of GL(N): IV, 4. |
| A(**G**) | algebraic hull of a formal group: IV, 5. |
| Int($s$), **Int**($s$) | interior automorphism and its formalization: IV, 7. |

# Introduction to the Theory of Formal Groups

CHAPTER I

# Definition of Formal Groups

## §1. *C*-groups and *C*-cogroups

**1.** *The notion of **C**-group.* Let **C** be a category which has a final object $e$ and in which the product of any two objects is defined. A **C**-*group* consists in a quadruplet (G, $m$, $\eta$, $i$) where G is an object of **C**, $m$: G × G → G, $\eta$: $e$ → G, and $i$: G → G three morphisms of **C**, with the following axioms:

1) The diagram

$$\begin{array}{ccc} G \times G \times G & \xrightarrow{m \times 1} & G \times G \\ {\scriptstyle 1 \times m}\downarrow & & \downarrow{\scriptstyle m} \\ G \times G & \xrightarrow[m]{} & G \end{array}$$

is commutative.

2) The two diagrams

$$\begin{array}{ccc} G \times e & \xrightarrow{1 \times \eta} & G \times G \\ & \searrow{\scriptstyle \sim} & \downarrow{\scriptstyle m} \\ & & G \end{array} \qquad \begin{array}{ccc} e \times G & \xrightarrow{\eta \times 1} & G \times G \\ & \searrow{\scriptstyle \sim} & \downarrow{\scriptstyle m} \\ & & G \end{array}$$

are commutative (the oblique arrow is the natural isomorphism).

3) The two diagrams

$$\begin{array}{ccc} G & \xrightarrow{(1,i)} & G \times G \\ \downarrow & & \downarrow m \\ e & \xrightarrow[\eta]{} & G \end{array} \qquad\qquad \begin{array}{ccc} G & \xrightarrow{(i,1)} & G \times G \\ \downarrow & & \downarrow m \\ e & \xrightarrow[\eta]{} & G \end{array}$$

are commutative.

When one takes as $\boldsymbol{C}$ the category of sets, the morphisms being arbitrary mappings, one obtains the usual definition of groups (also called "abstract" or "set-theoretic" groups). When one takes for $\boldsymbol{C}$ the category of topological spaces, the morphisms being continuous mappings, one obtains the topological groups (in that category, products are just the usual products of topological spaces, and the final object a space having only one point).

**2.** It is well-known that there is another way of defining $\boldsymbol{C}$-groups. For each object X in $\boldsymbol{C}$, consider the set of morphisms Mor(X, G) (or $\mathrm{Mor}_{\boldsymbol{C}}(X, G)$) of all morphisms (of $\boldsymbol{C}$) of X into G; the mapping $X \mapsto \mathrm{Mor}(X, G)$ is a *functor* from the dual category $\boldsymbol{C}^0$ of $\boldsymbol{C}$ into the category of sets: to each morphism $u: X \to X'$ in $\boldsymbol{C}$ is associated the mapping $f \mapsto f \circ u$ of $\mathrm{Mor}(X', G)$ into $\mathrm{Mor}(X, G)$. This is valid for an arbitrary object G in $\boldsymbol{C}$, but if G is a $\boldsymbol{C}$-group, it is possible to define on Mor(X, G) a *group structure* in a natural way, such that, for any morphism $u: X \to X'$, the mapping $f \mapsto f \circ u$ of $\mathrm{Mor}(X', G)$ into $\mathrm{Mor}(X, G)$ is a group homomorphism. To define the group law on Mor(X, G), observe that

$$\mathrm{Mor}(X, G) \times \mathrm{Mor}(X, G)$$

is naturally identified with $\mathrm{Mor}(X, G \times G)$ by definition of the product in $\boldsymbol{C}$; the group law is then simply

$$\mu_X : (f, g) \mapsto m \circ (f, g).$$

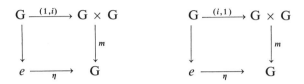

The associativity results from property 1); the neutral element is just the composite morphism $\varepsilon_X : X \to e \xrightarrow{\eta} G$, the fact that it is a neutral element following from property 2); finally the inverse is the mapping $f \mapsto i \circ f$, its properties following from property 3). We may thus say that a ***C***-group G defines in a natural way a *functor* $X \mapsto \mathrm{Mor}(X, G)$ from $\boldsymbol{C}^0$ to the *category of (abstract) groups* ***Gr***.

**3.** Conversely, if an object G of ***C*** is such that $X \mapsto \mathrm{Mor}(X, G)$ is a functor from $\boldsymbol{C}^0$ to ***Gr***, then there is a unique structure of ***C****-group* on G for which the group law $\mu_X$ on Mor(X, G) is deduced from that structure as in No. 2. Indeed, consider the object $G \times G$ of ***C***, and the corresponding group law $\mu_{G \times G}$ on $\mathrm{Mor}(G \times G, G)$; as

$$\mathrm{Mor}(G \times G, G) \times \mathrm{Mor}(G \times G, G)$$

is naturally identified with $\mathrm{Mor}(G \times G, G \times G)$, one can consider $\mu_{G \times G}$ as a mapping:

$$\mu_{G \times G} : \mathrm{Mor}(G \times G, G \times G) \to \mathrm{Mor}(G \times G, G);$$

however we have in the first set a privileged element $1_{G \times G}$; one then defines the morphism $m: G \times G \to G$ as equal to $\mu_{G \times G}(1_{G \times G})$; verification of the associativity diagram follows from the assumed associativity of the group law in Mor(X, G) for $X = G \times G \times G$. The morphism $\eta: e \to G$ is the neutral element $\varepsilon_e$ in Mor(*e*, G); the "inverse" morphism $i: G \to G$ is just the inverse (in the usual sense) of the identity morphism $1_G$ in the group Mor(G, G); verification of properties 2) and 3) is routine.

**4.** We now make ***C***-groups into a *category*, which will be denoted ***GC***; in order to do so we have to define the morphisms of that category. Given two ***C***-groups G, G′, we say that $u: G \to G'$ is a morphism of ***GC*** if it is a morphism in ***C***, and if, in addition, it makes the diagram

$$\begin{array}{ccc} G \times G & \xrightarrow{u \times u} & G' \times G' \\ \downarrow & & \downarrow \\ G & \xrightarrow[u]{} & G' \end{array}$$

commutative, the vertical arrows being the morphisms which define the ***C***-groups G and G′. It is equivalent to say that for each object $X \in \boldsymbol{C}$, the mapping $f \mapsto u \circ f$ of Mor(X, G) into Mor(X, G′) is a homomorphism of (abstract) groups.

**5.** *C-cogroups.* Suppose now that $\boldsymbol{C}$ is a category having an initial object $e$ and in which the *sum* $A \amalg B$ of any two objects A, B is defined. Then in the dual category $\boldsymbol{C}^0$, $e$ is a final object and the sum in $\boldsymbol{C}$ is the product in $\boldsymbol{C}^0$; we thus can define $\boldsymbol{C}^0$-*groups*. It is equivalent to consider $\boldsymbol{C}^0$-groups as objects of $\boldsymbol{C}$, called *C-cogroups*, and to say that such an object consists in a quadruplet $(B, c, \gamma, a)$ where B is an object of $\boldsymbol{C}$, $c\colon B \to B \amalg B$, $\gamma\colon B \to e$, $a\colon B \to B$ three morphisms of $\boldsymbol{C}$, with the following axioms (obtained by reversing the arrows in the axioms of No. 1, replacing products by sums, and $(G, m, \eta, i)$ by $(B, c, \gamma, a)$, respectively):

1) The diagram

$$\begin{array}{ccc} B & \xrightarrow{\ c\ } & B \amalg B \\ {\scriptstyle c}\downarrow & & \downarrow{\scriptstyle 1 \amalg c} \\ B \amalg B & \xrightarrow[c \amalg 1]{} & B \amalg B \amalg B \end{array}$$

is commutative.

2) The two diagrams

$$\begin{array}{ccc} B & & \\ {\scriptstyle c}\downarrow & \overset{\sim}{\searrow} & \\ B \amalg B & \xrightarrow[1 \amalg \gamma]{} & B \amalg e \end{array} \qquad \begin{array}{ccc} B & & \\ {\scriptstyle c}\downarrow & \overset{\sim}{\searrow} & \\ B \amalg B & \xrightarrow[\gamma \amalg 1]{} & e \amalg B \end{array}$$

are commutative.

3) The two diagrams

$$\begin{array}{ccc} B & \xrightarrow{\ \gamma\ } & e \\ {\scriptstyle c}\downarrow & & \downarrow \\ B \amalg B & \xrightarrow[(1,a)]{} & B \end{array} \qquad \begin{array}{ccc} B & \xrightarrow{\ \gamma\ } & e \\ {\scriptstyle c}\downarrow & & \downarrow \\ B \amalg B & \xrightarrow[(a,1)]{} & B \end{array}$$

are commutative (the definition of the sum $A \amalg B$ implies that, when $f\colon A \to C$ and $g\colon B \to C$ are morphisms, there is a unique morphism $u\colon A \amalg B \to C$ such that $f$ and $g$ factorize through $u$:

$$f: A \xrightarrow{j_1} A \amalg B \xrightarrow{u} C, \qquad g: B \xrightarrow{j_2} A \amalg B \xrightarrow{u} C$$

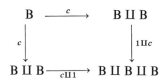

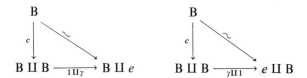

where $j_1$ and $j_2$ are the canonical morphisms; the morphism $u$ is also written $(f, g)$).
The alternative way of defining ***C***-cogroups (see No. 2) is to say that the functor $X \mapsto \text{Mor}(B, X)$ is, this time, a *functor* from ***C*** to the category ***Gr***. Finally, ***C***-cogroups are made into a category by defining a morphism of a ***C***-cogroup B into a ***C***-cogroup B′ as a morphism $u: B \to B'$ (in ***C***) satisfying the condition that it makes the diagram

$$\begin{array}{ccc} B & \xrightarrow{u} & B' \\ \downarrow & & \downarrow \\ B \amalg B & \xrightarrow[u \amalg u]{} & B' \amalg B' \end{array}$$

commutative.

**6.** *Example: affine group schemes.* Let $k$ be a commutative field, and denote by $\boldsymbol{Alg}_k$ the category whose objects consist in the *commutative k-algebras* (with unit) and the morphisms are the usual homomorphisms of $k$-algebras (sending unit into unit). It has an initial object, namely $k$, and sums: the sum of two commutative $k$-algebras A, B is simply the *tensor product over* $k$, $A \otimes_k B$, with its usual canonical homomorphisms $a \mapsto a \otimes 1$ and $b \mapsto 1 \otimes b$ of A and B into $A \otimes B$. Hence we can define the category $\boldsymbol{CAlg}_k$ of $\boldsymbol{Alg}_k$*-cogroups*: it will consist of objects $(B, c, \gamma, a)$ where B is a commutative $k$-algebra with unit, $e: B \to B \otimes_k B$, $\gamma: B \to k$ and $a: B \to B$, $k$-algebra homomorphisms satisfying the three conditions enumerated in No. 5. In the terminology of N. Bourbaki [5], such an object is called an *associative, coassociative, commutative bigebra* with *unit* $\varepsilon$ (the unique homomorphism $k \to B$), *counit* $\gamma$, and *antipodism* $a$ ($c$ is called the *coproduct* or *comultiplication* of the bigebra). It is interesting to describe explicitly the group structure on the set $\text{Hom}_{k\text{-alg.}}(B, X)$ defined by the $\boldsymbol{CAlg}_k$-cogroup B (cf. No. 2): to any pair $(u, v)$ of homomorphisms $u: B \to X$, $v: B \to X$, the product $u \cdot v$ in the group $\text{Hom}_{k\text{-alg.}}(B, X)$ is the composite homomorphism

(1) $$B \xrightarrow{c} B \otimes_k B \xrightarrow{u \otimes v} X \otimes_k X \xrightarrow{m} X$$

where $m$ is the homomorphism $x \otimes y \mapsto xy$.

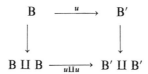

It is well known that the category $(\boldsymbol{Alg}_k)^0$ dual to $\boldsymbol{Alg}_k$ is the category of *affine k-schemes* $\boldsymbol{Aff}_k$, up to equivalence [28]; the category $\boldsymbol{CAlg}_k$ may therefore be identified with the category of *affine group schemes* $\boldsymbol{GAff}_k$ over $k$, the algebra associated to such a scheme being a bigebra of the type described above [16]. It is instructive to give a few explicit examples of such group schemes:

I. Let H be a *finite* (abstract) group, and consider the $k$-algebra $B = k^H$ of all mappings $f: H \to k$; $B \otimes_k B$ is naturally identified to $k^{H \times H}$, the tensor product $f \otimes g$ being identified to the mapping $(x, y) \mapsto f(x)g(y)$. One defines a comultiplication $c: B \to B \otimes_k B$ by defining $c(f)$ as the function of two variables $(x, y) \mapsto f(xy)$; a counit $\gamma: B \to k$ is defined by $\gamma(f) = f(e)$ where $e$ is the unit element of H; finally an antipodism $a: B \to B$ is defined by taking for $a(f)$ the mapping $x \mapsto f(x^{-1})$. It is easily checked that the axioms of No. 5 are satisfied. The corresponding affine group schemes are called *constant finite groups over k*.

II. The *additive group scheme* $\mathbf{G}_a = \mathrm{Spec}(k[T])$, where $B = k[T]$ is the algebra of polynomials in one indeterminate over $k$, is best defined by noting that for each commutative $k$-algebra X, a $k$-homomorphism $u: k[T] \to X$ is entirely determined by the element $x = u(T)$, and therefore, as a set, $\mathrm{Hom}_{k\text{-alg.}}(k[T], X)$ is identified with X; with this identification, we define the group structure on $\mathrm{Hom}_{k\text{-alg.}}(k[T], X)$ as being the *additive group structure* of the algebra X. It is easily verified that the coproduct $c: B \to B \otimes B$ is given by $c(T) = T \otimes 1 + 1 \otimes T$, the counit $\gamma: B \to k$ by $\gamma(T) = 0$, and the antipodism $a: B \to B$ by $a(T) = -T$.

III. *Multiplicative group schemes.* Let M be an arbitrary (set theoretic) *commutative group*, and consider its *group algebra* $B = k[M]$, which is commutative. For each commutative $k$-algebra X, a $k$-homomorphism $u: k[M] \to X$ is entirely determined by the elements $u(s)$ for $s \in M$, which of course must be such that the mapping $v = u|M: s \mapsto u(s)$ is a group homomorphism of M into the multiplicative group $X^*$ of invertible elements of X; the set $\mathrm{Hom}_{gr}(M, X^*)$ is itself a commutative group for the pointwise multiplication of homomorphisms, and this group structure is the one on $\mathrm{Hom}_{k\text{-alg.}}(k[M], X)$ which defines the structure of *affine group scheme* on $\mathrm{Spec}(k[M])$, which is written $\mathbf{D}_k(M)$. The comultiplication $c: B \to B \otimes B$ is here given by $c(s) = s \otimes s$ for any $s \in M$, the counit $\gamma: B \to k$ by $c(s) = 0$ for $s \in M$ and the antipodism $a: B \to B$ by $a(s) = s^{-1}$ for $s \in M$.

An interesting group scheme of this kind is given by the choice $M = \mathbf{Z}/n\mathbf{Z}$ (cyclic group of order $n$) for $m$; we have here

$$\mathbf{D}_k(M) = \mathrm{Spec}(k[T]/(T^n - 1)),$$

and $\mathbf{D}_k(\mathrm{M})$ is called the *group scheme of n-th roots of unity* over $k$. When $n$ is not a multiple of the characteristic $p$ of $k$, the algebra $\mathrm{B} = k[\mathrm{M}]$ is isomorphic to a direct sum of $n$ copies of $k$, and $\mathbf{D}_k(\mathrm{M})$ consists of $n$ points; on the other hand, if $n = p$,

$$\mathrm{B} = k[\mathrm{T}]/(\mathrm{T} - 1)^p,$$

hence $\mathbf{D}_k(\mathrm{M})$, as a scheme, has only *one point*, but the local ring at that point has nilpotent elements.

When B is a bigebra belonging to $\boldsymbol{CAlg}_k$, and $\mathrm{G} = \mathrm{Spec}(\mathrm{B})$ the corresponding affine group scheme, for any commutative $k$-algebra X, the elements of the group $\mathrm{Hom}_{k\text{-alg.}}(\mathrm{B}, \mathrm{X})$ are called the *points of* G *with values in* X.

## §2. Formal groups and their bigebras

**1.** *Linearly compact spaces.* Formal groups can be defined as co-groups in a category of a special type of *topological rings*. To define these rings we have to use the theory of linearly compact vector spaces; we will first rapidly recall the main notions and results of that theory [22, 36]. It is well known that the duality theory of vector spaces, which is a very useful tool when limited to finite-dimensional spaces, completely breaks down for infinite-dimensional spaces, the second dual of such a space being far too big and having no sensible relation to the original space. To restore the usefulness of duality theory, one has to introduce topological concepts. Let E be any vector space over an arbitrary commutative field $k$, and $\mathrm{E}^*$ its dual; we write as usual $\langle x, x^* \rangle$ for $x^*(x)$ when $x \in \mathrm{E}$ and $x^* \in \mathrm{E}^*$. For each $x \in \mathrm{E}$, the mapping $x^* \mapsto \langle x, x^* \rangle$ is a linear form on $\mathrm{E}^*$, which we denote by $\tilde{x}$. On $\mathrm{E}^*$ we put the topology $\sigma(\mathrm{E}^*, \mathrm{E})$ which is defined as the coarsest for which all the mappings $\tilde{x}$ of $\mathrm{E}^*$ into $k$ are continuous, when $x$ takes all values in E, and $k$ is given the discrete topology. A fundamental system of neighborhoods of 0 in $\mathrm{E}^*$ consists therefore of finite intersections of hyperplanes defined by equations $\langle x_j, x^* \rangle = 0$, for any finite sequence $(x_j)$ in E; they are of finite codimension in $\mathrm{E}^*$. A convenient way of describing this topology consists in choosing a basis $(e_\alpha)_{\alpha \in \mathrm{I}}$ of E; to each $\alpha \in \mathrm{I}$ we associate the corresponding linear "coordinate" form $e_\alpha^*$ on E such that $\langle e_\alpha, e_\beta^* \rangle = \delta_{\alpha\beta}$ (Kronecker delta), and we say that the family $(e_\alpha^*)_{\alpha\in I}$ is the *pseudobasis* of E*, *dual* to $(e_\alpha)_{\alpha\in I}$. The basis $(e_\alpha)$ defines a natural isomorphism $\varphi$ of E onto the vector space $k^{(I)}$, associating to each $e_\alpha$ the element $u_\alpha$ whose coordinates in $k^{(I)}$ are all 0 with the exception of the coordinate of index $\alpha$ which is equal to 1. It is immediate that the dual of the vector space $k^{(I)}$ is the product $k^I$; the transpose of $\varphi$ is an isomorphism of $k^I$ onto E* which maps $u_\alpha$ onto $e_\alpha^*$; the topology $\sigma(E^*, E)$ is obtained by *transporting* by this isomorphism the *product topology* of $k^I$ (where each factor is given the discrete topology). It should be observed that the subspace E′ of E* generated (algebraically) by the $e_\alpha^*$ is *dense* in E*, and E* may be considered as obtained by *completion* from E′, when E′ is given the topology for which a fundamental system of neighborhoods of 0 consists of the vector subspaces containing the $e_\alpha^*$ with the exception of at most a finite number.

Another way of defining the topological vector space E* is to say that it is the *inverse limit* of the discrete finite-dimensional vector spaces $E^*/V^0$ (dual to V), where V runs through the directed set of finite-dimensional subspaces of E.

It can be proved that the topological vector spaces E* thus defined (over the discrete field $k$) may be characterized by the property of *linear compactness*: they are vector spaces with a Hausdorff topology compatible with the additive group structure, with a fundamental system of neighborhoods of 0 consisting of vector subspaces, and any filter base consisting of affine linear varieties has a cluster point. We will not use this result.

**2.** With the same notations, any linear form on E* which is *continuous* for the topology $\sigma(E^*, E)$ is necessarily of type $x^* \mapsto \langle x, x^* \rangle$ for a unique $x \in E$; hence we get back E as *topological dual* of E*.

For any vector subspace M in E, the *orthogonal* subspace $M^0$ is the subspace of E* consisting of all $x^* \in E^*$ such that $\langle x, x^* \rangle = 0$ for every $x \in M$; it is a *closed* subspace of E*. Similarly, for any vector subspace M′ of E*, the *orthogonal* subspace $M'^0$ is the subspace of E consisting of the $x \in E$ such that $\langle x, x^* \rangle = 0$ for all $x^* \in M'$. If one takes a basis of E consisting of a basis of M and a basis of a supplementary subspace, it is immediately verified that $M^{00} = M$. Similarly, $M'^{00}$ is the *closure* $\overline{M'}$ of M′ in E*: indeed, as $M' \subset M'^{00}$ and the latter is closed, we have $\overline{M'} \subset M'^{00}$; on the other hand, if $x^* \notin \overline{M'}$, there exists in E* an open neighborhood V′ of 0 such that $(x^* + V') \cap M' = \varnothing$, hence $x^* \notin M' + V'$; but $M' + V'$ is open and contains 0, hence contains the intersection of a finite number of hyperplanes $H_i'$ defined by equations $\langle x_i, y^* \rangle = 0$ and containing M′; therefore one certainly has $\langle x_i, x^* \rangle \neq 0$ for some $i$, while $\langle x_i, y^* \rangle = 0$ for all $i$ and all $y^* \in \mathrm{M}'$; therefore $x^* \notin \mathrm{M}'^{00}$, which proves our assertion.

We thus define a bijection $\mathrm{M} \mapsto \mathrm{M}^0$ of the set of all vector subspaces of E onto the set of all *closed* vector subspaces of E*; it transforms finite sums into finite intersections and finite intersections into finite sums. In particular, every closed subspace M′ of E* admits a closed topological supplementary N′ (i.e., such that the mapping $(x^*, y^*) \mapsto x^* + y^*$ of $\mathrm{M}' \times \mathrm{N}'$ onto E* is bicontinuous); furthermore, if N′ is a closed supplementary subspace of M′, it is automatically a *topological* supplementary. Every finite-dimensional subspace of E* is closed.

It is clear that if $(e^*_\alpha)$ is a pseudobasis of E*, any subfamily of $(e^*_\alpha)$ is a pseudobasis of the closed subspace of E* it generates. One must beware of the fact that a family $(x^*_\alpha)$ in E* may generate E* as a closed subspace, and be such that for a family $(x_\alpha)$ in E having the same set of indices, $\langle x_\alpha, x^*_\beta \rangle = \delta_{\alpha\beta}$, *without being a pseudobasis of* E*. Take for instance E having a denumerable basis $(e_n)_{n \geqslant 0}$ and let

$$x^*_n = e^*_0 + e^*_n \qquad \text{for} \quad n \geqslant 1,$$

where $(e^*_n)_{n \geqslant 0}$ is the pseudobasis dual to $(e_n)_{n \geqslant 0}$. We have

$$\langle e_n, x^*_m \rangle = \delta_{nm} \qquad \text{for} \quad n \geqslant 1, m \geqslant 1,$$

and the $x^*_n$ generate E*, for if an element $x = \sum_{n=0}^{\infty} \xi_n e_n$ is orthogonal to all the $x^*_n$ for $n \geqslant 1$, one must have $\xi_n + \xi_0 = 0$ for all $n \geqslant 1$, which is impossible unless $x = 0$, since the $\xi_n$ can only be $\neq 0$ for finitely many indices.

**3.** Let E, F be any two vector spaces over $k$, E*, F* their dual spaces with the topologies $\sigma(\mathrm{E}^*, \mathrm{E})$ and $\sigma(\mathrm{F}^*, \mathrm{F})$. For any linear mapping $u\colon \mathrm{E} \to \mathrm{F}$, its *transposed* mapping ${}^t u\colon \mathrm{F}^* \to \mathrm{E}^*$ (defined by the identity

$$\langle u(x), y^* \rangle = \langle x, {}^t u(y^*) \rangle)$$

is *continuous*, and conversely, for any linear mapping $v\colon \mathrm{F}^* \to \mathrm{E}^*$ which is continuous, there exists a unique linear mapping $u\colon \mathrm{E} \to \mathrm{F}$ such that $v = {}^t u$; we write then $u = {}^t v$. For any subspace M of E

$$(u(\mathrm{M}))^0 = {}^t u^{-1}(\mathrm{M}^0) \qquad (1)$$

and for any subspace $N'$ of $F^*$,

$$(2) \qquad (v(N'))^0 = {}^t v^{-1}(N'^0);$$

in addition, if $N'$ is closed in $F^*$, then $v(N')$ is *closed* in $E^*$, hence equal to $({}^t v^{-1}(N'^0))^0$. For subspaces M of E and N of F, the relation $u(M) \subset N$ is equivalent to ${}^t u(N^0) \subset M^0$. The dual $M^*$ of M is topologically identified to $E^*/M^0$ and the dual $(E/M)^*$ to the subspace $M^0$ of $E^*$.

**4.** Let E, F be any two vector spaces over $k$; the tensor product $E^* \otimes_k F^*$ is naturally identified to a subspace of the dual $(E \otimes_k F)^*$ by the formula

$$(3) \qquad \langle x \otimes y, x^* \otimes y^* \rangle = \langle x, x^* \rangle \langle y, y^* \rangle.$$

In addition, this formula shows that if $(e_\alpha)$, $(f_\beta)$ are bases of E and F, respectively, $(e_\alpha^*)$ and $(f_\beta^*)$ the dual pseudobases in $E^*$ and $F^*$, then $(e_\alpha^* \otimes f_\beta^*)$ is the dual pseudobasis of $(e_\alpha \otimes f_\beta)$ in $(E \otimes_k F)^*$. This at once shows that $(E \otimes_k F)^*$ is the *completion* of the space $E^* \otimes_k F^*$ for a topology on that space (called the *tensor product topology*) for which a fundamental system of neighborhoods of 0 consist of the sets

$$V \otimes_k F^* + E^* \otimes_k W,$$

where V (resp. W) is an arbitrary element of a fundamental system of neighborhoods of 0 consisting in vector subspaces of $E^*$ (resp. $F^*$); this completion is also written $E^* \hat{\otimes}_k F^*$ and called the *completed tensor product* of $E^*$ and $F^*$; obviously the natural mapping $(x^*, y^*) \mapsto x^* \otimes y^*$ of $E^* \times F^*$ into $E^* \hat{\otimes}_k F^*$ is *continuous* for the topologies of these spaces.

When E *or* F has finite dimension, one has

$$E^* \hat{\otimes}_k F^* = E^* \otimes_k F^*.$$

If E is the direct sum of two subspaces M, N, so that $E^*$ is the topological direct sum of $M^* = N^0$ and $N^* = M^0$, then $E^* \hat{\otimes}_k F^*$ is the topological direct sum of $(N \otimes_k F)^0$, identified to $M^* \hat{\otimes}_k F^*$, and $(M \otimes_k F)^0$, identified to $N^* \hat{\otimes}_k F^*$.

Let

$$u : E \to E_1, \qquad v : F \to F_1$$

be two linear mappings; then ${}^t(u \otimes v)$ is the continuous extension to $E_1^* \hat{\otimes}_k F_1^*$ of the continuous mapping ${}^t u \otimes {}^t v$;

$$E_1^* \otimes_k F_1^* \to E^* \otimes_k F^*;$$

it is also noted ${}^t u \hat{\otimes} {}^t v$.

Let M (resp. N) be a subspace of E (resp. F), so that (tensor products being over a field) $M \otimes_k N$ is identified to a subspace of $E \otimes_k F$; taking suitable bases in E and F, it is immediately verified that the orthogonal of $M \otimes_k N$ in $E^* \hat{\otimes}_k F^*$ is

$$M^0 \hat{\otimes}_k F^* + E^* \hat{\otimes}_k N^0.$$

Similarly, if M′ (resp. N′) is a closed subspace of $E^*$ (resp. $F^*$), $M' \hat{\otimes}_k N'$ is identified to a closed subspace of $E^* \hat{\otimes}_k F^*$, whose orthogonal in $E \otimes_k F$ is

$$M'^0 \otimes_k F + E \otimes_k N'^0.$$

**5.** Let E be a vector space over $k$, and let K be an extension of $k$; the dual $(E_{(K)})^*$ of the vector subspace $E_{(K)} = E \otimes_k K$ over K is identified with the completion of the space

$$E^* \otimes_k K = (E^*)_{(K)},$$

with the topology having as fundamental system of neighborhoods of 0 the K-subspaces $V \otimes_k K$, where V is an arbitrary open vector subspace in $E^*$; this is immediately seen by taking a basis $(e_\alpha)$ of E, which gives a basis $(e_\alpha \otimes 1)$ of $E_{(K)}$ and a dual pseudobasis $(e_\alpha^* \otimes 1)$ in $(E_{(K)})^*$.

**6.** We will later consider semilinear mappings $u$: E → F, relative to an automorphism $\sigma$ of $k$; for each $y^* \in F^*$,

$$x \mapsto \langle u(x), y^* \rangle^{\sigma^{-1}}$$

is then a linear form on E, written ${}^t u(y^*)$; ${}^t u$ is thus a mapping of $F^*$ into $E^*$, which is semilinear relative to the automorphism $\sigma^{-1}$; we shall again say that ${}^t u$ is the *transposed* mapping of $u$, thus characterized by the identity

$$\langle u(x), y^* \rangle = \langle x, {}^t u(y^*) \rangle^\sigma;$$

it is clear that ${}^t u$ is *continuous*. Conversely, any continuous semilinear mapping $v$: $F^* \to E^*$ relative to $\sigma^{-1}$ has the form ${}^t u$, where $u$: $E \to F$ is semilinear relative to $\sigma$; we write again $u = {}^t v$. Let $F^\sigma$ be the $k$-vector space defined by keeping the same addition and taking as operation by scalars $(\lambda, y) \mapsto \lambda^\sigma \cdot y$; $u$ is then a linear mapping $E \to F^\sigma$; the bijection $y^* \mapsto y^{*\sigma^{-1}}$ transforms the dual $F^*$ into the dual $(F^\sigma)^*$, and allows one to identify the latter to the $k$-vector space $(F^*)^\sigma$ (defined by the operation $(\lambda, y^*) \mapsto \lambda^\sigma \cdot y^*$ by scalars); ${}^t u$ is then a continuous linear mapping $(F^*)^\sigma \to E^*$, or equivalently a continuous linear mapping $F^* \to (E^*)^{\sigma^{-1}}$.

**7.** *Linearly compact commutative algebras.* We will say that a commutative $k$-algebra A (with unit) is *linearly compact* if it has a topology for which, as a vector space over $k$, it is linearly compact, and in addition that topology is compatible with the ring structure of A.

**Proposition 1.** *In a linearly compact commutative algebra* A *over* $k$, *there exists a fundamental system of neighborhoods of* 0 *consisting of ideals of finite codimension.*

By definition, there is a fundamental system $\mathfrak{B}$ of neighborhoods of 0 consisting of vector subspaces of A of finite codimension. Let $V \in \mathfrak{B}$; the continuity of $(x, y) \mapsto xy$ at (0, 0) implies the existence of $W \in \mathfrak{B}$ such that $W \cdot W \subset V$. Let $(a_1, \ldots, a_r)$ be a basis over $k$ of a subspace of A supplementary to W; for each $j$, the continuity of $x \mapsto a_j x$ at 0 implies the existence of a neighborhood $W_j \subset W$ belonging to $\mathfrak{B}$ and such that $a_j W_j \subset V$. Let U be the intersection of the $W_j$ for $1 \leqslant j \leqslant r$; it is a vector subspace of finite codimension and a neighborhood of 0; furthermore, $a_j U \subset V$ for every $j$ and $WU \subset V$; hence $AU \subset V$; but as $U \subset AU$, AU is an ideal of finite codimension in A, which is a neighborhood of 0 and is contained in V. QED.

Let $(\mathfrak{a}_\alpha)$ be the set of all open ideals of A, ordered by the relation $\supset$, for which it is a directed set (we write $\alpha \geqslant \beta$ instead of $\mathfrak{a}_\alpha \subset \mathfrak{a}_\beta$); it follows from Proposition 1 that we may write

$$A = \varprojlim (A/\mathfrak{a}_\alpha) \tag{4}$$

so that A is an inverse limit of finite-dimensional commutative $k$-algebras, what is also called a *profinite* commutative $k$-algebra; conversely, it is clear that such an inverse limit is linearly compact (when the finite-dimensional algebras are given the discrete topology), so profinite and linearly compact commutative $k$-algebras are the same.

**8.** A *local* commutative linearly compact (or profinite) $k$-algebra is a linearly compact $k$-algebra A which is a local ring; its maximal ideal $\mathfrak{m}$ is then necessarily open and therefore of finite codimension, which means that the residual field $A/\mathfrak{m}$ is a *finite* extension of $k$.

**Proposition 2.** *A commutative linearly compact k-algebra* A *is isomorphic to a product* $\prod_{\xi \in X} A_\xi$ *of local linearly compact k-algebras.*

Let us denote by $(\mathfrak{m}_\xi)_{\xi \in X}$ the set of maximal ideals in A which are *open* (in other words, those which contain one of the ideals $\mathfrak{a}_\alpha$). For each $\alpha$, the quotients $\mathfrak{m}_\xi/\mathfrak{a}_\alpha$ for the open maximal ideals $\mathfrak{m}_\xi$ containing $\mathfrak{a}_\alpha$ are exactly the maximal ideals of the artinian ring $A/\mathfrak{a}_\alpha$; therefore [7], $A/\mathfrak{a}_\alpha$ is naturally isomorphic to the product of its local rings

$$(A/\mathfrak{a}_\alpha)_{\mathfrak{m}_\xi/\mathfrak{a}_\alpha} = A_{\mathfrak{m}_\xi}/(\mathfrak{a}_\alpha)_{\mathfrak{m}_\xi},$$

which we will denote by $A_{\xi,\alpha}$; furthermore, the natural homomorphism

$$f_{\alpha\beta} : A/\mathfrak{a}_\beta \to A/\mathfrak{a}_\alpha, \qquad \text{for} \quad \mathfrak{a}_\beta \subset \mathfrak{a}_\alpha,$$

gives a natural local homomorphism

$$f_{\alpha\beta}^\xi : A_{\xi,\beta} \to A_{\xi,\alpha} \qquad \text{for} \quad \mathfrak{m}_\xi \supset \mathfrak{a}_\alpha;$$

we write $f_{\alpha\beta}^\xi = 0$ if $\mathfrak{m}_\xi \supset \mathfrak{a}_\beta$, but $\mathfrak{m}_\xi \not\supset \mathfrak{a}_\alpha$. Let J be the set of pairs $(\xi, \alpha)$ such that $\mathfrak{a}_\alpha \subset \mathfrak{m}_\xi$, and consider the product $\prod_{(\xi,\alpha)\in J} A_{\xi,\alpha}$ of finite-dimensional $k$-algebras; each $A/\mathfrak{a}_\alpha$ is identified with the subalgebra of that product consisting of the partial product of the $A_{\xi,\alpha}$ for all $\xi$ such that $(\xi, \alpha) \in J$; and if $\alpha \leqslant \beta$, the homomorphism $f_{\alpha\beta}$ can be defined as having a restriction equal to $f_{\alpha\beta}^\xi$ for all $\xi$ such that $(\xi, \beta) \in J$. Now, A is the subalgebra of the product $\prod_\alpha (A/\mathfrak{a}_\alpha)$ consisting of all $(z_\alpha)$ such that

$$f_{\alpha\beta}(z_\beta) = z_\alpha \qquad \text{for} \quad \alpha \leqslant \beta;$$

it is therefore also the subalgebra of $\prod_{(\xi,\alpha)\in J} A_{\xi,\alpha}$ consisting of all $(x_{\xi,\alpha})$ such that

$$f_{\alpha\beta}^{\xi}(x_{\xi,\beta}) = x_{\xi,\alpha} \qquad \text{for} \quad \alpha \leqslant \beta.$$

But for each $\xi \in X$, the set $J_\xi$ of $\alpha$ such that $(\xi, \alpha) \in J$ is cofinal to the set of all indices $\alpha$; as we may write

$$\prod_{(\xi,\alpha)\in J} A_{\xi,\alpha} = \prod_{\xi\in X} \Big(\prod_{\alpha\in J_\xi} A_{\xi,\alpha}\Big),$$

we see that A is identified with the product algebra $\prod_{\xi\in X} A_\xi$, where $A_\xi = \varprojlim_\alpha A_{\xi,\alpha}$. It is clear that $A_\xi$ is a linearly compact $k$-algebra; furthermore, every ideal different from $A_\xi$ in that algebra projects in each $A_{\xi,\alpha}$ onto an ideal $\neq A_{\xi,\alpha}$, hence contained in the maximal ideal $\mathfrak{m}_{\xi,\alpha}$ of $A_{\xi,\alpha}$; the inverse image in $A_\xi$ of all the $\mathfrak{m}_{\xi,\alpha}$ is thus the unique maximal ideal of $A_\xi$, which is therefore a *local* algebra. Note that the product $\mathfrak{R} = \prod_{\xi\in X} \mathfrak{m}_\xi$ of the maximal ideals of the $A_\xi$ is the *radical* of A, for it is an intersection of maximal ideals in A (the inverse images of the $\mathfrak{m}_\xi$), and if $r = (r_\xi)_{\xi\in X}$ is in $\mathfrak{R}$, then $1 + r_\xi$ is invertible in $A_\xi$ for each $\xi$, which implies that $1 + r$ is invertible in A, and proves our contention.

**9.** Let $A_1$, $A_2$ be any two commutative linearly compact $k$-algebras; Proposition 1 shows that there exists in $A_1$ (resp. $A_2$) a fundamental system of neighborhoods of 0 consisting of finite-codimensional ideals $\mathfrak{K}$ (resp. $\mathfrak{L}$); as a linearly compact vector space, $A_1 \hat{\otimes}_k A_2$ is identified to

$$\varprojlim((A_1/\mathfrak{K}) \otimes_k (A_2/\mathfrak{L}));$$

but each $(A_1/\mathfrak{K}) \otimes_k (A_2/\mathfrak{L})$ has a natural structure of finite-dimensional $k$-algebra, the tensor product of the algebra structures of $A_1/\mathfrak{K}$ and $A_2/\mathfrak{L}$, and the inverse system $((A_1/\mathfrak{K}) \otimes_k (A_2/\mathfrak{L}))$ is an inverse system of algebras (and not only of vector spaces). The inverse limit of that system defines therefore $A_1 \hat{\otimes}_k A_2$ as a linearly compact commutative $k$-algebra, which we call the *completed tensor product* of $A_1$ and $A_2$; it can indeed be considered as the completion of the ordinary tensor product $A_1 \otimes_k A_2$ with the tensor product topology (No. 4).

**10.** We will denote by $\boldsymbol{Alc}_k$ the category whose objects are the commutative linearly compact $k$-algebras, and the morphisms the *continuous* $k$-homomorphisms of those algebras. It contains the category of all *finite dimensional* commutative $k$-algebras, which are the objects of $\boldsymbol{Alc}_k$ whose topology is *discrete*. In particular, $k$ is an object of $\boldsymbol{Alc}_k$, and it is obviously an initial object of that category. On the other hand, the completed tensor product $A_1 \hat{\otimes}_k A_2$ is the *sum* of $A_1$, $A_2$ in $\boldsymbol{Alc}_k$. Indeed, the natural mappings $x_1 \mapsto x_1 \otimes 1$ and $x_2 \mapsto 1 \otimes x_2$ of $A_1$, $A_2$ into

$$A_1 \otimes_k A_2 \subset A_1 \hat{\otimes}_k A_2$$

are continuous by definition of the tensor product topology (No. 4). Now let X be an arbitrary object in $\boldsymbol{Alc}_k$ and let $u_1 : A_1 \to X$, $u_2 : A_2 \to X$ be two continuous $k$-homomorphisms. There is a unique $k$-homomorphism $w: A_1 \otimes_k A_2 \to X$ such that $u_1(x_1) = w(x_1 \otimes 1)$ and $u_2(x_2) = w(1 \otimes x_2)$; we have only to prove that $w$ is *continuous*, for then it will extend uniquely to a continuous homomorphism of the completion of $A_1 \otimes_k A_2$ into X. But if $\mathfrak{M}$ is an open ideal in X, there is by assumption an open ideal $\mathfrak{K}$ (resp. $\mathfrak{L}$) in $A_1$ (resp. $A_2$) such that $u_1(\mathfrak{K}) \subset \mathfrak{M}$ (resp. $u_2(\mathfrak{L}) \subset \mathfrak{M}$); hence the image by $w$ of $\mathfrak{K} \otimes A_2 + A_1 \otimes \mathfrak{L}$ is still contained in $\mathfrak{M}$, which proves our assertion.

**11.** We can now apply to the category $\boldsymbol{Alc}_k$ the general construction of §1, No. 5 and define the category $\boldsymbol{CAlc}_k$ of $\boldsymbol{Alc}_k$-cogroups. An object in that category is therefore a commutative linearly compact $k$-algebra B together with a *coproduct*, which is a continuous homomorphism,

$$c : B \to B \hat{\otimes} B, \tag{5}$$

a *counit* (or *augmentation*), which is a continuous homomorphism,

$$\gamma : B \to k, \tag{6}$$

and an *antipodism* $a$, which is a continuous homomorphism,

$$a : B \to B \tag{7}$$

(all these homomorphisms sending unit onto unit); the diagrams described in §1, No. 5, where $e$ is replaced by $k$ and ∐ by $\hat{\otimes}$, must of course be commutative. A *morphism* in $\boldsymbol{CAlc}_k$ is a continuous homomorphism of algebras $u: B \to B'$ sending unit onto unit, and for which the diagram

$$\begin{array}{ccc} B & \xrightarrow{u} & B' \\ {\scriptstyle c}\downarrow & & \downarrow{\scriptstyle c'} \\ B \mathbin{\hat{\otimes}} B & \xrightarrow[u \hat{\otimes} u]{} & B' \mathbin{\hat{\otimes}} B' \end{array}$$

is commutative.

Observe that if $B^+$ is the closed "augmentation" ideal of B, kernel of $\gamma$, B is the (topological) direct sum $k \cdot 1 \oplus B^+$, and we may thus write

$$B \mathbin{\hat{\otimes}} B = (k \otimes k) \oplus (k \otimes B^+) \oplus (B^+ \otimes k) \oplus (B^+ \mathbin{\hat{\otimes}} B^+)$$
(topological direct sum).

For any element $x = \lambda + y$, where $\lambda \in k$ and $y \in B^+$, we may then write

$$c(x) = \lambda(1 \otimes 1) + 1 \otimes u + v \otimes 1 + w,$$

where $u$, $v$ are in $B^+$ and $w \in B^+ \mathbin{\hat{\otimes}} B^+$. The diagrams for the counit show that

$$(1 \mathbin{\hat{\otimes}} \gamma)\,(c(x)) = x \otimes 1 \qquad \text{and} \qquad (\gamma \mathbin{\hat{\otimes}} 1)\,(c(x)) = 1 \otimes x;$$

as $\gamma$ is continuous, one has

$$(1 \mathbin{\hat{\otimes}} \gamma)(w) = (\gamma \mathbin{\hat{\otimes}} 1)(w) = 0$$

for any $w \in B^+ \mathbin{\hat{\otimes}} B^+$, since this is true by definition for all $w \in B^+ \otimes B^+$; the preceding relations then prove that

$$v \otimes 1 = y \otimes 1, \qquad 1 \otimes u = 1 \otimes y,$$

or equivalently $u = v = y$, so that we may write, for any $y \in B^+$,

$$c(y) = 1 \otimes y + y \otimes 1 + c^+(y) \tag{8}$$

where $c^+(y) \in B^+ \mathbin{\hat{\otimes}} B^+$.

We may characterize an object of $\boldsymbol{CAlc}_k$ as a topological *commutative bigebra* over $k$, with unit, counit, and antipodism, which as a vector space over $k$ has a linearly compact topology for which it is a topological algebra, and the mappings $c$, $\gamma$, and $a$ are continuous.

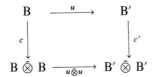

**12.** *Examples.* 1) Let $B = k[[T_1, \ldots, T_n]]$ be the formal power series algebra over $k$, in $n$ indeterminates; the elements of B are the families $(x_\alpha)_{\alpha \in \mathbf{N}^n}$, also written $\sum_\alpha x_\alpha T^\alpha$, the product being defined as for convergent power series; as a vector space over $k$, B is isomorphic to $k^{\mathbf{N}^n}$, and the product topology on that space is compatible with the structure of algebra of B. It is easily verified that $B \hat{\otimes}_k B$ is naturally identified to the algebra of formal power series in $2n$ indeterminates, the tensor product $f \otimes g$ of two elements of B being identified to the formal power series

$$f(T_1, \ldots, T_n)g(U_1, \ldots, U_n)$$

in $2n$ indeterminates. Let us take as counit $\gamma$ for B the natural "augmentation" which to each formal power series $\sum_\alpha x_\alpha T^\alpha$ associates its constant term $x_0$, so that the kernel $B^+$ of $\gamma$ is the ideal of all power series without constant terms. Then it follows from (8) that, for $1 \leqslant i \leqslant n$, one must have for the coproduct on B

$$c(T_i) = T_i + U_i + \sum_{|\alpha| > 1, |\beta| > 1} b_{\alpha\beta} T^\alpha U^\beta. \tag{9}$$

And conversely, given arbitrarily $n$ formal power series of such form, there is a unique continuous $k$-homomorphism $c: B \to B \hat{\otimes} B$ which takes these values for the elements $T_i$ of $B^+$. Furthermore, the coassociativity diagram for $c$ boils down to the "formal associativity" condition

$$\boldsymbol{\varphi}(T, \boldsymbol{\varphi}(U, V)) = \boldsymbol{\varphi}(\boldsymbol{\varphi}(T, U), V) \tag{10}$$

where $\boldsymbol{\varphi}(T, U)$ stands for the system $(\varphi_i(T, U))_{1 \leqslant i \leqslant n}$ of the right-hand sides of (9). We will see later (II, §2, No. 8) that, when such a system of $n$ formal power series in $2n$ indeterminates over $k$ is given, there exists automatically one and only one antipodism $a: B \to B$ defining B as an object of $\boldsymbol{CAlc}_k$. We will refer to the formal groups being obtained in this way as the "naive formal Lie groups of dimension $n$ over $k$" [3].

2) From any *algebraic group* G over an algebraically closed field $k$, one deduces a "naive formal Lie group over $k$" by the following natural "localization and completion" process. If $e$ is the neutral element of G and $\mathfrak{o}_e$ (resp. $\mathfrak{o}_{(e,e)}$) the local ring of G (resp. $G \times G$) at the neutral element, to the multiplication morphism $G \times G \to G$ there corresponds a local homomorphism of local rings $u: \mathfrak{o}_e \to \mathfrak{o}_{(e,e)}$, hence by completion the continuous extension of $u$, $\hat{u}: \hat{\mathfrak{o}}_e \to \hat{\mathfrak{o}}_{(e,e)}$. But $\mathfrak{o}_e$ is a regular ring of dimension

$n$, hence its completion $\hat{\mathfrak{o}}_e$ is isomorphic to $k[[T_1, T_2, \ldots, T_n]]$, and similarly $\hat{\mathfrak{o}}_{(e,e)}$ is isomorphic to $k[[T_1, \ldots, T_n, U_1, \ldots, U_n]]$.
If we express the associativity property of the multiplication in G, we find that $\hat{u}$ is a coproduct satisfying conditions (9) and (10) of example 1). We will say that the formal group thus obtained is the *formal Lie group associated to* G (cf. Chapter IV). When $G = \mathrm{Spec}(A)$ is an affine group over $k$, $\hat{\mathfrak{o}}_e$ is just the $\mathfrak{m}$-*adic completion* of the bigebra A (see §1, No. 6) for the maximal ideal $\mathfrak{m}$ of A corresponding to $e$. For instance, if $G = \mathbf{G}_a = \mathrm{Spec}(k[T])$, $\mathfrak{m}$ is the principal ideal (T), and $\hat{\mathfrak{o}}_e$ the $k$-algebra $k[[T]]$ of formal power series in one indeterminate.

The *multiplicative group* $\mathbf{G}_m$ over $k$ is the special case of §1, No. 6, Example III, for the choice of $M = \mathbf{Z}$ (additive group); then $\mathbf{G}_m = \mathrm{Spec}(k[T, T^{-1}])$ (the basis of $k[T, T^{-1}]$ over $k$, consisting of the $T^n$ for $n \in \mathbf{Z}$, being identified with the elements of the group $\mathbf{Z}$). The maximal ideal $\mathfrak{m}$ corresponding to the neutral element is here the principal ideal generated by $T - 1 = t$; as $k[T, T^{-1}]$ is contained in the local ring of the polynomial ring $k[t]$ at the maximal ideal $(t)$, it follows that $\hat{\mathfrak{o}}_e$ is again the $k$-algebra $k[[t]]$ of formal power series in one indeterminate, but here the comultiplication is defined by $c(t) = t + u + tu$.

**13.** *Cartier duality.* A linearly compact $k$-algebra $A \in \boldsymbol{Alc}_k$ may also be described as a linearly compact vector space over $k$, equipped in addition with a continuous multiplication (associative and commutative) and a unit. If $m_0 \colon A \times A \to A$ is the multiplication, it factorizes as

$$A \times A \to A \otimes_k A \xrightarrow{m} A$$

where $m$ is linear; in addition $m$ is *continuous* for the tensor product topology, for if $\mathfrak{J}$ is an open ideal in A, the image by $m$ of $A \otimes \mathfrak{J} + \mathfrak{J} \otimes A$ is contained in $\mathfrak{J}$; hence $m$ may be extended by continuity to a continuous linear mapping again written $m \colon A \hat{\otimes}_k A \to A$. Conversely, such a continuous linear mapping defines, by composition with the natural mapping $A \times A \to A \otimes_k A$, a multiplication on A, for which A becomes a commutative $k$-algebra, provided the associativity and commutativity conditions are satisfied, and that means that the diagrams

(11)
$$\begin{array}{ccc} A \hat{\otimes}_k A \hat{\otimes}_k A & \xrightarrow{m \hat{\otimes} 1} & A \hat{\otimes}_k A \\ \Big\downarrow{\scriptstyle 1 \hat{\otimes} m} & & \Big\downarrow{\scriptstyle m} \\ A \hat{\otimes}_k A & \xrightarrow[m]{} & A \end{array}$$

and

$$
(12)\qquad
\begin{array}{ccc}
A \mathbin{\hat{\otimes}_k} A & \xrightarrow{\ \sigma\ } & A \mathbin{\hat{\otimes}_k} A \\
 & {\scriptstyle m}\searrow \quad \swarrow{\scriptstyle m} & \\
 & A &
\end{array}
$$

are commutative, $\sigma$ being the natural isomorphism of $A \mathbin{\hat{\otimes}_k} A$ onto itself such that $\sigma(x \otimes y) = y \otimes x$.

We also note that, if $\varepsilon: k \to A$ is the unique $k$-linear mapping which sends the unit element of $k$ onto that of A, the diagram

$$
(13)\qquad
\begin{array}{ccc}
A \mathbin{\hat{\otimes}_k} k & \xrightarrow{\ 1 \hat{\otimes} \varepsilon\ } & A \mathbin{\hat{\otimes}_k} A \\
 & \searrow{\scriptstyle \sim} & \downarrow{\scriptstyle m} \\
 & & A
\end{array}
$$

is commutative (the oblique arrow being the isomorphism which sends $x \otimes \lambda$ to $\lambda x$ for $x \in A$ and $\lambda \in k$).

Observe now, that, as a vector space, A can be written as the *dual* $C^*$ of its topological dual C, and furthermore that $(C \otimes_k C)^*$ is identified to $A \mathbin{\hat{\otimes}_k} A$ (No. 4). If $\Delta = {}^t m$ is the transposed mapping of $m$ (No. 3), it is a linear mapping $\Delta: C \to C \otimes_k C$, and the commutative diagrams obtained from (11), (12), and (13) by taking the transposed of the linear mappings expresses that $\Delta$ is the *coproduct* for a structure of *coassociative* and *cocommutative cogebra* over $k$ on C, for which $\eta = {}^t\varepsilon: C \to k$ is a *counit*. Furthermore, if $u: A \to A'$ is a continuous homomorphism of commutative linearly compact $k$-algebras, it makes the diagram

$$
\begin{array}{ccc}
A \mathbin{\hat{\otimes}_k} A & \xrightarrow{\ u \hat{\otimes} u\ } & A' \mathbin{\hat{\otimes}_k} A' \\
{\scriptstyle m}\downarrow & & \downarrow{\scriptstyle m'} \\
A & \xrightarrow[\ u\ ]{} & A'
\end{array}
$$

commutative; this implies that if $A' = C'^*$, ${}^t u: C' \to C$ is a morphism of cogebras.

In other words, if to each $A \in \boldsymbol{Alc}_k$ we associate its dual cogebra C, and to each morphism $u: A \to A'$ of $\boldsymbol{Alc}_k$ the transposed ${}^t u: C' \to C$, we have established an *equivalence* of the dual category $(\boldsymbol{Alc}_k)^0$ and of the category $\boldsymbol{Cog}_k$ of *coassociative, cocommutative cogebras over k with counit.* It is clear that $k$ is the final object in $\boldsymbol{Cog}_k$, and that any two objects C′, C″ in that category, with coproducts $\Delta'$, $\Delta''$ and counits $\eta'$, $\eta''$, have a *product* in $\boldsymbol{Cog}_k$, which as a vector space over $k$ is $\mathrm{C}' \otimes_k \mathrm{C}''$, the coproduct $\Delta$ being defined by the composite

$$\mathrm{C}' \otimes \mathrm{C}'' \xrightarrow{\Delta' \otimes \Delta''} \mathrm{C}' \otimes \mathrm{C}' \otimes \mathrm{C}'' \otimes \mathrm{C}'' \xrightarrow{\tau} \mathrm{C}' \otimes \mathrm{C}'' \otimes \mathrm{C}' \otimes \mathrm{C}''$$

where $\tau$ is the natural linear mapping such that

$$\tau(x \otimes y \otimes z \otimes t) = x \otimes z \otimes y \otimes t.$$

Similarly, the counit $\eta$ in $\mathrm{C}' \otimes_k \mathrm{C}''$ is merely $\eta' \otimes \eta''$, when $k \otimes_k k$ is identified naturally with $k$.

We may therefore consider the category $\boldsymbol{GCog}_k$ of groups in the category $\boldsymbol{Cog}_k$ (§1, No. 1); it follows from No. 10 that an object in that category is a coassociative and cocommutative cogebra $\mathfrak{G}$ with counit, whose dual $\mathfrak{G}^* = \mathrm{B}$ belongs to $\boldsymbol{CAlc}_k$; by transposition, the mappings $c$, $\gamma$, and $a$ for B give linear mappings

$${}^t c = \mu : \mathfrak{G} \otimes_k \mathfrak{G} \to \mathfrak{G}, \qquad {}^t\gamma = e : k \to \mathfrak{G}, \qquad {}^t a = \alpha : \mathfrak{G} \to \mathfrak{G}.$$

Transposing the diagrams of §1, No. 5 for $c$ and $\gamma$ shows at once that $\mu$ defines on $\mathfrak{G}$ a structure of *associative k-algebra with unit* $e(1)$ (not commutative in general). In addition, the fact that $c$ is an *algebra* homomorphism of B into $\mathrm{B} \hat{\otimes} \mathrm{B}$ yields, by duality, the fact that $\mu$ must be a *cogebra* homomorphism of $\mathfrak{G} \otimes \mathfrak{G}$ into $\mathfrak{G}$; this means, by definition, that the coproduct $\Delta: \mathfrak{G} \to \mathfrak{G} \otimes \mathfrak{G}$ and the product $\mu$ must make the diagram

$$\begin{array}{ccccc}
\mathfrak{G} \otimes \mathfrak{G} & \xrightarrow{\mu} & \mathfrak{G} & \xrightarrow{\Delta} & \mathfrak{G} \otimes \mathfrak{G} \\
{\scriptstyle \Delta\otimes\Delta}\downarrow & & & & \uparrow{\scriptstyle \mu\otimes\mu} \\
\mathfrak{G} \otimes \mathfrak{G} \otimes \mathfrak{G} \otimes \mathfrak{G} & & \xleftrightarrow{\tau} & & \mathfrak{G} \otimes \mathfrak{G} \otimes \mathfrak{G} \otimes \mathfrak{G}
\end{array}$$

commutative; but the diagram also expresses the fact that $\Delta$ is an *algebra* homomorphism for the structure of algebra defined on $\mathfrak{G}$ by $\mu$ and its tensor product by itself on $\mathfrak{G} \otimes \mathfrak{G}$. In other words, the mappings $\mu$, $\Delta$, $e$, and $\eta$ define on $\mathfrak{G}$ a structure of coassociative, cocommutative *bigebra* with unit and counit. Furthermore, the commutative diagrams expressing that $a\colon \mathrm{B} \to \mathrm{B}$ is an antipodism (§1, No. 5) may be written in more detail as

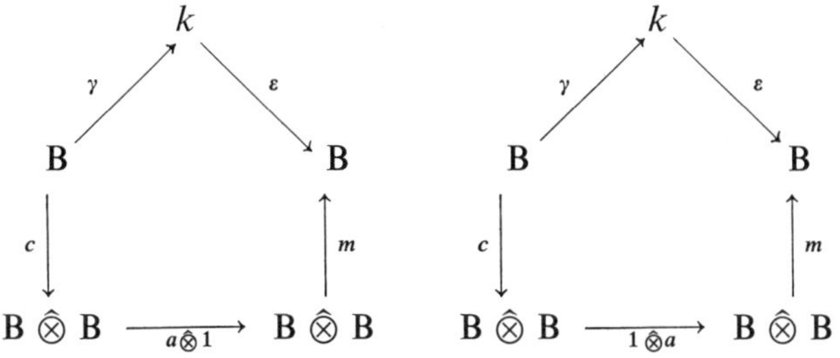


where $m$ and $\varepsilon$ are the multiplication and unit in B; by duality, they give the commutative diagrams

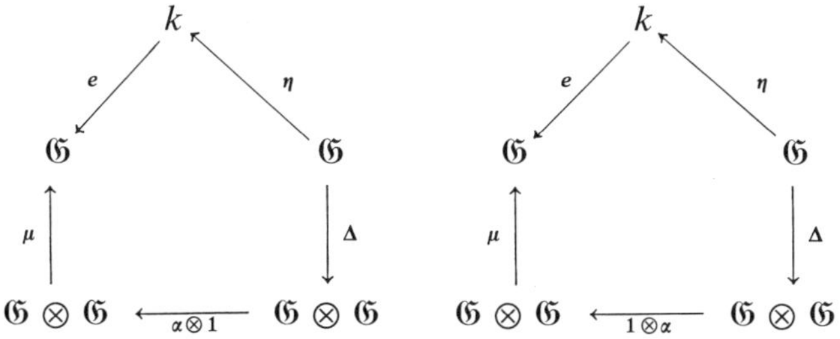


But it is clear that these diagrams express the fact that $\alpha$ is an *antipodism* for the bigebra $\mathfrak{G}$. Observe in passing that from its functorial definition it follows that $a^2 = \mathrm{Id.}$ and similarly $\alpha^2 = \mathrm{Id.}$; the antipodisms are *bijective* mappings.

The kernel $\mathfrak{G}^+$ of the counit $\eta\colon \mathfrak{G} \to k$ of the bigebra $\mathfrak{G}$ is a two-sided ideal in the algebra $\mathfrak{G}$, such that $\mathfrak{G} = k \cdot 1 \oplus \mathfrak{G}^+$, called the *augmentation ideal* of $\mathfrak{G}$. The same argument as in No. 11 proves the relation

(14)

$$\Delta(Z) = Z \otimes 1 + 1 \otimes Z + \Delta^+(Z) \quad \text{for} \quad Z \in \mathfrak{G}^+, \quad \text{with} \quad \Delta^+(Z) \in \mathfrak{G}^+ \otimes \mathfrak{G}^+.$$

Finally, the morphisms in $\boldsymbol{CAlc}_k$ are the continuous linear mappings $\mathrm{B} \to \mathrm{B}'$ which are *bigebra* homomorphisms, that is, homomorphisms for both the algebra and cogebra structures; this implies at once that the morphisms in $\boldsymbol{GCog}_k$ are also the *bigebra* homomorphisms $\mathfrak{G} \to \mathfrak{G}'$. The equivalence thus obtained between the categories $\boldsymbol{GCog}_k$ and the dual category $(\boldsymbol{CAlc}_k)^0$ is called *Cartier duality.*

**14.** To summarize what we have done up to now, let us agree that whenever we speak of a bigebra, it will be understood that it is associative, coassociative and has a unit and a counit. We have defined essentially *three* categories:

I) The category $\boldsymbol{GCog}_k$ of (discrete) *cocommutative bigebras with antipodism*, where the morphisms are the bigebra homomorphisms.

II) The category $\boldsymbol{CAlc}_k$ of *linearly compact commutative bigebras with antipodism*, where the morphisms are the continuous bigebra homomorphisms.

III) The category of *functors*

$$\mathbf{G} : \mathrm{A} \mapsto \mathrm{Hom}_{\boldsymbol{Alc}_k}(\mathrm{B}, \mathrm{A})$$

from $\boldsymbol{Alc}_k$ to the category of groups $\boldsymbol{Gr}$, where B belongs to $\boldsymbol{CAlc}_k$; the morphisms of that category are the *functorial morphisms* $\mathbf{u}: \mathbf{G} \to \mathbf{G}'$, defined as associating to each $\mathrm{A} \in \boldsymbol{Alc}_k$ the group homomorphism

$$\mathrm{Hom}_{\boldsymbol{Alc}_k}(\mathrm{B}, \mathrm{A}) \to \mathrm{Hom}_{\boldsymbol{Alc}_k}(\mathrm{B}', \mathrm{A}) \quad \text{equal to } \mathrm{Hom}({}^t u, 1_{\mathrm{A}}),$$

where ${}^t u: \mathrm{B}' \to \mathrm{B}$ is a continuous bigebra homomorphism.

The first and third of these categories are equivalent, and also equivalent to the dual of the second. Nevertheless, it will be useful to work with all three categories at the same time.

To fix a simple terminology, we shall say that the objects of the third category are *formal groups over* $k$, and the morphisms of that category *formal homomorphisms.* To each formal group $\mathbf{G}$ there is associated a bigebra $\mathfrak{G} \in \boldsymbol{GCog}_k$, which we shall call the *covariant bigebra* of $\mathbf{G}$, and its Cartier dual $\mathfrak{G}^* \in \boldsymbol{CAlc}_k$, which we shall call the *contravariant bigebra* of $\mathbf{G}$; to each formal homomorphism $\mathbf{u}: \mathbf{G} \to \mathbf{G}'$ is associated a bigebra homomorphism $u: \mathfrak{G} \to \mathfrak{G}'$ of the covariant bigebras of $\mathbf{G}$ and $\mathbf{G}'$, and its transposed ${}^t u: \mathfrak{G}'^* \to \mathfrak{G}^*$ which is a continuous bigebra homomorphism.

**15.** It is clear that for any bigebra $\mathfrak{G} \in \boldsymbol{GCog}_k$ and any extension K of $k$, we may deduce a bigebra $\mathfrak{G}_{(\mathrm{K})} = \mathfrak{G} \otimes_k \mathrm{K}$ in $\boldsymbol{GCog}_{\mathrm{K}}$ by defining as usual the multiplication on $\mathfrak{G}_{(\mathrm{K})}$ and taking for comultiplication $\Delta \otimes 1$ after identification of $\mathfrak{G}_{(\mathrm{K})} \otimes_{\mathrm{K}} \mathfrak{G}_{(\mathrm{K})}$ with $(\mathfrak{G} \otimes_k \mathfrak{G}) \otimes_k \mathrm{K}$; the counit and the antipodism are extended trivially.

## §3. Elementary theory of formal groups

**1.** *Points of formal groups.* Let $\mathbf{G}$ be a formal group over $k$, $\mathfrak{G}$ and $\mathfrak{G}^*$ its covariant and contravariant bigebras. Using the same terminology as for affine group schemes (§1, No. 6) we shall say that, for any linearly compact commutative $k$-algebra A, an element of the group $\mathbf{G}_{\mathrm{A}} = \mathrm{Hom}_{\boldsymbol{Alc}_k}(\mathfrak{G}^*, \mathrm{A})$, in other words, a continuous $k$-algebra homomorphism $x: \mathfrak{G}^* \to \mathrm{A}$, is *a point of* $\mathbf{G}$ *with values in* A. Let A′ be a second $k$-algebra in $\boldsymbol{Alc}_k$, and let $f: \mathrm{A} \to \mathrm{A}'$ be a continuous $k$-algebra homomorphism; then $f \circ x: \mathfrak{G}^* \to \mathrm{A}'$ is a point of $\mathbf{G}$ with values in A′, which we call the *specialization of* $x$ *by* $f$; the mapping $x \mapsto f \circ x$ of $\mathbf{G}_{\mathrm{A}}$ into $\mathbf{G}_{\mathrm{A}'}$, which is a homomorphism of groups, is called the *specialization* defined by $f$.

Let now $\mathbf{u}: \mathbf{G} \to \mathbf{G}'$ be a formal homomorphism, ${}^t u: \mathfrak{G}'^* \to \mathfrak{G}^*$ the corresponding continuous bigebra homomorphism; then $x \circ {}^t u$ is a point of $\mathbf{G}'$ with values in A, which we denote by $\mathbf{u}_{\mathrm{A}}(x)$, and call the *image* of $x$ by $\mathbf{u}_{\mathrm{A}}$ (or $\mathbf{u}$); the mapping $x \mapsto \mathbf{u}_{\mathrm{A}}(x)$ is a homomorphism of the group $\mathbf{G}_{\mathrm{A}}$ into the group $\mathbf{G}'_{\mathrm{A}}$; with the same notations as above, one has

$$f \circ \mathbf{u}_{\mathrm{A}}(x) = \mathbf{u}_{\mathrm{A}'}(f \circ x).$$

**2.** We will say that a point $x \in \mathbf{G}_{\mathrm{A}}$ is *generic* if for any $\mathrm{A}' \in \boldsymbol{Alc}_k$, *any* point of $\mathbf{G}_{\mathrm{A}'}$ is a specialization of $x$. For instance, the *identity mapping* $1_{\mathfrak{G}^*}: \mathfrak{G}^* \to \mathfrak{G}^*$ is a generic point of $\mathbf{G}$ (called the *canonical point* in $\mathbf{G}_{\mathfrak{G}^*}$).

Several generic points $x_1, \ldots, x_n$ of $\mathbf{G}$ with values in A are said to be *independent* if for any sequence $(y_i)_{1 \leqslant i \leqslant n}$ of points of $\mathbf{G}$ with values in an algebra $\mathrm{A}' \in \boldsymbol{Alc}_k$, there exists a continuous $k$-algebra homomorphism $f: \mathrm{A} \to \mathrm{A}'$ such that $y_i = f \circ x_i$ for $1 \leqslant i \leqslant n$.

The existence of such systems of independent generic points, for any $n$, is easily established: more generally, if $\mathbf{G}_1, \ldots, \mathbf{G}_n$ are formal groups, and if

$$\mathfrak{G}^* = \mathfrak{G}_1^* \mathbin{\hat{\otimes}} \mathfrak{G}_2^* \mathbin{\hat{\otimes}} \cdots \mathbin{\hat{\otimes}} \mathfrak{G}_n^*,$$

if for each $i$ we denote by $x_i$ the natural homomorphism

$$t_i \mapsto 1 \otimes \cdots \otimes 1 \otimes t_i \otimes 1 \otimes \cdots \otimes 1 \qquad (t_i \text{ in the } i\text{-th slot})$$

of $\mathfrak{G}_i^*$ into $\mathfrak{G}^*$, the universal property of completed tensor products (§2, No. 10) proves the existence, for any system of homomorphisms $y_i: \mathfrak{G}_i^* \to A'$, of a continuous homomorphism $f: \mathfrak{G}^* \to A'$ such that $y_i = f \circ x_i$ for $1 \leqslant i \leqslant n$. The $x_i$ are called the *canonical points* of the $\mathbf{G}_i$ with values in $\mathfrak{G}^*$.

In particular, for any formal group $\mathbf{G}$, we obtain two independent *canonical* generic points $x_1$, $x_2$ of $\mathbf{G}$ with values in $\mathfrak{G}^* \hat{\otimes} \mathfrak{G}^*$; it follows at once from the definitions that $x_1 \otimes x_2$ is the mapping sending $t_1 \otimes t_2$ to $(t_1 \otimes 1) \otimes (1 \otimes t_2)$, and therefore (§1, formula (1)), $x_1 x_2$ is simply the *coproduct* $c: \mathfrak{G}^* \to \mathfrak{G}^* \hat{\otimes} \mathfrak{G}^*$.

This last result enables one to characterize the continuous algebra homomorphisms ${}^t u: \mathfrak{G}'^* \to \mathfrak{G}^*$ which correspond to formal homomorphisms $\mathbf{u}: \mathbf{G} \to \mathbf{G}'$. One must have by definition

$$(x_1 x_2) \circ {}^t u = (x_1 \circ {}^t u)\,(x_2 \circ {}^t u);$$

conversely, that condition is sufficient, for if $y_1, y_2$ are two points of $\mathbf{G}$ with values in $A \in \boldsymbol{Alc}_k$, there is a continuous $k$-algebra homomorphism $f: \mathfrak{G}^* \hat{\otimes} \mathfrak{G}^* \to A$ such that $y_i = f \circ x_i$ for $i = 1, 2$, and therefore

$$\begin{aligned}(y_1 y_2) \circ {}^t u &= f \circ (x_1 x_2) \circ {}^t u = f \circ ((x_1 \circ {}^t u)\,(x_2 \circ {}^t u)) \\ &= (f \circ x_1 \circ {}^t u)\,(f \circ x_2 \circ {}^t u) = (y_1 \circ {}^t u)\,(y_2 \circ {}^t u),\end{aligned}$$

which proves that $y \mapsto y \circ {}^t u$ is a homomorphism of $\mathbf{G}_A$ into $\mathbf{G}'_A$.

**3.** A *subcogebra* of $\mathfrak{G}$ is a vector subspace C such that $\Delta(C) \subset C \otimes C$, so that the restriction of $\Delta$ to C defines on C a structure of cogebra; an equivalent condition is that the orthogonal $C^0$ be a *closed ideal* in the commutative $k$-algebra $\mathfrak{G}^*$ (§2, No. 4); as an arbitrary intersection of closed ideals in $\mathfrak{G}^*$ is a closed ideal, an arbitrary sum of subcogebras in $\mathfrak{G}$ is a subcogebra. Similarly, the closure of an arbitrary sum of closed ideals in $\mathfrak{G}^*$ is a closed ideal, hence an arbitrary intersection of subcogebras of $\mathfrak{G}$ is a subcogebra.

A *coideal* of $\mathfrak{G}$ is a vector subspace L such that $\Delta(L) \subset \mathfrak{G} \otimes L + L \otimes \mathfrak{G}$; passing to the quotient shows that $\Delta$ defines on $\mathfrak{G}/L$ a structure of cocommutative cogebra. An equivalent condition is that the orthogonal $L^0$ be a *closed subalgebra* of $\mathfrak{G}^*$ (§2, No. 4); intersections of closed subalgebras of $\mathfrak{G}^*$ being closed subalgebras, we deduce that sums of coideals of $\mathfrak{G}$ are coideals.

Similarly, to say that A is a *subalgebra* of $\mathfrak{G}$ is equivalent to saying that $A^0$ is a *closed coideal* of $\mathfrak{G}^*$ (i.e., $c(A^0) \subset A^0 \hat{\otimes} \mathfrak{G}^* + \mathfrak{G}^* \hat{\otimes} A^0$): to say that I is a *two-sided ideal* of G is equivalent to saying that $I^0$ is a *closed subcogebra* of $\mathfrak{G}^*$ (i.e., $c(I^0) \subset I^0 \hat{\otimes} I^0$).

A *subbigebra* of $\mathfrak{G}$ is both a subalgebra and a subcogebra B; the restriction of $\mu$ to $B \otimes B$ and of $\Delta$ to B then define on B a structure of cocommutative bigebra, and the restriction to B of the counit $\eta$ is a counit for B. If C is a subcogebra of $\mathfrak{G}$, the subalgebra of $\mathfrak{G}$ generated by C is a subcogebra, since $\Delta(xy) = \Delta(x)\Delta(y)$, hence it is a *subbigebra.* This implies that if A is a subalgebra of $\mathfrak{G}$, the largest subcogebra contained in A (i.e., the sum of all subcogebras contained in A) is a subbigebra. We will say that a subbigebra $\mathfrak{H} \subset \mathfrak{G}$ is a *covariant subbigebra* of $\mathfrak{G}$ if in addition it contains the unit of $\mathfrak{G}$, and the antipodism $\alpha$ of $\mathfrak{G}$ is such that $\alpha(\mathfrak{H}) \subset \mathfrak{H}$; it is then clear that the restriction of $\alpha$ to $\mathfrak{H}$ is an antipodism (§2, No. 13), and therefore $\mathfrak{H} \in \boldsymbol{GCog}_k$. It is equivalent to say that $\mathfrak{H}^0$ is a *closed biideal* (i.e., both ideal and coideal) of $\mathfrak{G}^*$, stable under the antipodism *a* of $\mathfrak{G}^*$, and contained in the "augmentation" biideal $\mathfrak{G}^{*+}$, kernel of the counit $\gamma$ of $\mathfrak{G}^*$ ($\mathfrak{G}^{*+}$ is thus the largest (stable under $\alpha$) closed nontrivial biideal of $\mathfrak{G}^*$, the orthogonal of the smallest covariant bigebra $k \cdot 1$ of $\mathfrak{G}$); such biideals will be called *contravariant biideals*; by passage to the quotient, $\mathfrak{G}^*/\mathfrak{H}^0$ becomes a contravariant bigebra in $\boldsymbol{CAlc}_k$, which is naturally isomorphic to the dual $\mathfrak{H}^*$ of $\mathfrak{H}$.

A *biideal* of $\mathfrak{G}$ is both a two-sided ideal and a coideal I; this is equivalent to saying that its orthogonal $I^0$ is a *closed subbigebra* of $\mathfrak{G}^*$. We will say that $I^0$ is a *contravariant subbigebra* of $\mathfrak{G}^*$ if in addition it contains the unit of $\mathfrak{G}^*$ and is stable under the antipodism *a* of $\mathfrak{G}^*$; this is equivalent to saying that I is stable under the antipodism $\alpha$ of $\mathfrak{G}$ and is contained in the biideal $\mathfrak{G}^+$, kernel of the counit of $\mathfrak{G}$ (and which is obviously the largest such biideal of $\mathfrak{G}$, the orthogonal of the smallest contravariant subbigebra $k \cdot 1$ of $\mathfrak{G}^*$); by passage to the quotient, $\mathfrak{G}/I$ becomes a covariant bigebra in $\boldsymbol{GCog}_k$, whose dual is naturally isomorphic to $I^0$.

**4.** *Formal subgroups and quotient groups.* Let us now consider a formal homomorphism $\mathbf{u}: \mathbf{G} \to \mathbf{G}'$ of formal groups; it is clear from the definition of a bigebra homomorphism that the image $\mathfrak{H} = u(\mathfrak{G})$ is a subbigebra of $\mathfrak{G}'$, containing the unit; in addition, the fact that, for any $A \in \boldsymbol{Alc}_k$,

$$\mathbf{u}_A(x^{-1}) = (\mathbf{u}_A(x))^{-1} \qquad \text{for any point} \quad x \in \mathbf{G}_A,$$

implies that $u$ commutes with the antipodisms of $\mathfrak{G}$ and $\mathfrak{G}'$, and therefore $\mathfrak{H}$ is a *covariant subbigebra* of $\mathfrak{G}'$. For any algebra $A \in \boldsymbol{Alc}_k$, there is a natural identification of the image $\mathbf{u}_A(\mathbf{G}_A)$ in $\mathbf{G}_A$ to a subgroup of the group

$$\mathbf{H}_A = \mathrm{Hom}_{\boldsymbol{Alc}_k}(\mathfrak{H}^*, A);$$

indeed, the homomorphism $u$ factorizes into $\mathfrak{G} \xrightarrow{v} \mathfrak{H} \xrightarrow{j} \mathfrak{G}'$, where $v$ is surjective and $j$ injective: the dual $\mathfrak{H}^*$ is naturally identified to the quotient bigebra $\mathfrak{G}'^*/\mathfrak{H}^0$, ${}^tj\colon \mathfrak{G}'^* \to \mathfrak{G}'^*/\mathfrak{H}^0$ being the natural surjective homomorphism, and ${}^tv\colon \mathfrak{G}'^*/\mathfrak{H}^0 \to \mathfrak{G}^*$ an injective homomorphism. The image $\mathbf{u}_A(\mathbf{G}_A)$ consists of the continuous homomorphisms $\mathfrak{G}'^* \to A$ which can be written $f \circ {}^tu$, where $f$ is a continuous homomorphism of $\mathfrak{G}^*$ into A; such a homomorphism vanishes on ${}^tu^{-1}(0) = \mathfrak{H}^0$, hence is naturally identified with an algebra homomorphism of $\mathfrak{H}^*$ into A. Whether $\mathbf{u}_A(\mathbf{G}_A)$ is equal to $\mathbf{H}_A$ or not depends on the possibility of extending a continuous algebra homomorphism of $\mathfrak{H}^* = \mathfrak{G}'^*/\mathfrak{H}^0$ into A to a continuous homomorphism of $\mathfrak{G}^*$ into A ($\mathfrak{G}'^*/\mathfrak{H}^0$ being identified to a subbigebra of $\mathfrak{G}^*$) (cf. II, §3, No. 4, Proposition 8). If $u = j$ is injective, $\mathfrak{G}'^*/\mathfrak{H}^0$ is equal to $\mathfrak{G}^*$, so the preceding extension is trivial. In other words, if $\mathfrak{G}$ is a covariant subbigebra of $\mathfrak{G}'$, then $\mathbf{G}_A$ is (functorially) a subgroup of $\mathbf{G}'_A$; we are therefore justified is defining a *subgroup* of a formal group $\mathbf{G}'$ as a formal group $\mathbf{G}$ corresponding to a *covariant subbigebra* $\mathfrak{G}$ of $\mathfrak{G}'$. With the preceding notations, we will say that the subgroup $\mathbf{H}$ corresponding to the subbigebra $\mathfrak{H} = u(\mathfrak{G})$ of $\mathfrak{G}'$ is the *image* of $\mathbf{G}$ by $\mathbf{u}$, and that $\mathbf{u}$ is *surjective* if $u(\mathfrak{G}) = \mathfrak{G}'$.

**5.** Keeping the same notations as in No. 4, let us now consider the *kernel* of $\mathbf{u}_A$: it consists of the elements $f \in \mathbf{G}_A$ such that $f \circ {}^tu = e'_A$, neutral element of $\mathbf{G}'_A$. That relation means that $f({}^tu(x')) = 0$ for all elements in the "augmentation" biideal $\mathfrak{G}'^{*+}$ of $\mathfrak{G}'^*$. Let ${}^tu(\mathfrak{G}'^*) = I^0$, which is a closed contravariant subbigebra of $\mathfrak{G}^*$, orthogonal to the biideal I, kernel of the bigebra homomorphism $u$; the intersection $I^0 \cap \mathfrak{G}'^{*+}$ is the augmentation ideal $I^{0+}$ of the bigebra $I^0$. Let $\mathfrak{N}^0$ be the closed ideal in $\mathfrak{G}'^*$ generated by $I^{0+}$; it is readily verified that $\mathfrak{N}^0$ is a contravariant biideal; then, to say that $f$ vanishes on $I^{0+}$ is equivalent to saying that it vanishes in $\mathfrak{N}^0$; in other words, there is a (functorial) isomorphism of the kernel of $\mathbf{u}_A$ onto the group $\mathrm{Hom}_{\boldsymbol{Alc}_k}(\mathfrak{G}'^*/\mathfrak{N}^0, A)$. Now $\mathfrak{N}^0$ is the orthogonal of a covariant subbigebra $\mathfrak{N}$ of $\mathfrak{G}$, which may be characterized as the *largest* covariant subbigebra such that $\mathfrak{N}^+$ is *contained in the kernel* I of $u$. We are therefore justified in saying that the formal group $\mathbf{N}$ corresponding to the bigebra $\mathfrak{N}$ is the *kernel* of the formal homomorphism $\mathbf{u}$; $\mathbf{N}_A$ is then identified to the kernel of $\mathbf{u}_A$. If we now return to the image $\mathfrak{H} = u(\mathfrak{G})$, we observe that it is a covariant subbigebra of $\mathfrak{G}'$ isomorphic to the quotient bigebra $\mathfrak{G}/I$ of $\mathfrak{G}$; we will say that the formal group corresponding to $\mathfrak{G}/I$ is the *quotient* of the formal group $\mathbf{G}$ by the formal subgroup $\mathbf{N}$, kernel of $\mathbf{u}$, and write it $\mathbf{G}/\mathbf{N}$. The covariant bigebra corresponding to $\mathbf{G}/\mathbf{N}$ is $\mathfrak{G}/I$, and its contravariant bigebra is $I^0$. To say that $\mathbf{u}_A$ is injective for any $A \in \boldsymbol{Alc}_k$ means that $I = \{0\}$, or equivalently, $\mathfrak{N} = k \cdot 1$; we then say that the formal homomorphism $\mathbf{u}$ is *injective*. With the same notations, for an arbitrary formal homomorphism $\mathbf{u}: \mathbf{G} \to \mathbf{G}'$, there is a canonical factorization

$$\mathbf{G} \xrightarrow{\mathbf{p}} \mathbf{G}/\mathbf{N} \xrightarrow{\mathbf{g}} \mathbf{H} \xrightarrow{\mathbf{j}} \mathbf{G}'$$

where $\mathbf{p}$ is surjective, $\mathbf{g}$ an isomorphism, and $\mathbf{j}$ injective.

**6.** The fact that a subbigebra $\mathfrak{N}$ of $\mathfrak{G}$ is such that the corresponding formal group $\mathbf{N}$ is the kernel of a formal homomorphism $\mathbf{u}: \mathbf{G} \to \mathbf{G}'$ implies that, for any $A \in \boldsymbol{Alc}_k$, $\mathbf{N}_A = \mathrm{Hom}(\mathfrak{N}^*, A)$ is an *invariant* subgroup of $\mathbf{G}_A$. This can be expressed in a different way, using the concept of generic points (No. 2): let $\mathbf{i}$ be the natural formal homomorphism $\mathbf{N} \to \mathbf{G}$, corresponding to the natural injection $i$ of $\mathfrak{N}$ into $\mathfrak{G}$, and let $x$ and $t$ be the canonical points of $\mathbf{G}$ and $\mathbf{N}$ in the algebra $A_0 = \mathfrak{G}^* \hat{\otimes} \mathfrak{N}^*$; as $\mathbf{i}(t)$ is a point of $\mathbf{G}$ with values in $\mathfrak{G}^* \hat{\otimes} \mathfrak{N}^*$, $x\mathbf{i}(t)x^{-1}$ belongs to the invariant subgroup $\mathbf{N}_{A_0}$ of $\mathbf{G}_{A_0}$. Conversely, if this condition is satisfied, $\mathbf{N}_A$ is an invariant subgroup of $\mathbf{G}_A$ for *any* algebra $A \in \boldsymbol{Alc}_k$, for if $y \in \mathbf{G}_A$, $z \in \mathbf{N}_A$, there is a continuous homomorphism $f: A_0 \to A$ such that $y = f \circ x, z = f \circ t$, hence $\mathbf{i}_A(z) = f \circ \mathbf{i}(t)$, and $y\mathbf{i}_A(z)y^{-1}$ is the specialization by $f$ of $x\mathbf{i}(t)x^{-1}$, hence belongs to $\mathbf{N}_A$, which is therefore an invariant subgroup of $\mathbf{G}_A$.

We will say that a covariant subbigebra $\mathfrak{N}$ of $\mathfrak{G}$ is *invariant* in $\mathfrak{G}$ if the corresponding formal subgroup $\mathbf{N}$ is such that $\mathbf{N}_A$ is invariant in $\mathbf{G}_A$ for any $A \in \boldsymbol{Alc}_k$ (in which case we will also say that $\mathbf{N}$ is *invariant* in $\mathbf{G}$). These subbigebras may also be defined as the largest subbigebras contained in vector spaces of the type $k \cdot 1 \oplus I$, where I is a covariant biideal in $\mathfrak{G}$.

The condition $x\mathbf{i}(t)x^{-1} \in \mathbf{N}_{A_0}$ implies the following property: *the left and right ideals of the algebra $\mathfrak{G}$ generated by the augmentation ideal $\mathfrak{N}^+$ of $\mathfrak{N}$ are identical.* Indeed, the assumption means that there exists a continuous algebra endomorphism $f$ of $A_0 = \mathfrak{G}^* \hat{\otimes} \mathfrak{N}^*$, such that

$$f \circ \mathbf{i}(t) = x\mathbf{i}(t)x^{-1},$$

or equivalently

$$x\mathbf{i}(t) = (f \circ \mathbf{i}(t))x.$$

Now it is readily verified that $x = {}^t\rho$, $\mathbf{i}(t) = {}^t\sigma$, where

$$\rho(Z \otimes U) = \eta(U)Z \quad \text{and} \quad \sigma(Z \otimes U) = \eta(Z)U \quad \text{for } Z \in \mathfrak{G} \text{ and } U \in \mathfrak{N}.$$

If $f = {}^tg$, where $g$ is an endomorphism of the cogebra $\mathfrak{G} \otimes \mathfrak{N}$, the preceding relation is equivalent to

$$\mu \circ (\sigma \otimes \rho) \circ \Delta = \mu \circ (\rho \otimes (\sigma \circ g)) \circ \Delta,$$

where $\Delta$ is the coproduct in the cogebra $\mathfrak{G} \otimes \mathfrak{N}$. From Eq. (14) in §2, No. 13, and the definition of the coproduct in a tensor product of cogebras (§2, No. 13) it follows immediately that, for $Z \in \mathfrak{G}^+$ and $U \in \mathfrak{N}^+$, the only term in $\Delta(Z \otimes U)$ which gives a nonzero value in $(\sigma \otimes \rho)(\Delta(Z \otimes U))$ is $1 \otimes U \otimes Z \otimes 1$, and

$$\mu((\sigma \otimes \rho)(\Delta(Z \otimes U))) = UZ.$$

Similarly the image of $Z \otimes U$ by $\mu \circ (\rho \otimes (\sigma \circ g)) \circ \Delta$ has the form $\sum_i Z_i U_i$, with $Z_i \in \mathfrak{G}$ and $U_i \in \mathfrak{N}^+$, which proves that the right ideal generated by $\mathfrak{N}^+$ is contained in the left ideal generated by $\mathfrak{N}^+$; the converse is seen by changing $x$ to $x^{-1}$.

It is readily verified that the two-sided ideal thus obtained in $\mathfrak{G}$ is in fact a covariant biideal contained in I, but it is not obvious that it should be identical to I. We will show that such is the case for some types of formal groups (Chapter II, §3, No. 3).

**7.** Let $\mathfrak{H}$ and $\mathfrak{H}'$ be two covariant subbigebras in $\mathfrak{G}$, $\mathbf{H}$ and $\mathbf{H}'$ the corresponding formal subgroups of $\mathbf{G}$. Then we have $\mathbf{H}_A \subset \mathbf{H}'_A$ (subgroups of $\mathbf{G}_A$) for every $A \in \boldsymbol{Alc}_k$ if and only if $\mathfrak{H} \subset \mathfrak{H}'$. Indeed, that condition is trivially sufficient. Conversely, if $\mathfrak{H}$ is not contained in $\mathfrak{H}'$, then $\mathfrak{H}'^0$ is not contained in $\mathfrak{H}^0$; taking $A = \mathfrak{H}^* = \mathfrak{G}^*/\mathfrak{H}^0$, the identity mapping of $\mathfrak{H}^*$

(identified to the natural mapping of $\mathfrak{G}^*$ onto $\mathfrak{G}^*/\mathfrak{H}^0$) is not 0 on $\mathfrak{H}'^0$, and therefore does not belong to $\mathbf{H}'_A$.

If we define the order relation $\mathbf{H} \subset \mathbf{H}'$ for formal subgroups of $\mathbf{G}$ as meaning $\mathbf{H}_A \subset \mathbf{H}'_A$ for any $A \in \mathbf{Alc}_k$, we thus see that the natural bijection of the set of covariant subbigebras of $\mathfrak{G}$ onto the set of formal subgroups of $\mathbf{G}$ is order preserving. As the covariant subbigebras of $\mathfrak{G}$ form a complete lattice for the inclusion relation (No. 3), the same is true for formal subgroups of $\mathbf{G}$, and we may therefore speak of $\inf(\mathbf{H}_\lambda)$ and $\sup(\mathbf{H}_\lambda)$ for any family of formal subgroups $\mathbf{H}_\lambda$ of $\mathbf{G}$. Furthermore, we have, for any $A \in \mathbf{Alc}_k$

$$(1) \qquad (\inf(\mathbf{H}_\lambda))_A = \bigcap_\lambda (\mathbf{H}_\lambda)_A.$$

Indeed, if $\mathfrak{H}_\lambda$ is the covariant subbigebra of $\mathfrak{G}$ corresponding to $\mathbf{H}_\lambda$, $\mathfrak{H} = \bigcap_\lambda \mathfrak{H}_\lambda$ is a covariant subbigebra and corresponds to $\inf(\mathbf{H}_\lambda)$; $\mathfrak{H}^0$ is therefore the closed *vector subspace* of $\mathfrak{G}^*$ generated by the union of the subspaces $\mathfrak{H}_\lambda^0$; the intersection of the groups $(\mathbf{H}_\lambda)_A = \mathrm{Hom}_{\mathbf{Alc}_k}(\mathfrak{G}^*/\mathfrak{H}_\lambda^0, A)$ consists of the continuous homomorphisms $\mathfrak{G}^* \to A$ which vanish on each $\mathfrak{H}_\lambda^0$, hence also on $\mathfrak{H}^0$; as the converse is trivial, this proves (1). It follows at once from (1) and from the characterization of invariant formal subgroups (No. 6) that if $(\mathbf{N}_\lambda)$ is a family of invariant formal subgroups of $\mathbf{G}$, $\inf(\mathbf{N}_\lambda)$ is also an invariant formal subgroup of $\mathbf{G}$.

I do not know if $(\sup(\mathbf{H}_\lambda))_A$ is the subgroup of $\mathbf{G}_A$ generated by the $(\mathbf{H}_\lambda)_A$ for all $A \in \mathbf{Alc}_k$.

**8.** Let $\mathbf{H}'$, $\mathbf{H}''$ be two formal subgroups of a formal group $\mathbf{G}$, and let $\mathfrak{H}'$, $\mathfrak{H}''$ be the corresponding subbigebras in the covariant bigebra $\mathfrak{G}$ of $\mathbf{G}$. Assume that for any $A \in \mathbf{Alc}_k$, the subgroups $\mathbf{H}'_A$, $\mathbf{H}''_A$ of the group $\mathbf{G}_A$ are such that $\mathbf{H}'_A\mathbf{H}''_A = \mathbf{H}'_A\mathbf{H}''_A$. Then, if we denote by $\mathfrak{H}'\mathfrak{H}''$ the vector subspace of $\mathfrak{G}$ generated by the products $Z'Z''$ with $Z' \in \mathfrak{H}'$ and $Z'' \in \mathfrak{H}''$, we have $\mathfrak{H}'\mathfrak{H}'' = \mathfrak{H}''\mathfrak{H}'$, and that vector subspace $\mathfrak{H}$ is in fact a *covariant subbigebra* of $\mathfrak{G}$, equal to $\sup(\mathfrak{H}', \mathfrak{H}'')$. To see this, let $i'$ and $i''$ be the natural injections of $\mathfrak{H}'$, $\mathfrak{H}''$ into $\mathfrak{G}$; let $x'$ and $x''$ be the canonical points of $\mathbf{H}'$ and $\mathbf{H}''$ in the algebra $A_0 = \mathfrak{H}'^* \hat{\otimes} \mathfrak{H}''^*$ (No. 2); then any point in $\mathbf{H}'_A\mathbf{H}''_A$ is a specialization of the point $\mathbf{i}'(x')\mathbf{i}''(x'')$ of $\mathbf{G}_{A_0}$, and in particular this is true for the point $\mathbf{i}''(x'')\mathbf{i}'(x')$ of $\mathbf{H}''_{A_0}\mathbf{H}'_{A_0} = \mathbf{H}'_{A_0}\mathbf{H}''_{A_0}$. This means that there is a continuous endomorphism $f$ of $A_0$ such that

$$f \circ (\mathbf{i}'(x')\mathbf{i}''(x'')) = \mathbf{i}''(x'')\mathbf{i}'(x');$$

now we have, as in No. 6, $x' = {}^t\rho$, $x'' = {}^t\sigma$, where

$$\rho(Z' \otimes Z'') = \eta(Z'')Z' \quad \text{and} \quad \sigma(Z' \otimes Z'') = \eta(Z')Z'' \quad \text{for} \quad Z' \in \mathfrak{H}', \quad Z'' \in \mathfrak{H}'';$$

if we write $f = {}^t g$, where $g$ is an endomorphism of the cogebra $\mathfrak{H}' \otimes \mathfrak{H}''$, the previous relation is equivalent to

$$\mu \circ (\sigma \otimes \rho) \circ \Delta = (\mu \circ (\rho \otimes \sigma) \circ \Delta) \circ g.$$

Computing the values of both sides for $Z' \otimes Z''$, with $Z' \in \mathfrak{H}'^{+}$ and $Z'' \in \mathfrak{H}''^{+}$, we find a relation of the form

$$Z''Z' = \sum_j Z_j'Z_j'' \quad \text{with} \quad Z_j' \in \mathfrak{H}'^{+} \quad \text{and} \quad Z_j'' \in \mathfrak{H}''^{+};$$

this proves that $\mathfrak{H}''\mathfrak{H}' \subset \mathfrak{H}'\mathfrak{H}''$, and the converse inclusion is proved similarly. This proves that the vector space $\mathfrak{H} = \mathfrak{H}'\mathfrak{H}''$ is a *subalgebra* of $\mathfrak{G}$; furthermore, $\mathfrak{H}' + \mathfrak{H}''$, as a sum of cogebras, is a cogebra, and the *algebra* it generates (equal to $\mathfrak{H}$) is a covariant subbigebra of $\mathfrak{G}$ (No. 3), which is obviously the smallest one containing both $\mathfrak{H}'$ and $\mathfrak{H}''$. In this case, it is clear that

$$\mathbf{H}_A = \mathbf{H}'_A\mathbf{H}''_A = \sup(\mathbf{H}'_A, \mathbf{H}''_A).$$

**9.** An important case of the situation considered in No. 8 is that of a subgroup $\mathbf{H}$ and an *invariant* subgroup $\mathbf{N}$ of $\mathbf{G}$, since in that case $\mathbf{H}_A\mathbf{N}_A = \mathbf{N}_A\mathbf{H}_A$ for every A, $\mathbf{N}_A$ being an invariant subgroup of $\mathbf{G}_A$. Let $\mathbf{p}$ be the natural formal homomorphism $\mathbf{G} \to \mathbf{G}/\mathbf{N}$ and $\mathbf{u}$ its restriction to $\mathbf{H}$; the kernel of $\mathbf{u}_A$ is then $\mathbf{H}_A \cap \mathbf{N}_A = (\inf(\mathbf{H}, \mathbf{N}))_A$, so the kernel of $\mathbf{u}$ is the invariant subgroup $\inf(\mathbf{H}, \mathbf{N})$ of $\mathbf{H}$; if I is the kernel of the bigebra homomorphism $p$, $\mathrm{I} \cap \mathfrak{H}$ is the kernel of the bigebra homomorphism $u$, and therefore $\mathfrak{H} \cap \mathfrak{N}$ is the largest subbigebra of $\mathfrak{H}$ contained in

$$k \cdot 1 + (\mathrm{I} \cap \mathfrak{H}).$$

The same argument applied to $\sup(\mathbf{H}, \mathbf{N})$ instead of $\mathbf{H}$ shows that $\mathbf{N}$ is the kernel of the restriction of $\mathbf{p}$ to $\sup(\mathbf{H}, \mathbf{N})$; furthermore $p(\mathfrak{H}\mathfrak{N}) = p(\mathfrak{H})$, so there are natural isomorphisms of $\mathbf{H}/\inf(\mathbf{H}, \mathbf{N})$ and $\sup(\mathbf{H}, \mathbf{N})/\mathbf{N}$ onto the subgroup of $\mathbf{G}/\mathbf{N}$ corresponding to the subbigebra $p(\mathfrak{H}) = p(\mathfrak{H}\mathfrak{N})$, extending to formal groups the well-known "first isomorphism theorem" of the classical theory. I do not know, however, if *all* formal subgroups of $\mathbf{G}/\mathbf{N}$ correspond to covariant subbigebras of $\mathfrak{G}$ containing $\mathfrak{N}$. By duality, covariant subbigebras $\mathfrak{H}'$ of $\mathfrak{G}/\mathrm{I}$ (which correspond to subgroups of $\mathbf{G}/\mathbf{N}$) are orthogonal to closed contravariant biideals $\mathfrak{H}'^0$ of the bigebra $\mathrm{I}^0$; to say that the subbigebra $\mathfrak{H}'$ has the form $(\mathfrak{H} + \mathrm{I})/\mathrm{I}$, where $\mathfrak{H}$ is the largest subbigebra contained in the subalgebra $p^{-1}(\mathfrak{H}')$ of $\mathfrak{G}$, means that the closed ideal $\mathfrak{H}^0$ of $\mathfrak{G}^*$ generated by $\mathfrak{H}'^0$ is such that $\mathfrak{H}^0 \cap \mathrm{I}^0 = \mathfrak{H}'^0$. We will later prove that this is true for some types of formal groups (II, §3, No. 4).

When $\mathbf{H}$ itself is an invariant subgroup of $\mathbf{G}$, then sup($\mathbf{H}$, $\mathbf{N}$) is also an invariant subgroup of $\mathbf{G}$, since $\mathbf{H}_{\mathrm{A}}\mathbf{N}_{\mathrm{A}}$ is then an invariant subgroup of $\mathbf{G}_{\mathrm{A}}$ for all $\mathrm{A} \in \boldsymbol{Alc}_k$. If in addition $\mathbf{H} \supset \mathbf{N}$ and J is the corresponding biideal of $\mathfrak{G}$ (so that $\mathfrak{H}$ is the largest subbigebra contained in $k \cdot 1 \oplus \mathrm{J}$) then $\mathfrak{G}/\mathrm{J}$ is isomorphic to the quotient bigebra $(\mathfrak{G}/\mathrm{I})/(\mathrm{J}/\mathrm{I})$ of $\mathfrak{G}/\mathrm{I}$ by the biideal $\mathrm{J}/\mathrm{I}$; the image $(\mathfrak{H} + \mathrm{I})/\mathrm{I}$ of $\mathfrak{H}$ in $\mathfrak{G}/\mathrm{I}$ is a subbigebra contained in $k \cdot 1 \oplus (\mathrm{J}/\mathrm{I})$, but I do not know if it is always the largest one. This is equivalent to knowing if the biideal $\mathfrak{H}^0 \cap \mathrm{I}^0$ *of the bigebra* $\mathrm{I}^0$ is generated by the augmentation ideal of the subbigebra $\mathrm{J}^0$ (which generates $\mathfrak{H}^0$ *in* $\mathfrak{G}^*$).

**10.** *Centralizers.* Let $\mathbf{H}'$ and $\mathbf{H}''$ be two formal subgroups of a formal group $\mathbf{G}$, $\mathfrak{H}'$ and $\mathfrak{H}''$ the corresponding subbigebras of the covariant bigebra $\mathfrak{G}$ of $\mathbf{G}$. The following properties are equivalent:

1) Every element of $\mathfrak{H}'$ commutes with every element of $\mathfrak{H}''$.

2) For any algebra $\mathrm{A} \in \boldsymbol{Alc}_k$, every element of the subgroup $\mathbf{H}'_{\mathrm{A}}$ commutes with every element of the subgroup $\mathbf{H}''_{\mathrm{A}}$ (in other words, $\mathbf{H}'_{\mathrm{A}}$ and $\mathbf{H}''_{\mathrm{A}}$ centralize each other in the group $\mathbf{G}_{\mathrm{A}}$).

With the same notations as in No. 8, condition 2) is equivalent to the relation

$$\mathbf{i}'(x')\mathbf{i}''(x'') = \mathbf{i}''(x'')\mathbf{i}'(x'),$$

since every point of $\mathbf{H}'_{\mathrm{A}}$ (resp. $\mathbf{H}''_{\mathrm{A}}$) is a specialization of $\mathbf{i}'(x')$ (resp. $\mathbf{i}''(x'')$) by the same homomorphism $\mathrm{A}_0 \to \mathrm{A}$ (No. 2). The same argument as in No. 8 shows that this relation is equivalent to

$$\mu \circ (\sigma \otimes \rho) \circ \Delta = \mu \circ (\rho \otimes \sigma) \circ \Delta;$$

applying both sides to $Z' \otimes Z''$ for $Z' \in \mathfrak{H}'^+$, $Z'' \in \mathfrak{H}''^+$, yields the relation $Z''Z' = Z'Z''$ and this proves our proposition.

When the equivalent conditions 1) and 2) are satisfied, we say that the formal subgroups $\mathbf{H}'$ and $\mathbf{H}''$ of $\mathbf{G}$ *centralize* each other.

**11.** For any formal subgroup **H** of **G**, there is a *largest* formal subgroup $\mathscr{Z}(\mathbf{H})$ of **G** which centralizes **H** (and is called the *centralizer* of **H** in **G**): indeed, one has only to argue on the corresponding subbigebras; the set of all elements of $\mathfrak{G}$ commuting with every element of $\mathfrak{H}$ is a *subalgebra* $\mathfrak{A}$ of $\mathfrak{G}$, stable under the antipodism. Therefore the largest subcogebra in $\mathfrak{A}$ is the largest covariant subbigebra of $\mathfrak{G}$ consisting of elements commuting with all elements of $\mathfrak{H}$ (No. 3).

The centralizer $\mathscr{Z}(\mathbf{G})$ is called the *center* of **G**; **G** is said to be *commutative* when $\mathscr{Z}(\mathbf{G}) = \mathbf{G}$, which is equivalent to saying that the algebra $\mathfrak{G}$ is *commutative*. For any formal group **G**, the center $\mathscr{Z}(\mathbf{G})$ is commutative, and it follows from No. 6 that it is an *invariant* subgroup of **G**.

**12.** *Direct products and semidirect products of formal groups.* Let $\mathbf{G}_1$, $\mathbf{G}_2$ be two formal subgroups, $\mathfrak{G}_1$, $\mathfrak{G}_2$ their covariant bigebras. The *product* of $\mathbf{G}_1$ and $\mathbf{G}_2$ exists in the category of formal groups. It is of course enough to prove that, in the category $\boldsymbol{CAlc}_k$, the *sum* of $\mathfrak{G}_1^*$ and $\mathfrak{G}_2^*$ exist; but the sum of $\mathfrak{G}_1^*$ and $\mathfrak{G}_2^*$ in the category $\boldsymbol{Alc}_k$ is equal to $\mathfrak{G}_1^* \hat{\otimes} \mathfrak{G}_2^*$, with the natural maps

$$x_1 : t_1 \mapsto t_1 \otimes 1 \qquad \text{and} \qquad x_2 : t_2 \mapsto 1 \otimes t_2.$$

Now it is readily verified that these maps are not only algebra homomorphisms but also bigebra homomorphisms; hence the only thing to verify is that, for any two bigebra homomorphisms ${}^t v_1 : \mathfrak{G}_1^* \to \mathfrak{H}^*$, ${}^t v_2 : \mathfrak{G}_2^* \to \mathfrak{H}^*$, the corresponding algebra homomorphism

$${}^t w : \mathfrak{G}_1^* \hat{\otimes} \mathfrak{G}_2^* \to \mathfrak{H}^*$$

is also a cogebra homomorphism. This is equivalent to saying that the cogebra homomorphism $w : \mathfrak{H} \to \mathfrak{G}_1 \otimes \mathfrak{G}_2$ corresponding to the bigebra homomorphisms $v_1 : \mathfrak{H} \to \mathfrak{G}_1$ and $v_2 : \mathfrak{H} \to \mathfrak{G}_2$ is also an algebra homomorphism. But by definition $w$ is the composite

$$\mathfrak{H} \xrightarrow{\ \Delta\ } \mathfrak{H} \otimes \mathfrak{H} \xrightarrow{v_1 \otimes v_2} \mathfrak{G}_1 \otimes \mathfrak{G}_2,$$

and both $\Delta$ and $v_1 \otimes v_2$ are algebra homomorphisms, hence the conclusion.

We will write $\mathbf{G}_1 \times \mathbf{G}_2$ the product of the formal groups $\mathbf{G}_1$, $\mathbf{G}_2$; its covariant bigebra is therefore $\mathfrak{G}_1 \otimes \mathfrak{G}_2$, its contravariant bigebra $\mathfrak{G}_1^* \hat{\otimes} \mathfrak{G}_2^*$, and for any $\mathrm{A} \in \boldsymbol{Alc}_k$

$$(\mathbf{G}_1 \times \mathbf{G}_2)_A = \mathbf{G}_{1A} \times \mathbf{G}_{2A}.$$

The canonical point of $\mathbf{G}_1 \times \mathbf{G}_2$ (i.e., the identity mapping of $\mathfrak{G}_1^* \hat{\otimes} \mathfrak{G}_2^*$) is the pair $(x_1, x_2)$.

The natural projections

$$\mathbf{p}_1 : \mathbf{G}_1 \times \mathbf{G}_2 \to \mathbf{G}_1, \qquad \mathbf{p}_2 : \mathbf{G}_1 \times \mathbf{G}_2 \to \mathbf{G}_2$$

correspond to the bigebra homomorphisms

$$p_1 : \mathfrak{G}_1 \otimes \mathfrak{G}_2 \to \mathfrak{G}_1, \qquad p_2 : \mathfrak{G}_1 \otimes \mathfrak{G}_2 \to \mathfrak{G}_2$$

such that

$$p_1(Z_1 \otimes Z_2) = \eta_2(Z_2)Z_1, \qquad p_2(Z_1 \otimes Z_2) = \eta_1(Z_1)Z_2$$

where $\eta_1, \eta_2$ are the counits in $\mathfrak{G}_1, \mathfrak{G}_2$; one has $x_1 = {}^t p_1$, $x_2 = {}^t p_2$. It is readily verified, by taking in $\mathfrak{G}_1$ (resp. $\mathfrak{G}_2$) a basis consisting of 1 and a basis of $\mathfrak{G}_1^+$ (resp. $\mathfrak{G}_2^+$) that the kernel of $p_1$ (resp. $p_2$) is $\mathfrak{G}_1 \otimes \mathfrak{G}_2^+$ (resp. $\mathfrak{G}_1^+ \otimes \mathfrak{G}_2$). A supplement to that kernel consists of the image of the injection $j_1 : Z_1 \mapsto Z_1 \otimes 1$ (resp. $j_2 : Z_2 \mapsto 1 \otimes Z_2$) of $\mathfrak{G}_1$ (resp. $\mathfrak{G}_2$) into $\mathfrak{G}_1 \otimes \mathfrak{G}_2$; these two subbigebras

$$j_1(\mathfrak{G}_1) = \mathfrak{G}_1 \otimes k \qquad \text{and} \qquad j_2(\mathfrak{G}_2) = k \otimes \mathfrak{G}_2$$

centralize each other, and we have

$$j_1(\mathfrak{G}_1) \cap j_2(\mathfrak{G}_2) = k \qquad \text{and} \qquad j_1(\mathfrak{G}_1) j_2(\mathfrak{G}_2) = \mathfrak{G}_1 \otimes \mathfrak{G}_2.$$

The corresponding formal subgroups $\mathbf{j}_1(\mathbf{G}_1)$, $\mathbf{j}_2(\mathbf{G}_2)$ are invariant subgroups of $\mathbf{G}_1 \times \mathbf{G}_2$ which centralize each other and are, respectively, the kernels of $\mathbf{p}_2$ and $\mathbf{p}_1$.

**13.** With the same notations, let $\mathbf{u}_1, \mathbf{u}_2$ be formal homomorphisms of a formal group $\mathbf{G}$ into $\mathbf{G}_1$ and $\mathbf{G}_2$, respectively; then, by definition, there is a unique formal homomorphism $\mathbf{u}$ of $\mathbf{G}$ into $\mathbf{G}_1 \times \mathbf{G}_2$, written $(\mathbf{u}_1, \mathbf{u}_2)$, such that $\mathbf{u}_A = (\mathbf{u}_{1A}, \mathbf{u}_{2A})$ for any $A \in \boldsymbol{Alc}_k$, the corresponding homomorphism $u$ of $\mathfrak{G}$ into $\mathfrak{G}_1 \times \mathfrak{G}_2$ being the composite

$$\mathfrak{G} \xrightarrow{\Delta} \mathfrak{G} \otimes \mathfrak{G} \xrightarrow{u_1 \otimes u_2} \mathfrak{G}_1 \otimes \mathfrak{G}_2.$$

If $\mathbf{N}$, $\mathbf{N}_1$, $\mathbf{N}_2$ are the invariant formal subgroups of $\mathbf{G}$, kernels of $\mathbf{u}$, $\mathbf{u}_1$, and $\mathbf{u}_2$, one has $\mathbf{N}_A = \mathbf{N}_{1A} \cap \mathbf{N}_{2A}$ for any $A \in \boldsymbol{Alc}_k$, hence $\mathbf{N} = \inf(\mathbf{N}_1, \mathbf{N}_2)$.

**14.** Let $\mathbf{G}$ be a formal group, $\mathbf{G}_1$, $\mathbf{G}_2$ two formal subgroups of $\mathbf{G}$ which centralize each other and are such that $\inf(\mathbf{G}_1, \mathbf{G}_2) = \mathbf{e}$ (formal unit, corresponding to the subbigebra $k \cdot 1$) and $\sup(\mathbf{G}_1, \mathbf{G}_2) = \mathbf{G}$. Then there is a unique isomorphism of $\mathbf{G}_1 \times \mathbf{G}_2$ onto $\mathbf{G}$, transforming $\mathbf{j}_1(\mathbf{G}_1)$ into $\mathbf{G}_1$ and $\mathbf{j}_2(\mathbf{G}_2)$ into $\mathbf{G}_2$. We have only to prove the corresponding properties for the covariant bigebras: the subbigebras $\mathfrak{G}_1$, $\mathfrak{G}_2$ centralize each other, and the algebra $\mathfrak{G}$ is equal to $\mathfrak{G}_1\mathfrak{G}_2$ (No. 8); as $\mathfrak{G}_1 \cap \mathfrak{G}_2 = k \cdot 1$, $\mathfrak{G}$ is naturally identified as an *algebra* with the tensor product $\mathfrak{G}_1 \otimes \mathfrak{G}_2$ of the algebras $\mathfrak{G}_1$, $\mathfrak{G}_2$, the tensor product $Z_1 \otimes Z_2$ being identified with $Z_1Z_2 = Z_2Z_1$. Now we may write

$$\Delta(Z_1) = \sum_i X_i' \otimes Y_i' \qquad \text{with} \quad X_i', Y_i' \text{ in } \mathfrak{G}_1$$

and

$$\Delta(Z_2) = \sum_j X_j'' \otimes Y_j'' \qquad \text{with} \quad X_j'', Y_j'' \text{ in } \mathfrak{G}_2;$$

we have

$$\Delta(Z_1Z_2) = \Delta(Z_1)\Delta(Z_2) = \sum_{i,j} (X_i'X_j'') \otimes (Y_i'Y_j'')$$

and on the other hand

$$\tau(\Delta(Z_1) \otimes \Delta(Z_2)) = \sum_{i,j} (X_i' \otimes X_j'') \otimes (Y_i' \otimes Y_j'')$$

Taking into account the previous identification, we see that $\mathfrak{G}$ is also identified to $\mathfrak{G}_1 \otimes \mathfrak{G}_2$ as a bigebra, proving the proposition.

**15.** Let $\mathbf{G}$ be a formal group, $\mathbf{H}$ a formal subgroup, and $\mathbf{N}$ an *invariant* formal subgroup of $\mathbf{G}$. If $\inf(\mathbf{H}, \mathbf{N}) = \mathbf{e}$ and $\sup(\mathbf{H}, \mathbf{N}) = \mathbf{G}$, we say that $\mathbf{G}$ is a *semidirect product* of $\mathbf{H}$ and $\mathbf{N}$; the restriction to $\mathbf{H}$ of the natural homomorphism $\mathbf{p}: \mathbf{G} \to \mathbf{G}/\mathbf{N}$ is then an isomorphism of $\mathbf{H}$ onto $\mathbf{G}/\mathbf{N}$. The bigebras $\mathfrak{H}$, $\mathfrak{N}$ of $\mathbf{H}$ and $\mathbf{N}$, respectively, are then such that $\mathfrak{H} \cap \mathfrak{N} = k \cdot 1$ and $\mathfrak{H}\mathfrak{N} = \mathfrak{N}\mathfrak{H} = \mathfrak{G}$, from which it follows that, as an *algebra* (but *not* as a bigebra) $\mathfrak{G}$ is isomorphic to $\mathfrak{H} \otimes \mathfrak{N}$.

If I is the biideal of $\mathfrak{G}$ corresponding to the invariant subbigebra $\mathfrak{N}$, the fact that the restriction to $\mathfrak{H}$ of the natural homomorphism $\mathfrak{G} \to \mathfrak{G}/\mathrm{I}$ is bijective implies that $\mathfrak{G}$ is the direct sum $\mathfrak{H} \oplus \mathrm{I}$ as a vector space, and conversely, this implies that **G** is the semidirect product of **H** and **N**.

**16.** *Commutator subgroup of a formal group.* Let **G** be a formal group, $\mathbf{N}_1$, $\mathbf{N}_2$ two *invariant* formal subgroups of **G**. Then there is a *smallest invariant formal subgroup* **N** of **G** such that the natural images of $\mathbf{N}_1$ and $\mathbf{N}_2$ in $\mathbf{G}/\mathbf{N}$ centralize each other.

Indeed, suppose that **M** is an invariant formal subgroup of **G** such that the images of $\mathbf{N}_1$ and $\mathbf{N}_2$ in $\mathbf{G}/\mathbf{M}$ centralize each other; if J is the biideal of $\mathfrak{G}$ corresponding to **M**, the covariant subbigebras of $\mathfrak{G}/\mathrm{J}$ corresponding to the images of $\mathbf{N}_1$ and $\mathbf{N}_2$ are the natural images of $\mathfrak{N}_1$ and $\mathfrak{N}_2$, and to say that they centralize each other means (No. 10) that, for any $Z_1 \in \mathfrak{N}_1$ and any $Z_2 \in \mathfrak{N}_2$, $Z_1Z_2 - Z_2Z_1$ belongs to J. We therefore obtain the *smallest* possible invariant subgroup **N** by taking as its corresponding biideal I the *smallest biideal of* $\mathfrak{G}$ containing all elements $Z_1Z_2 - Z_2Z_1$ for $Z_1 \in \mathfrak{N}_1$, $Z_2 \in \mathfrak{N}_2$.

Observe that if $\mathrm{I}_1$, $\mathrm{I}_2$ are the two biideals of $\mathfrak{G}$ corresponding to $\mathbf{N}_1$, $\mathbf{N}_2$, respectively, then

$$Z_1Z_2 - Z_2Z_1 \in \mathrm{I}_1 \cap \mathrm{I}_2;$$

this implies that $\mathbf{N} \subset \inf(\mathbf{N}_1, \mathbf{N}_2)$.

The invariant subgroup **N** may be characterized differently: let

$$\mathbf{j}_1 : \mathbf{N}_1 \to \mathbf{G}, \qquad \mathbf{j}_2 : \mathbf{N}_2 \to \mathbf{G}$$

be the natural formal injections, and let $x_1$, $x_2$ be the canonical points of $\mathbf{N}_1$ and $\mathbf{N}_2$ with values in the algebra $\mathrm{A}_0 = \mathfrak{N}_1^* \hat{\otimes} \mathfrak{N}_2^*$ (No. 2). Then **N** may be defined as the *smallest invariant formal subgroup* of **G** such that $\mathbf{N}_{\mathrm{A}_0}$ contains the point

$$\mathbf{j}_1(x_1)\mathbf{j}_2(x_2)\mathbf{j}_1(x_1)^{-1}\mathbf{j}_2(x_2)^{-1}.$$

Indeed, suppose **M** is an invariant formal subgroup of **G** such that $\mathbf{M}_{\mathrm{A}_0}$ contains that point; if $\mathbf{g}: \mathbf{G} \to \mathbf{G}/\mathbf{M}$ is the natural homomorphism, this condition is equivalent to the relation

$$\mathbf{g}(\mathbf{j}_1(x_1))\mathbf{g}(\mathbf{j}_2(x_2)) = \mathbf{g}(\mathbf{j}_2(x_2))\mathbf{g}(\mathbf{j}_1(x_1)). \tag{2}$$

On the other hand, if $x_i'$ ($i = 1, 2$) is the canonical point of the formal subgroup $\mathbf{N}_i' = \mathbf{g}(\mathbf{N}_i)$ of $\mathbf{G}/\mathbf{M}$, with values in $\mathrm{A}_0' = \mathfrak{N}_1'^* \hat{\otimes} \mathfrak{N}_2'^*$, and $\mathbf{j}_i'$ ($i = 1, 2$) the natural injection $\mathbf{N}_i' \to \mathbf{G}/\mathbf{M}$, then $\mathbf{j}_i'(x_i')$ is the image of $\mathbf{g}(\mathbf{j}_i(x_i))$ by the natural injection $\mathrm{A}_0' \to \mathrm{A}_0$. Relation (2) is thus equivalent to

$$\mathbf{j}_1'(x_1')\mathbf{j}_2'(x_2') = \mathbf{j}_2'(x_2')\mathbf{j}_1'(x_1'),$$

which means that $\mathbf{N}_1'$ and $\mathbf{N}_2'$ centralize each other in $\mathbf{G}/\mathbf{M}$ (No. 10).

We will write $\mathbf{N} = [\mathbf{N}_1, \mathbf{N}_2]$ and say that $[\mathbf{N}_1, \mathbf{N}_2]$ is the *commutator subgroup* of the invariant formal subgroups $\mathbf{N}_1$, $\mathbf{N}_2$. It follows from the preceding property that for any algebra $\mathrm{A} \in \boldsymbol{Alc}_k$, one has

$$(3) \qquad (\mathbf{N}_{1\mathrm{A}}, \mathbf{N}_{2\mathrm{A}}) \subset [\mathbf{N}_1, \mathbf{N}_2]_\mathrm{A}$$

where the left-hand side is the usual commutator subgroup; I do not know if there is always equality. We will write $[\mathfrak{N}_1, \mathfrak{N}_2]$ for the subbigebra corresponding to $[\mathbf{N}_1, \mathbf{N}_2]$.

**17.** The definition of the biideal corresponding to $[\mathbf{N}_1, \mathbf{N}_2]$ shows immediately that, if $\mathbf{u}\colon \mathbf{G} \to \mathbf{G}'$ is a *surjective* homomorphism of formal groups, then

$$(4) \qquad [\mathbf{u}(\mathbf{N}_1), \mathbf{u}(\mathbf{N}_2)] = \mathbf{u}([\mathbf{N}_1, \mathbf{N}_2]).$$

Similarly, if $\mathbf{G}'$ is a formal subgroup of $\mathbf{G}$, $\mathbf{N}_1'$, $\mathbf{N}_2'$ two formal invariant subgroups *of* $\mathbf{G}'$ such that $\mathbf{N}_1' \subset \mathbf{N}_1$, $\mathbf{N}_2' \subset \mathbf{N}_2$, then

$$(5) \qquad [\mathbf{N}_1', \mathbf{N}_2'] \subset [\mathbf{N}_1, \mathbf{N}_2].$$

**18.** For a formal group $\mathbf{G}$, we will note $\mathscr{D}(\mathbf{G})$ the invariant formal subgroup $[\mathbf{G}, \mathbf{G}]$ and call it the *commutator subgroup* or *derived group* of $\mathbf{G}$; it is the smallest invariant formal subgroup $\mathbf{M}$ of $\mathbf{G}$ such that $\mathbf{G}/\mathbf{M}$ is commutative.

The biideal J of $\mathfrak{G}$ corresponding to $\mathscr{D}(\mathbf{G})$ is the *left* (or *right*) ideal of the algebra $\mathfrak{G}$ generated by the alternants $[U, V]$ for $U$, $V$ in $\mathfrak{G}$. Indeed, the Jacobi identity

$$(6) \qquad U[V, W] - [V, W]U = [V, [U, W]] - [W, [U, V]]$$

first shows that the left and right ideals generated by the $[U, V]$ are the same, hence J is a two-sided ideal; one need only prove that it is also a *coideal.* As

$$\Delta([U, V]) = [\Delta(U), \Delta(V)],$$

all we have to verify is that $[U \otimes V, U' \otimes V']$ belongs to $\mathfrak{G} \otimes \mathrm{J} + \mathrm{J} \otimes \mathfrak{G}$; but this is immediate, since

$$[U \otimes V, U' \otimes V'] = [U, U'] \otimes VV' + UU' \otimes [V, V'].$$

If $\mathbf{N}$ is an invariant formal subgroup of G, it is clear by definition that $\mathscr{D}(\mathbf{N}) \subset [\mathbf{N}, \mathbf{N}]$, but I do not know if the two subgroups are equal, or, equivalently, if $\mathscr{D}(\mathbf{N})$ is an invariant subgroup *of* $\mathbf{G}$.

**19.** With the help of the concept of commutator subgroup, the criterion on No. 14 on the decomposition of a formal group into a direct product of two subgroups may be given a slightly different formulation: one has only to assume that $\mathbf{G}_1$, $\mathbf{G}_2$ are *invariant* subgroups of $\mathbf{G}$, and

$$\inf(\mathbf{G}_1, \mathbf{G}_2) = \mathbf{e}, \qquad \sup(\mathbf{G}_1, \mathbf{G}_2) = \mathbf{G}.$$

Indeed, one need prove only that $\mathbf{G}_1$, $\mathbf{G}_2$ centralize each other. But this follows from the relation $[\mathbf{G}_1, \mathbf{G}_2] \subset \inf(\mathbf{G}_1, \mathbf{G}_2)$ and the definition of $[\mathbf{G}_1, \mathbf{G}_2]$.

CHAPTER II

# Infinitesimal Formal Groups

## §1. The decomposition theorem

**1.** *Structure of linearly compact algebras over a perfect field.* Let A be a *local* linearly compact algebra over a field $k$; it is by definition *complete* for its topology; in addition, if $\mathfrak{m}$ is the maximal ideal of A, the filter base of the $\mathfrak{m}^n$ ($n$ integer $> 0$) *converges to* 0 in A: indeed, if $\mathfrak{a}$ is an open ideal in A, the $k$-algebra $A/\mathfrak{a}$ is an artinian local ring with maximal ideal $\mathfrak{m}/\mathfrak{a}$, hence $(\mathfrak{m}/\mathfrak{a})^n = 0$ for $n$ large enough, which means that $\mathfrak{m}^n \subset \mathfrak{a}$, and proves our assertion.

Suppose now that the field $k$ is *perfect*; then there is in A a unique subfield K which is such that $A = K \oplus \mathfrak{m}$, so that K, isomorphic to $A/\mathfrak{m}$, will be of *finite* degree over $k$. Indeed, the field $A/\mathfrak{m}$ is a finite extension of $k$, hence generated by a single element $\bar{x}$, since it is separable over $k$; let P(X) be the minimal polynomial of $\bar{x}$ in $k$[X]; as P is separable over $k$, $P'(\bar{x}) \neq 0$. It follows then, from the fact that $(\mathfrak{m}^n)$ tends to 0 and that A is complete, that there is a unique root $x \in A$ of P(X) in the class $\bar{x}$ ([7], Chapter III, §4, No. 5, Corollary 1 of Theorem 2); the field $K = k(x)$ satisfies the required conditions, for if $Q(x) \in \mathfrak{m}$ for a polynomial $Q \in k[X]$, one has $Q(\bar{x}) = 0$ in $A/\mathfrak{m}$, hence Q is a multiple of P in $k$[X], and $Q(x) = 0$.

One may add the following remark: if an irreducible polynomial $Q \in k[X]$ has roots in A, they must belong to K; for any such root $y$ has a class $\bar{y}$ in $A/\mathfrak{m}$ which is a root of Q in that field, and the isomorphism of K and $A/\mathfrak{m}$ show that there is a root $y_1 \in K$ of Q belonging to the class $\bar{y}$, hence necessarily equal to $y$ by *loc. cit.* above, the polynomial Q being separable over $k$.

**2.** Turning now to arbitrary linearly compact algebras A, let us recall that one has $A = \prod_\xi A_\xi$, where $A_\xi$ is a local linearly compact algebra, and if $\mathfrak{m}_\xi$ is the maximal ideal of $A_\xi$, $\mathfrak{R}_A = \prod_\xi \mathfrak{m}_\xi$ is the radical of A (I, §2, No. 8). Supposing *k perfect*, each $A_\xi = K_\xi \times \mathfrak{m}_\xi$ for a finite extension $K_\xi$ of $k$, hence one may write $A = A_e \times \mathfrak{R}_A$, where $A_e = \prod_\xi K_\xi$. We will say that A is *etale* over $k$ if it is a product of finite extensions of $k$, and for any linearly compact algebra A, $A_e$ will be called the *etale* part of A. Let $B = B_e \times \mathfrak{R}_B$ be a second linearly compact $k$-algebra. Then, for any continuous homomorphism $u: A \to B$, one has

$$u(\mathfrak{R}_A) \subset \mathfrak{R}_B, \qquad u(A_e) \subset B_e. \tag{1}$$

The first relation is immediate, since $\mathfrak{R}_A$ is the set of $x \in A$ such that $\lim_{n\to\infty} x^n = 0$. To prove the second, it will be enough (by projection of B on its local components) to consider the case when B is *local*, so $B = K \times \mathfrak{m}$, where K is a finite extension of $k$ and $\mathfrak{m}$ the maximal ideal. On the other hand, as $u$ is continuous, it is enough to prove that, for each $\xi$, $u(K'_\xi) \subset K$, where $K'_\xi$ is the subalgebra of $A_e$ consisting of the elements $x = (x_\eta)$ where $x_\eta = 0$ for $\eta \neq \xi$ and $x_\xi \in K_\xi$. We may of course suppose $x_\xi \neq 0$; let P(X) be its minimal polynomial in $k$[X]; then it is clear that $xP(x) = 0$, hence, if we put $y = u(x) \in B$, $yP(y) = 0$. If $P(y) = 0$, we have seen in No. 1 that $y \in K$; on the other hand, if $y \in \mathfrak{m}$, $P(y)$ is the sum of the constant term of P (which is not 0) and an element of $\mathfrak{m}$, hence $P(y)$ is invertible in B and we get $y = 0$; finally, if $P(y) \neq 0$ and $y \notin \mathfrak{m}$, we cannot have $P(y) \in \mathfrak{m}$, for that would imply $P(y) = 0$ (No. 1), hence again $P(y)$ is invertible and $y = 0$, proving the second relation (1).

**3.** With the same notations (and always supposing $k$ perfect), let $C = A \hat{\otimes} B$; one has then (I, §2, No. 4)

$$C = (A_e \hat{\otimes} B_e) \oplus (A_e \hat{\otimes} \mathfrak{R}_B) \oplus (\mathfrak{R}_A \hat{\otimes} B_e) \oplus (\mathfrak{R}_A \hat{\otimes} \mathfrak{R}_B).$$

Now one has $A \hat{\otimes} \mathfrak{R}_B \subset \mathfrak{R}_C$: indeed, for any open ideal $\mathfrak{b}$ in B, there is an integer $n$ such that

$$(\mathrm{A} \otimes \mathfrak{N}_{\mathrm{B}})^n \subset \mathrm{A} \otimes \mathfrak{b} \subset \mathrm{A} \hat{\otimes} \mathfrak{b},$$

hence also by continuity $(\mathrm{A} \hat{\otimes} \mathfrak{N}_{\mathrm{B}})^n \subset \mathrm{A} \hat{\otimes} \mathfrak{b}$, and therefore $\lim_{n\to\infty} z^n = 0$ for any $z \in \mathrm{A} \hat{\otimes} \mathfrak{N}_{\mathrm{B}}$ by definition of the topology of C (I, §2, No. 4); similarly $\mathfrak{N}_{\mathrm{A}} \hat{\otimes} \mathfrak{B} \subset \mathfrak{N}_{\mathrm{C}}$. On the other hand, let

$$\mathrm{A}_e = \prod_{\xi} \mathrm{K}_{\xi}, \qquad \mathrm{B}_e = \prod_{\eta} \mathrm{L}_{\eta},$$

where the $\mathrm{K}_{\xi}$ and $\mathrm{L}_{\eta}$ are finite extensions of $k$; the topology of $\mathrm{A}_e$ (resp. $\mathrm{B}_e$) can be defined by taking as a fundamental system of neighborhoods of 0 the products of all the $\mathrm{K}_{\xi}$ (resp. $\mathrm{L}_{\eta}$) except for a finite number of indices, where $\mathrm{K}_{\xi}$ (resp. $\mathrm{L}_{\eta}$) is replaced by 0; $\mathrm{A}_e \hat{\otimes} \mathrm{B}_e$ is then the completion of $(\bigoplus_{\xi} \mathrm{K}_{\xi}) \otimes (\bigoplus_{\eta} \mathrm{L}_{\eta})$ for the tensor product topology (I, §2, No. 4). But

$$(\bigoplus_{\xi} \mathrm{K}_{\xi}) \otimes (\bigoplus_{\eta} \mathrm{L}_{\eta}) = \bigoplus_{\xi,\eta} (\mathrm{K}_{\xi} \otimes \mathrm{L}_{\eta}),$$

and as $k$ is perfect, $\mathrm{K}_{\xi} \otimes \mathrm{L}_{\eta}$ is a direct sum of finitely many extensions of finite degree of $k$ ([6], §7, No. 3, Corollary 1 of Theorem 1); hence one may write

$$(\bigoplus_{\xi} \mathrm{K}_{\xi}) \otimes (\bigoplus_{\eta} \mathrm{L}_{\eta}) = \bigoplus_{\zeta} \mathrm{M}_{\zeta}$$

where the $\mathrm{M}_{\zeta}$ are finite extensions of $k$, and the tensor product topology is readily verified to be defined from the $\mathrm{M}_{\zeta}$ as the topologies of $\prod_{\xi} \mathrm{K}_{\xi}$ and $\prod_{\eta} \mathrm{L}_{\eta}$ from the $\mathrm{K}_{\xi}$ and $\mathrm{L}_{\eta}$. Hence $\mathrm{A}_e \hat{\otimes} \mathrm{B}_e$ is identified with the product $\prod_{\zeta} \mathrm{M}_{\zeta}$ (discrete topologies on the $\mathrm{M}_{\zeta}$). We may therefore summarize the preceding arguments in the relations

$$(2) \quad \mathrm{C}_e = \mathrm{A}_e \hat{\otimes} \mathrm{B}_e, \qquad \mathfrak{N}_{\mathrm{C}} = (\mathrm{A}_e \hat{\otimes} \mathfrak{N}_{\mathrm{B}}) \oplus (\mathfrak{N}_{\mathrm{A}} \hat{\otimes} \mathrm{B}_e) \oplus (\mathfrak{N}_{\mathrm{A}} \hat{\otimes} \mathfrak{N}_{\mathrm{B}}).$$

**4.** We will say that a formal group **G** over $k$ is *etale* if its contravariant bigebra $\mathfrak{G}^*$ is etale (in other words, a product of finite extensions of $k$), and that **G** is *infinitesimal* if $\mathfrak{G}^*$ is a *local* linearly compact algebra with $k$ as its residual field. We will now prove the

*Decomposition theorem* (Cartier-Gabriel). *If $k$ is a perfect field, any formal group $\mathbf{G}$ over $k$ is the semidirect product of an infinitesimal invariant subgroup $\mathbf{G}_i$ and of an etale subgroup $\mathbf{G}_e$, both of which are uniquely determined. In addition, for any formal homomorphism $\mathbf{u}: \mathbf{G} \to \mathbf{G}'$ of formal groups over $k$, $\mathbf{u}(\mathbf{G}_i)$ is a subgroup of $\mathbf{G}'_i$ and $\mathbf{u}(\mathbf{G}_e)$ a subgroup of $\mathbf{G}'_e$.*

By No. 2, we may write

$$\mathfrak{G}^* = \mathfrak{G}_e^* \oplus \mathfrak{N}_{\mathfrak{G}^*}$$

for the contravariant bigebra of $\mathbf{G}$. The comultiplication

$$c: \mathfrak{G}^* \to \mathfrak{G}^* \hat{\otimes} \mathfrak{G}^*$$

being a continuous algebra homomorphism, it follows from (1) and (2) that we have

$$c(\mathfrak{G}_e^*) \subset \mathfrak{G}_e^* \hat{\otimes} \mathfrak{G}_e^* \qquad \text{and} \qquad c(\mathfrak{N}_{\mathfrak{G}^*}) \subset \mathfrak{G}^* \hat{\otimes} \mathfrak{N}_{\mathfrak{G}^*} + \mathfrak{N}_{\mathfrak{G}^*} \hat{\otimes} \mathfrak{G}^*:$$

similarly, for the antipodism $a$ of $\mathfrak{G}^*$, one derives from (1) that

$$a(\mathfrak{G}_e^*) \subset \mathfrak{G}_e^* \qquad \text{and} \qquad a(\mathfrak{N}_{\mathfrak{G}^*}) \subset \mathfrak{N}_{\mathfrak{G}^*}.$$

In other words, $\mathfrak{G}_e^*$ is a *contravariant subbigebra* of $\mathfrak{G}^*$ and $\mathfrak{N}_{\mathfrak{G}^*}$ a *contravariant biideal* of $\mathfrak{G}^*$. For the covariant bigebra $\mathfrak{G}$ of $\mathbf{G}$, we have therefore the decomposition in direct sum $\mathfrak{G} = \mathrm{I} \oplus \mathfrak{G}_e$, where $\mathfrak{G}_e$ is a covariant subbigebra such that $\mathfrak{G}_e^0 = \mathfrak{N}_{\mathfrak{G}^*}$ (so that the dual of $\mathfrak{G}_e$ is isomorphic to $\mathfrak{G}^*/\mathfrak{N}_{\mathfrak{G}^*}$, hence to $\mathfrak{G}_e^*$, which justifies the notation), and I is a covariant biideal such that $\mathrm{I}^0 = \mathfrak{G}_e^*$.

Now consider the counit $\gamma: \mathfrak{G}^* \to k$ of $\mathfrak{G}^*$, which is a continuous $k$-algebra homomorphism; it is clear that $\gamma(\mathfrak{N}_{\mathfrak{G}^*}) = 0$, no element of $k$ other than 0 being nilpotent. On the other hand, we have $\mathfrak{G}_e^* = \prod_{\xi \in \mathrm{X}} k_\xi$, where each $k_\xi$ is a finite extension of $k$; as $\gamma(1) = 1$, $\gamma$ may not be 0 on all $k_\xi$; on the other hand, the restriction of $\gamma$ to the product of any two of the $k_\xi$ must be 0 on at least one of them. The conclusion is that there is a *single* element $\nu \in \mathrm{X}$ such that $\gamma(k_\nu) \neq 0$, hence $\gamma(k_\nu) = k$; this is only possible if $k_\nu$ is isomorphic to $k$.

We now apply the general method of I, §3, No. 5 to determine the largest covariant subbigebra $\mathfrak{G}_i$ of $\mathfrak{G}$ contained in $k.1 + \mathrm{I}$. We have to consider the augmentation ideal $\mathfrak{G}_e^{*+}$ of the subbigebra $\mathfrak{G}_e^*$, which is of course the intersection of $\mathfrak{G}_e^*$ and of the augmentation ideal $\mathfrak{G}^{*+}$ of $\mathfrak{G}^*$. It follows immediately from the preceding remarks that $\mathfrak{G}_e^{*+}$ is just the product of all $k_\eta$ for $\eta \neq \nu$. The orthogonal $\mathfrak{G}_i^0$ of $\mathfrak{G}_i$ will be the ideal in $\mathfrak{G}^*$ generated by $\mathfrak{G}_e^{*+}$. If $\mathfrak{G}^* = \prod_{\xi \in \mathrm{X}} \mathrm{A}_\xi$, where the $\mathrm{A}_\xi$ are local linearly compact rings with residual fields $k_\xi$, it is immediate that this ideal is just the product of the $\mathrm{A}_\xi$ for $\xi \neq \nu$. As $\mathfrak{G}$ is the direct sum $\bigoplus_{\xi \in \mathrm{X}} \mathrm{C}_\xi$, where $\mathrm{C}_\xi$ is the cogebra dual to the algebra $\mathrm{A}_\xi$ (and orthogonal to the product of the $\mathrm{A}_\eta$ with $\eta \neq \xi$), we see that $\mathfrak{G}_i = \mathrm{C}_\nu$, which we know is also a bigebra.

We will denote by $\mathbf{G}_i$ and $\mathbf{G}_e$ the formal subgroups of the formal group $\mathbf{G}$ having as covariant subbigebras $\mathfrak{G}_i$ and $\mathfrak{G}_e$. It is then clear (I, §3, No. 15) that $\mathbf{G}$ is the semidirect product of $\mathbf{G}_i$ and $\mathbf{G}_e$, $\mathbf{G}_e$ being etale by definition, and $\mathbf{G}_i$ infinitesimal, since $k_\nu = k$.

**5.** The remainder of this book will be devoted exclusively, from the next paragraph on (with the exception of III, §1), to the study of *infinitesimal* formal groups. Let us here mention only a few properties of the *etale* formal groups. Suppose first that $\mathfrak{G}^*$ has the form $k^{\mathrm{X}}$, in other words, all the extensions $k_\xi$ are equal to $k$ (which is always the case if $k$ is algebraically closed). We still write $k_\xi$ for the factor of that product of index $\xi$, so that $k_\xi = k \cdot e_\xi^*$, the $e_\xi^*$ forming a pseudobasis of $\mathfrak{G}^*$, and $e_\xi^*$ being the unit element of $k_\xi$. The set $\mathrm{Hom}_{Alc_k}(\mathfrak{G}^*, k)$ is then in natural one to one correspondence with X; to each $\xi \in \mathrm{X}$ there corresponds the homomorphism $\mathfrak{G}^* \to k$ sending $e_\xi^*$ onto 1 and the $e_\eta^*$ with $\eta \neq \xi$ to 0; there are no other homomorphisms. As, by definition, this set of homomorphisms has a group structure, we see that X is thus equipped with a *group structure*, the element $\nu$ being the neutral element. It is then easy to determine completely, from that structure, the bigebra structure of the covariant bigebra $\mathfrak{G}$. Let $(e_\xi)$ be the basis of $\mathfrak{G}$, dual to the pseudobasis $(e_\xi^*)$, so that $\mathfrak{G}$ is the direct sum of the cogebras $\mathrm{C}_\xi = k \cdot e_\xi$ of dimension 1, and the homomorphisms $\mathfrak{G}^* \to k$ described above are the transposed of the cogebra homomorphisms $u_\xi \colon k \to \mathrm{C}_\xi$ such that $u_\xi(1) = e_\xi$. As $u_{\xi\eta}$ is given as the composite mapping

$$k \to k \otimes k \xrightarrow{u_\xi \otimes u_\eta} \mathrm{C}_\xi \otimes \mathrm{C}_\eta \to \mathrm{C}_{\xi\eta}$$

we see at once that $e_\xi e_\eta = e_{\xi\eta}$, in other words, the algebra structure of $\mathfrak{G}$

makes it into the *group algebra* of X over $k$. Furthermore, if we write that $u_\xi$ is a cogebra homomorphism, we must express that the diagram

$$\begin{array}{ccc} k & \xrightarrow{u_\xi} & C_\xi \\ \downarrow & & \downarrow{\scriptstyle\Delta} \\ k \otimes k & \xrightarrow[u_\xi \otimes u_\xi]{} & C_\xi \otimes C_\xi \end{array}$$

is commutative, and this gives $\Delta(e_\xi) = e_\xi \otimes e_\xi$ for every $\xi \in X$.

Consider now the general case, and let $\bar{k}$ be the algebraic closure of $k$. Then $\mathfrak{G} \otimes_k \bar{k}$ is a covariant bigebra over $\bar{k}$, a direct sum of the cogebras $C_\xi \otimes_k \bar{k}$, which are dual to the algebras $k_\xi \otimes_k \bar{k}$ over $\bar{k}$. Now the algebra $k_\xi \otimes_k \bar{k}$ is the product of fields $\bar{k}_\zeta$ isomorphic to $\bar{k}$, where $\zeta$ runs through a finite set $Z(\xi)$ having a number of elements equal to the degree $[k_\xi : k]$. If Z is the union of the $Z(\xi)$ for $\xi \in X$, then, as already noted, Z has a natural group structure, and $\mathfrak{G} \otimes_k \bar{k}$ is the group algebra of the group Z, with the comultiplication defined by

$$\Delta(e_\zeta) = e_\zeta \otimes e_\zeta \qquad \text{for every} \quad \zeta \in Z.$$

Furthermore, the Galois group $\Gamma$ of $\bar{k}$ over $k$ acts on each $k_\xi \otimes_k \bar{k}$ by its usual action on the second factor, and it is well known that this action amounts to permuting among themselves the fields $\bar{k}_\zeta$ for $\zeta \in Z(\xi)$. Hence we may consider that we have an action of $\Gamma$ on the group Z, each $\sigma \in \Gamma$ acting as an automorphism of the group Z, and each orbit of Z under $\Gamma$ being *finite*. Conversely, it may be shown that such "Galois modules" form a category that is equivalent to the category of etale formal groups over $k$ (P. Cartier).

## §2. The structure theorem for stable infinitesimal formal groups

From now on, we will always assume that the basic field $k$ is *perfect* and has a *characteristic* $p > 0$; we only briefly indicate in No. 5 what happens when $k$ has characteristic 0 (a relatively uninteresting case for us).

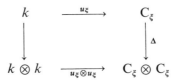

**1.** *Notations.* Let I be an arbitrary nonempty set; we will have to consider the mappings $\alpha: i \mapsto \alpha(i)$ defined in I with values in the set $\mathbf{N}$ of integers $\geqslant 0$, having *finite* support ("multiindices"). They form a lattice-ordered set $\mathbf{N}^{(\mathrm{I})}$, the relation $\alpha \leqslant \beta$ meaning that $\alpha(i) \leqslant \beta(i)$ for every $i \in \mathrm{I}$; the *degree* of $\alpha$ is by definition $|\alpha| = \sum_{i \in \mathrm{I}} \alpha(i)$, which is strictly increasing. The sum $\alpha + \beta$, difference $\beta - \alpha$ (when $\alpha \leqslant \beta$) and product $m \cdot \alpha$ by an integer $m \geqslant 0$ are defined as usual; one has

$$|\alpha + \beta| = |\alpha| + |\beta|.$$

We define the *height* $\mathrm{ht}(\alpha)$ as the smallest integer $k \geqslant 0$ such that $\alpha(i) < p^{k+1}$ for every $i \in \mathrm{I}$; each $\alpha$ of height $k$ can be written uniquely as

$$\alpha_0 + p\alpha_1 + \cdots + p^k \alpha_k \tag{1}$$

where the $\alpha_j$ all have height 0; we define the *factorial* $\alpha!$ in the following way: if $\mathrm{ht}(\alpha) = 0$,

$$\alpha! = \prod_{i \in \mathrm{I}} (\alpha(i))!,$$

and if $\alpha$ is given by (1),

$$\alpha! = \alpha_0!\alpha_1! \cdots \alpha_k!;$$

the element $\alpha!$ is therefore always $\neq 0$ in $k$.

The only elements $\alpha \in \mathbf{N}^{(\mathrm{I})}$ such that $|\alpha| = 1$ are the functions $\varepsilon_i: j \mapsto \delta_{ij}$ (Kronecker symbol). Any $\alpha \in \mathbf{N}^{(\mathrm{I})}$ can be written

$$\alpha = \sum_{i \in \mathrm{I}} \alpha(i)\varepsilon_i.$$

**2.** *The fundamental filtrations.* Our object of study, from now on, will be a commutative linearly compact *local* algebra $\mathfrak{G}^* = k \oplus \mathfrak{R}$, where $\mathfrak{R}$ is the radical, which is made into a bigebra by a continuous coassociative coproduct $c: \mathfrak{G}^* \to \mathfrak{G}^* \hat{\otimes} \mathfrak{G}^*$, and a counit $\gamma: \mathfrak{G}^* \to k$, which is 0 in $\mathfrak{R}$ and such that $\gamma(1) = 1$. We explicitly *do not* assume the existence of an antipodism, and in fact we shall prove later (No. 8) that the previous assumptions *imply* the existence of a unique antipodism of $\mathfrak{G}^*$, so that $\mathfrak{G}^*$ will indeed be the contravariant bigebra of an infinitesimal formal group **G** over $k$. Together with $\mathfrak{G}^*$, we shall of course consider its Cartier dual $\mathfrak{G}$, which is a (discrete) bigebra whose comultiplication $\Delta: \mathfrak{G} \to \mathfrak{G} \otimes \mathfrak{G}$ is the transpose of the multiplication in $\mathfrak{G}^*$ and the multiplication the transpose of the comultiplication $c$; we have $\mathfrak{G} = k \oplus \mathfrak{G}^+$, where the augmentation ideal $\mathfrak{G}^+$, kernel of the counit $\eta$, is the orthogonal of the subspace $k$ of $\mathfrak{G}^*$. *These assumptions and notations will be in force during the whole chapter and will not be repeated.* We note that formulas (8) of I, §2, No. 12 and (14) of I, §2, No. 13 are of course valid in $\mathfrak{G}^*$ and $\mathfrak{G}$.

By abuse of notation, we shall write $\mathfrak{R}^n$, for any integer $n > 0$, to mean *not* the usual product of $n$ ideals identical to the radical $\mathfrak{R}$, but the *closure* of that product. It is clear that (as $\mathfrak{R}$ is the largest ideal $\neq \mathfrak{G}^*$ in $\mathfrak{G}^*$) the closed ideals $\mathfrak{R}^n$ form the *largest decreasing filtration* of the algebra $\mathfrak{G}^*$ consisting of closed ideals, by which I mean that for any such other filtration

$$\mathfrak{J}_1 \supset \mathfrak{J}_2 \supset \cdots \supset \mathfrak{J}_n \supset \cdots$$

of that type, one has necessarily $\mathfrak{J}_n \subset \mathfrak{R}^n$.

For any such filtration, the $C_n = \mathfrak{J}^0_{n-1}$, consisting of subcogebras of $\mathfrak{G}$, is a *cogebra* (increasing) filtration of the cogebra $\mathfrak{G}$, i.e., satisfies the relation

$$\Delta(C_n) \subset C_0 \otimes C_n + C_1 \otimes C_{n-1} + \cdots + C_n \otimes C_0. \tag{2}$$

Indeed, from the relation $\mathfrak{J}_h\mathfrak{J}_k \subset \mathfrak{J}_{h+k}$ satisfied by the $\mathfrak{J}_n$, which can be written $m(\mathfrak{J}_h \hat{\otimes} \mathfrak{J}_k) \subset \mathfrak{J}_{h+k}$, we deduce (I, §2, No. 4)

$$\Delta(C_{h+k-1}) \subset C_{h-1} \otimes \mathfrak{G} + \mathfrak{G} \otimes C_{k-1} \tag{3}$$

for all indices $h, k$ (taking $C_{-1} = 0$ by convention). Let $E_n$ be a supplementary subspace to $C_{n-1}$ in $C_n$, and $C'_n$ a supplementary subspace to $C_n$ in G. Writing

$$G = C_n \oplus C'_n, \quad C_n = C_{h-1} \oplus E_h \oplus E_{h+1} \oplus \cdots \oplus E_n =$$
$$C_{k-1} \oplus E_k \oplus E_{k+1} \oplus \cdots \oplus E_n \qquad \text{for} \quad n = h + k - 1,$$

we see from (3) that the components of $\Delta(C_n)$ in the following summands

$$C'_n \otimes C'_n, \quad E_h \otimes C'_n, \quad C'_n \otimes E_k, \quad E_h \otimes E_k \qquad \text{for} \quad h + k \geqslant n$$

are all 0; but if we develop the tensor product

$$\mathfrak{G} \otimes \mathfrak{G} = (\mathrm{E}_0 \oplus \mathrm{E}_1 \oplus \cdots \oplus \mathrm{E}_n \oplus \mathrm{C}'_n) \otimes (\mathrm{E}_0 \oplus \mathrm{E}_1 \oplus \cdots \oplus \mathrm{E}_n \oplus \mathrm{C}'_n)$$

into a direct sum of subspaces, we find that when we delete from that sum the preceding summands, what remains is exactly the right-hand side of (2). Conversely, if (2) holds for every $n$, it implies at once (3) for every pair of indices, hence the $\mathfrak{J}_n = \mathrm{C}^0_{n+1}$ form a decreasing filtration of $\mathfrak{G}^*$ consisting of closed ideals.

We shall write $\mathrm{M}_n = (\mathfrak{N}^{n+1})^0$, so that the $\mathrm{M}_n$ form the *smallest increasing filtration* of $\mathfrak{G}$ consisting of *cogebras*, with $\mathrm{M}_0 = k \cdot 1$. We may write $\mathrm{M}_n = k \cdot 1 \oplus \mathrm{P}_n$, with $\mathrm{P}_n = \mathfrak{G}^+ \cap \mathrm{M}_n$, so that the subspace $\mathrm{P}_n$ is in natural correspondence with $\mathrm{M}_n/\mathrm{M}_0$, hence in natural duality with $\mathfrak{N}/\mathfrak{N}^{n+1}$. Taking into account the relation

$$\Delta(Z) - 1 \otimes Z - Z \otimes 1 \in \mathfrak{G}^+ \otimes \mathfrak{G}^+ \qquad \text{for any} \quad Z \in \mathfrak{G}^+,$$

we see that relation (2), applied to the cogebras $\mathrm{M}_n$, gives

(4) $$\Delta(Z) = 1 \otimes Z + Z \otimes 1 \qquad \text{for} \quad Z \in \mathrm{P}_1$$

(5) $$\Delta(Z) - 1 \otimes Z - Z \otimes 1 \in \mathrm{P}_1 \otimes \mathrm{P}_{n-1} + \mathrm{P}_2 \otimes \mathrm{P}_{n-2} + \cdots + \mathrm{P}_{n-1} \otimes \mathrm{P}_1 \qquad \text{for} \quad Z \in \mathrm{P}_n.$$

Conversely, the $\mathrm{P}_n$ may be *defined* inductively by these conditions. Indeed, it is easily verified by induction on $n$ that they form an increasing sequence and that $k \cdot 1 \oplus \mathrm{P}_n$ is a cogebra (i.e.,

$$\Delta(k \cdot 1 \oplus \mathrm{P}_n) \subset (k \cdot 1 \oplus \mathrm{P}_n) \otimes (k \cdot 1 \oplus \mathrm{P}_n));$$

as obviously $k \cdot 1 \oplus \mathrm{P}_n$ is contained in $\mathrm{C}_n$ for any increasing filtration of $\mathfrak{G}$ consisting of cogebras and satisfying (2), this proves our contention.

The elements of $\mathrm{P}_1$, i.e., those which satisfy (4), are called the *primitive elements* of the bigebra $\mathfrak{G}$. If $X$, $Y$ are primitive, then

$$\begin{aligned} \Delta(XY) &= (1 \otimes X + X \otimes 1)(1 \otimes Y + Y \otimes 1) \\ &= 1 \otimes XY + Y \otimes X + X \otimes Y + XY \otimes 1 \\ \Delta(YX) &= (1 \otimes Y + Y \otimes 1)(1 \otimes X + X \otimes 1) \\ &= 1 \otimes YX + X \otimes Y + Y \otimes X + YX \otimes 1 \end{aligned}$$

hence $\Delta([X, Y]) = 1 \otimes [X, Y] + [X, Y] \otimes 1$, i.e., $[X, Y]$ is primitive.

Similarly, as $1 \otimes X$ and $X \otimes 1$ commute in $\mathfrak{G} \otimes \mathfrak{G}$,

$$\Delta(X^p) = (\Delta(X))^p = (1 \otimes X + X \otimes 1)^p = 1 \otimes X^p + X^p \otimes 1,$$

i.e., $X^p$ is also primitive. We thus may say that the primitive elements of $\mathfrak{G}$ form a *p-Lie algebra* over $k$, which we shall denote by $\mathfrak{g}_0$.

We may add that the $\mathrm{M}_n$ form an increasing *algebra* filtration of $\mathfrak{G}$, i.e.,

(6) $$\mathrm{M}_m \mathrm{M}_n \subset \mathrm{M}_{m+n}$$

or, what amounts to the same,

(7) $$\mathrm{P}_m \mathrm{P}_n \subset \mathrm{P}_{m+n}.$$

This is obvious if $n = 0$ and is proved by induction on $m$ and $n$: if $Z \in \mathrm{P}_m$, $Z' \in \mathrm{P}_n$, the relation

$$\Delta(ZZ') = \Delta(Z)\Delta(Z')$$

and relations (5) applied to $Z$ and $Z'$ show that

$$\Delta(ZZ') - 1 \otimes ZZ' - ZZ' \otimes 1$$

is a sum of terms of the following forms:

$$Z \otimes Z' \in \mathrm{P}_m \otimes \mathrm{P}_n; \qquad Z' \otimes Z \in \mathrm{P}_n \otimes \mathrm{P}_m;$$

$ZU'_j \otimes V'_{n-j}$ where $U'_j \in \mathrm{P}_j$, $V'_{n-j} \in \mathrm{P}_{n-j}$, and $1 \leqslant j \leqslant n-1$, hence that element belongs to $\mathrm{P}_{m+j} \otimes \mathrm{P}_{n-j}$ by the induction assumption; $U'_j \otimes ZV'_{n-j}$, which similarly belongs to $\mathrm{P}_j \otimes \mathrm{P}_{m+n-j}$; $U_i Z' \otimes V_{m-i}$ where $U_i \in \mathrm{P}_i$, $V_{m-i} \in \mathrm{P}_{m-i}$, and $1 \leqslant i \leqslant m-1$, hence that element belongs to $\mathrm{P}_{n+i} \otimes \mathrm{P}_{m-i}$; $U_i \otimes V_{m-i}Z'$, which similarly belongs to $\mathrm{P}_i \otimes \mathrm{P}_{m+n-i}$; and finally $U_i U'_j \otimes V_{m-i} V'_{n-j}$ which belongs to $\mathrm{P}_{i+j} \otimes \mathrm{P}_{m+n-i-j}$.

Finally, we have seen (§1, No. 1) that the intersection of the $\mathfrak{R}^n$ is reduced to 0, hence $\mathfrak{G}$ is the union of the $\mathrm{M}_n$ (one says that the filtration $(\mathrm{M}_n)$ is *exhaustive*).

**3.** *The Frobenius subbigebras.* For any integer $r \in \mathbf{Z}$, we will define on the additive group $\mathfrak{G}^*$ a new scalar multiplication by elements of $k$, by putting $\lambda \cdot x = \lambda^{p^r} x$ (on the right-hand side the scalar multiplication is that of $\mathfrak{G}^*$). Remember that as $k$ is perfect, $\xi \mapsto \xi^{p^r}$ is an automorphism of $k$ for positive or negative integers $r$; hence we thus define on $\mathfrak{G}^*$ a new structure of linearly compact vector space over $k$ (I, §2, No. 6), and it is clear that the product, coproduct, and counit of $\mathfrak{G}^*$, together with that new vector space structure, define on $\mathfrak{G}^*$ a new structure of bigebra; we will write $\mathfrak{G}^{*(r)}$ for that bigebra. Its Cartier dual is the discrete bigebra $\mathfrak{G}^{(r)}$ defined from $\mathfrak{G}$ in the same manner.

Now, $\mathfrak{G}^*$ being commutative and $k$ of characteristic $p$, the mapping $\mathrm{F}: x \mapsto x^p$ is a continuous *ring* homomorphism of $\mathfrak{G}^*$, but as

$$(\lambda x)^p = \lambda^p x^p \qquad \text{for} \quad \lambda \in k,$$

it is an *algebra* homomorphism of $\mathfrak{G}^*$ into $\mathfrak{G}^{*(1)}$; we have

$$c(x^p) = (c(x))^p$$

since $c$ is a ring homomorphism of $\mathfrak{G}^*$ into $\mathfrak{G}^* \hat{\otimes} \mathfrak{G}^*$, and $\mathfrak{G}^* \hat{\otimes} \mathfrak{G}^*$ is commutative, hence the mapping $\mathrm{F} \otimes \mathrm{F}$ of $\mathfrak{G}^* \otimes \mathfrak{G}^*$ into itself is just the mapping $z \mapsto z^p$ and by continuity the same is true of $\mathrm{F} \hat{\otimes} \mathrm{F}$ in $\mathfrak{G}^* \hat{\otimes} \mathfrak{G}^*$. We therefore see that F is a continuous *bigebra* homomorphism of $\mathfrak{G}^*$ into $\mathfrak{G}^{*(1)}$, which we shall call the *Frobenius homomorphism.*

It is clearly equivalent to say that a subset $\mathfrak{H}^*$ of $\mathfrak{G}^*$ is a closed subbigebra of $\mathfrak{G}^*$ or of any of the $\mathfrak{G}^{*(r)}$. For any integer $r > 0$, $\mathrm{F}^r(\mathfrak{G}^*)$ is therefore a closed subbigebra of $\mathfrak{G}^*$ (I, §2, No. 3) equal to $k \cdot 1 \oplus \mathrm{F}^r(\mathfrak{R})$; observe that by definition we have

$$\mathrm{F}^r(\mathfrak{R}) \subset \mathfrak{R}^{p^r}. \tag{8}$$

Let $\mathfrak{B}_r$ be the closed ideal generated by $\mathrm{F}^r(\mathfrak{R})$ in $\mathfrak{G}^*$; it is a biideal, hence its orthogonal $\mathfrak{s}_{r-1}$ in $\mathfrak{G}$ is a subbigebra (also written $\mathfrak{s}_{r-1}(\mathfrak{G})$), which is the largest subbigebra such that $\mathfrak{s}_{r-1}^+$ is contained in the biideal orthogonal to $\mathrm{F}^r(\mathfrak{R})$. We will say that the subbigebras $\mathrm{F}^r(\mathfrak{G}^*)$ and $\mathfrak{s}_{r-1}$ are the *Frobenius subbigebras* of $\mathfrak{G}^*$ and $\mathfrak{G}$, respectively; the dual of $\mathfrak{s}_{r-1}$ is $\mathfrak{G}^*/\mathfrak{B}_r$. We will say that $\mathfrak{G}$ and $\mathfrak{G}^*$ have *finite height h* if $h$ is the smallest integer $\geqslant 0$ such that $\mathfrak{s}_h = \mathfrak{G}$, or equivalently $\mathrm{F}^{h+1}(\mathfrak{R}) = 0$; if no integer satisfies these relations, we will say that $\mathfrak{G}$ and $\mathfrak{G}^*$ have *infinite height.* For any $r$ smaller or equal to the height of $\mathfrak{G}$, $\mathfrak{s}_r$ is a bigebra of height $r$.

**4.** *Bigebras of height* 0. The structure of bigebras of height 0 is a very perspicuous one, compared with the more complicated situation in bigebras of finite height $> 0$ (see No. 6). Consider a *basis* $(X_i)_{i \in \mathrm{I}}$ of the

$k$-vector space $\mathfrak{g}_0$ of primitive elements of a bigebra $\mathfrak{G}$ of height 0. Take on I an arbitrary *total order*, and for any $\alpha: i \mapsto \alpha(i)$ of $\mathbf{N}^{(\mathrm{I})}$ of *height* 0 (i.e., $\alpha(i) < p$ for every $i$) let us write $X_\alpha$ the product

$$X_{i_1}^{\alpha(i_1)} X_{i_2}^{\alpha(i_2)} \cdots X_{i_m}^{\alpha(i_m)}$$

where the $i_k$ are the values of $i$ for which $\alpha(i) > 0$, labeled in such a way as to form an increasing sequence $i_1 < \cdots < i_m$ in I. The unit element of $\mathfrak{G}$ is written $X_0$.

**Proposition 1.** *The elements $Z_\alpha = X_\alpha/\alpha!$, where $\alpha$ runs through all elements of $\mathbf{N}^{(\mathrm{I})}$ of height 0, form a basis of $\mathfrak{G} = \mathfrak{s}_0$ over $k$, so that $\mathfrak{s}_0$ is the algebra generated by $\mathfrak{g}_0$; furthermore, the comultiplication in $\mathfrak{G}$ is given by*

$$\Delta(Z_\alpha) = \sum_{\beta+\gamma=\alpha} Z_\beta \otimes Z_\gamma. \tag{9}$$

Relation (9) is evident by definition for $|\alpha| = 1$, since the $X_i$ are primitive elements; we will first prove it for any $\alpha$ by induction on the degree $|\alpha|$. We may always write

$$X_\alpha = X_i^m X_\beta \qquad \text{with} \quad m = \alpha(i) > 0,$$

the multiindex $\beta$ being such that $\beta(j) = 0$ for all indices $j \leqslant i$; by definition of $\alpha!$, one has then

$$\alpha! = m!\beta!, \qquad Z_\alpha = \frac{1}{m!} X_i^m Z_\beta,$$

and therefore, from the inductive assumption,

$$\begin{aligned}\Delta(Z_\alpha) &= \frac{1}{m!}(1 \otimes X_i + X_i \otimes 1)^m \Big(\sum_{\gamma+\delta=\beta} Z_\gamma \otimes Z_\delta\Big)\\ &= \Big(\sum_{r+s=m} Z_{r\varepsilon_i} \otimes Z_{s\varepsilon_i}\Big)\Big(\sum_{\gamma+\delta=\beta} Z_\gamma \otimes Z_\delta\Big)\end{aligned}$$

taking into account the binomial theorem and the fact that $m < p$. Now every multiindex $\lambda \leqslant \alpha$ can be written uniquely in the form $r\varepsilon_i + \mu$ with $r \leqslant m$ and $\mu \leqslant \beta$, and by definition

$$Z_\lambda = Z_{r\varepsilon_i} Z_\mu;$$

this proves formula (9) for $Z_\alpha$.

We next prove that the $Z_\alpha$ are linearly independent; by definition the $Z_\alpha$ for $|\alpha| = 1$ are linearly independent; we prove by induction on $m$ that this is still true for the $Z_\alpha$ such that $|\alpha| \leqslant m$. If this were not the case we would have a relation

$$\sum_{|\alpha|=m} c_\alpha Z_\alpha = \sum_{|\beta|<m} d_\beta Z_\beta \tag{10}$$

where the $c_\alpha$ are not all zero. Observe that by assumption the $Z_\lambda \otimes Z_\mu$ for $|\lambda| < m$ and $|\mu| < m$ are linearly independent in $\mathfrak{G} \otimes \mathfrak{G}$. If we call $Z$ the common value of both sides of (10) and replace $Z$ in

$$\Delta(Z) - 1 \otimes Z - Z \otimes 1$$

successively by the left-hand side and the right-hand side of (10), we obtain for the left-hand side a combination of $Z_\lambda \otimes Z_\mu$ with $|\lambda| + |\mu| = m$, $\lambda \neq 0$, $\mu \neq 0$, and on the right-hand side a combination of similar products but with $|\lambda| + |\mu| < m$, hence both sides must be 0; but there are no terms in $Z_\lambda \otimes Z_\mu$ for the same indices coming from $\Delta(Z_\alpha)$ for two distinct multiindices $\alpha$, since $\lambda + \mu = \alpha$. Hence one must have $c_\alpha = 0$ for all $\alpha$ such that $|\alpha| = m$, contrary to assumption.

Finally, we prove that the $Z_\alpha$ form a basis of $\mathfrak{G}$ by showing, by induction on $n$, that the $Z_\alpha$ for $|\alpha| \leqslant n$ form a basis of the $n$-th cogebra $M_n$ of the fundamental filtration of $\mathfrak{G}$ (No. 2). Again this follows from the definitions if $n = 1$; suppose the assertion has been proved for $n$; then the $Z_\alpha \otimes Z_\beta$ with $|\alpha| \leqslant r$, $|\beta| \leqslant s$, $\alpha \neq 0$, $\beta \neq 0$, and $r + s \leqslant n$ constitute a basis for $P_r \otimes P_s$, and $Z_\alpha \in P_{n+1}$ for $|\alpha| = n + 1$ by (9) and (5). Hence for any $Z \in P_{n+1}$, it follows from (5) that

$$\Delta(Z) - 1 \otimes Z - Z \otimes 1$$

can be written in only one way as

$$\sum_{\alpha,\beta} b_{\alpha\beta} Z_\alpha \otimes Z_\beta \qquad \text{with} \quad \alpha \neq 0, \quad \beta \neq 0, \quad |\alpha| + |\beta| \leqslant n + 1.$$

I claim that if $\alpha + \beta = \gamma + \delta$, one has necessarily $b_{\alpha\beta} = b_{\gamma\delta}$. This follows from the coassociativity condition

$$(\Delta \otimes 1) \circ \Delta = (1 \otimes \Delta) \circ \Delta,$$

which yields the relation

$$\sum_{\substack{\alpha,\beta \\ \rho \leqslant \alpha}} b_{\alpha\beta} Z_\rho \otimes Z_{\alpha-\rho} \otimes Z_\beta = \sum_{\substack{\alpha,\beta \\ \sigma \leqslant \beta}} b_{\alpha\beta} Z_\alpha \otimes Z_\sigma \otimes Z_{\beta-\sigma}$$

and that relation can also be written

$$\sum_{|\lambda|+|\mu|+|\nu| \leqslant n+1} b_{\lambda+\mu,\nu} Z_\lambda \otimes Z_\mu \otimes Z_\nu = \sum_{|\lambda|+|\mu|+|\nu| \leqslant n+1} b_{\lambda,\mu+\nu} Z_\lambda \otimes Z_\mu \otimes Z_\nu$$

hence $b_{\lambda+\mu,\nu} = b_{\lambda,\mu+\nu}$. But if $\alpha + \beta = \gamma + \delta$, there are 4 elements $\lambda$, $\mu$, $\gamma$, $\pi$, in $\mathbf{N}^{(I)}$ such that

$$\alpha = \lambda + \mu, \quad \beta = \nu + \pi, \quad \gamma = \lambda + \nu, \quad \delta = \mu + \pi,$$

and we conclude that $b_{\alpha\beta} = b_{\gamma\delta}$. We may thus write

$$\Delta(Z) - 1 \otimes Z - Z \otimes 1 = \sum_{|\alpha| \leqslant n+1} b_\alpha(\Delta(Z_\alpha) - 1 \otimes Z_\alpha - Z_\alpha \otimes 1),$$

hence, if $Y = Z - \sum_{|\alpha| \leqslant n+1} b_\alpha Z_\alpha$,

$$\Delta(Y) - 1 \otimes Y - Y \otimes 1 = 0;$$

but this implies that $Y \in \mathrm{P}_1 = \mathfrak{g}_0$, hence $Z$ is a linear combination of the $Z_\alpha$ with $|\alpha| \leqslant n + 1$. Q.E.D.

It can be shown that $\mathfrak{G} = \mathfrak{s}_0$ is the *universal restricted enveloping algebra* of the $p$-Lie algebra $\mathfrak{g}_0$, in the following sense: any linear mapping $f\colon \mathfrak{g}_0 \to \mathrm{B}$ of $\mathfrak{g}_0$ into an associative algebra B over $k$, such that

$$f([X, Y]) = f(X)f(Y) - f(Y)f(X) \qquad \text{and} \qquad f(X^p) = (f(X))^p$$

can be extended to an algebra homomorphism $\bar{f}\colon \mathfrak{s}_0 \to \mathrm{B}$. This may be proved by defining $\bar{f}(X_\alpha)$ as equal to

$$(f(X_{i_1}))^{\alpha(i_1)} \cdots (f(X_{i_m}))^{\alpha(i_m)},$$

and then proving that

$$\bar{f}(X_\alpha X_\beta) = \bar{f}(X_\alpha)\bar{f}(X_\beta)$$

by induction on the degrees $|\alpha|$ and $|\beta|$. As we will not need this result, we leave the details to the reader.

Let $(e_\alpha^*)$ be the pseudobasis of $\mathfrak{G}^*$ dual to $(Z_\alpha)$ (I, §2, No. 1); as the multiplication $m: \mathfrak{G}^* \hat{\otimes} \mathfrak{G}^* \to \mathfrak{G}^*$ in $\mathfrak{G}^*$ is the transpose of $\Delta$, we have, by (9),

$$\langle Z_\alpha, m(e_\lambda^* \otimes e_\mu^*)\rangle = \langle \Delta(Z_\alpha), e_\lambda^* \otimes e_\mu^*\rangle = \sum_{\beta+\gamma=\alpha} \langle Z_\beta, e_\lambda^*\rangle\langle Z_\gamma, e_\mu^*\rangle$$

and the sum is 0 unless $\lambda + \mu = \alpha$, in which case it is 1; therefore (I, §2, No. 4)

$$m(e_\lambda^* \otimes e_\mu^*) = e_{\lambda+\mu}^*$$

for any two multiindices $\lambda$, $\mu$ of height 0 in $\mathbf{N}^{(\mathrm{I})}$ whose sum is still of height 0, i.e., such that

$$\lambda(i) + \mu(i) < p \qquad \text{for } \textit{all} \quad i \in \mathrm{I};$$

if this condition is not satisfied, $m(e_\lambda^* \otimes e_\mu^*) = 0$. Putting

$$x_i = e_{\varepsilon_i}^* \qquad \text{for} \quad i \in \mathrm{I},$$

we may therefore write $e_\alpha^* = x^\alpha$ with the usual convention

$$x^\alpha = \prod_{i\in\mathrm{I}} x_i^{\alpha(i)}$$

(the order here is irrelevant since $\mathfrak{G}^*$ is commutative), for any $\alpha$ of height 0, and we have in addition $x_i^p = 0$ for every $i \in \mathrm{I}$. We may therefore say that, as a linearly compact $k$-algebra, $\mathfrak{G}^*$ is isomorphic to *the quotient of the algebra of formal power series* $k[[(\mathrm{T}_i)]]_{i\in\mathrm{I}}$ *by the closed ideal generated by the* $\mathrm{T}_i^p$; as $\langle x_i, X_j\rangle = \delta_{ij}$ by definition and $\langle x^\alpha, X_j\rangle = 0$ for $|\alpha| > 1$, we also see that the *classes* $\bar{x}_i$ of the $x_i$ modulo $\mathfrak{N}^2$ form a pseudobasis of $\mathfrak{N}/\mathfrak{N}^2$, which is *dual* to the basis $(X_i)$ of $\mathfrak{g}_0$ in the natural duality between these vector spaces (No. 2).

**5.** *The case of characteristic* 0. Before we study bigebras of arbitrary finite height, let us consider briefly the corresponding situation when $k$ has characteristic 0, $\mathfrak{G}^*$ being always a local algebra with residual field $k$. The concepts of height and of Frobenius homomorphism of course disappear; the arguments of No. 2 carry over without change; $X_\alpha$ is defined as above, but for *every* $\alpha \in \mathbf{N}^{(I)}$, and so is $Z_\alpha$ (but now $\alpha! = \prod_{i \in I} (\alpha(i))!$, which is always $\neq 0$). The argument in Proposition 1 is then repeated without change, except for the suppression of the condition $m < p$. We thus get a basis $(Z_\alpha)$ of $\mathfrak{G}$ satisfying (9), where now $\alpha$ runs through the *whole* set $\mathbf{N}^{(I)}$; it can be shown that $\mathfrak{G}$ is the *universal enveloping algebra* of the Lie algebra $\mathfrak{g}_0$, and $\mathfrak{G}^*$ is isomorphic to the *algebra of formal power series* $k[[(T_i)]]_{i \in I}$; we are back to the example 1) in I, §2, No. 12, except that no restriction is placed on the set of indices of the indeterminates.

**6.** *Bigebras of finite height.* We are now going to generalize proposition 1 to bigebras $\mathfrak{G}$ of finite height. To do so, we must first introduce a descending sequence of subspaces of the Lie algebra $\mathfrak{g}_0$, linked to the existence of the Frobenius homomorphism. Consider, for any integer $r \geqslant 0$, the iterated Frobenius homomorphism $F^r$, which sends $\mathfrak{N}$ into $\mathfrak{N}^{p^r}$, and therefore $\mathfrak{N}^2$ into $\mathfrak{N}^{2p^r} \subset \mathfrak{N}^{p^r+1}$. By passage to quotients, we deduce therefore from $F^r$ a semilinear mapping of vector spaces

$$\overline{F}_r : \mathfrak{N}/\mathfrak{N}^2 \to \mathfrak{N}^{p^r}/\mathfrak{N}^{p^r+1}.$$

Let $N_r$ be the kernel of that homomorphism, which is a subspace of $\mathfrak{N}/\mathfrak{N}^2$; the sequence of subspaces $(N_r)$ is *increasing*, since if $F^r(x) \in \mathfrak{N}^{p^r+1}$, then

$$F^{r+1}(x) \in \mathfrak{N}^{p^{r+2}} \subset \mathfrak{N}^{p^{r+1}+1}.$$

Recall that $\mathfrak{g}_0$ and $\mathfrak{N}/\mathfrak{N}^2$ are in natural duality; therefore, the subspaces

$$\mathfrak{g}'_r = N_r^0 \tag{11}$$

orthogonal to the $N_r$ in $\mathfrak{g}_0$ form a descending sequence, such that $\mathfrak{g}'_0 = \mathfrak{g}_0$.

We may observe here that when we replace $\mathfrak{G}$ by any Frobenius subbigebra $\mathfrak{s}_h$, the definition of the mappings $\overline{F}_r$ for $r \leqslant h$ does not change (up to the standard identification of a quotient P/Q of subspaces of $\mathfrak{G}^*$ containing $\mathfrak{B}_{h+1}$ and such that $P \supset Q$ with the quotient $(P/\mathfrak{B}_{h+1})/(Q/\mathfrak{B}_{h+1})$ in $\mathfrak{G}^*/\mathfrak{B}_{h+1}$); nor do, therefore, the definitions of the subspaces $\mathfrak{g}'_r$ for $r \leqslant h$.

Now suppose that $\mathfrak{G} = \mathfrak{s}_r$ has finite height $r$; then $F^{r+1}(\mathfrak{N}) = 0$

and therefore $N_{r+1} = \mathfrak{R}/\mathfrak{R}^2$, and $\mathfrak{g}'_{r+1} = 0$. We are going to choose a basis $(X_i)_{i\in I}$ of $\mathfrak{g}_0$ having the property that there exists a decreasing sequence $(I_k)$ of subsets of I such that $I_0 = I$, $I_{r+1} = \varnothing$, and the $X_i$ such that $i \in I_k$ form a basis of $\mathfrak{g}'_k$ for $0 \leqslant k \leqslant r$. We will then say that a basis $(Z_\alpha)$ of the vector space $\mathfrak{G}$ is *privileged* with respect to $(X_i)_{i\in I}$ if it satisfies the following conditions:

A) *The indices* $\alpha$ *run through the subset* $L_r$ *of* $\mathbf{N}^{(I)}$ *consisting of the multiindices satisfying the conditions* $\alpha(i) < p^{k+1}$ *for* $i \in I_k - I_{k+1}$ *and* $0 \leqslant k \leqslant r$ (it is clear that if $\alpha \in L_r$ and $\beta \leqslant \alpha$, then $\beta \in L_r$).

B) *For any* $\alpha \in L_r$,

$$\Delta(Z_\alpha) = \sum_{\beta+\gamma=\alpha} Z_\beta \otimes Z_\gamma. \tag{12}$$

C) *For* $\alpha = \varepsilon_i$, $i \in I$, *we have* $Z_\alpha = X_i$.

It is clear that if $(Z_\alpha)$ is such a basis (with the convention $Z_0 = 1$), then, for any integer $h \geqslant 0$, the cogebra $M_h$ of the minimal filtration of $\mathfrak{G}$ (No. 2) is just the vector subspace having as a basis the $Z_\alpha$ such that $|\alpha| \leqslant h$; this follows at once from condition C) for $h = 1$, and is proved by induction on $h$ exactly as in the proof of Proposition 1.

We will use the following lemmas:

**Lemma 1.** *Let* $(x_i)_{i\in I}$ *be a family of elements of* $\mathfrak{R}$, *and for each multiindex* $\alpha \in \mathbf{N}^{(I)}$ *write*

$$x^\alpha = \prod_{i\in I} x_i^{\alpha(i)}.$$

*If* $|\alpha| = h$, *then we have for the coproduct* $c$ *in* $\mathfrak{G}$

$$c(x^\alpha) - \sum_{(\beta,\gamma)} \frac{\alpha!}{\beta!\,\gamma!} x^\beta \otimes x^\gamma \in \mathfrak{R} \mathbin{\hat{\otimes}} \mathfrak{R}^h + \mathfrak{R}^2 \mathbin{\hat{\otimes}} \mathfrak{R}^{h-1} + \cdots \tag{13}$$
$$\cdots + \mathfrak{R}^{h-1} \mathbin{\hat{\otimes}} \mathfrak{R}^2 + \mathfrak{R}^h \mathbin{\hat{\otimes}} \mathfrak{R}$$

*where, in the summation,* $(\beta, \gamma)$ *runs through the set of finite pairs of multiindices satisfying the following condition: if*

$$\alpha = \alpha_0 + p\alpha_1 + \cdots + p^m\alpha_m$$

*is the natural decomposition* (1) *of the multiindex* $\alpha$,

$$\beta = \beta_0 + p\beta_1 + \cdots + p^m\beta_m, \qquad \gamma = \gamma_0 + p\gamma_1 + \cdots + p^m\gamma_m$$

*the similar decompositions of* $\beta$ *and* $\gamma$, *then one must have*

$$(14) \qquad \beta_j + \gamma_j = \alpha_j \qquad \text{for every index } j \text{ such that } 0 \leqslant j \leqslant m.$$

*In particular, for* $\alpha = p^m\varepsilon_i$, *the sum on the left-hand side of* (13) *reduces to* $1 \otimes x^\alpha + x^\alpha \otimes 1$.

The lemma easily follows from I, §2, formula (8): by induction on $h$, one may write

$$c(x^\alpha) = \prod_{i \in \mathrm{I}} (c(x_i))^{\alpha(i)} = \mathrm{P}_\alpha + \mathrm{Q}_\alpha,$$

where $\mathrm{Q}_\alpha$ is an element in the right-hand side of (13), and

$$\mathrm{P}_\alpha = \prod_{i \in \mathrm{I}} (1 \otimes x_i + x_i \otimes 1)^{\alpha(i)} = \prod_{j=0}^{m} (\prod_{i \in \mathrm{I}} (1 \otimes x_i^{p^j} + x_i^{p^j} \otimes 1)^{\alpha_j(i)}).$$

Using the fact that each of the integers $\alpha_j(i)$ is $< p$ by definition and the definition of the factorial $\alpha!$ for a multiindex (No. 1), we readily obtain for $\mathrm{P}_\alpha$, by the binomial theorem, the sum in the left-hand side of (13).

**Lemma 2.** *Let* $(x_i)_{i \in \mathrm{I}}$ *be a family of elements of* $\mathfrak{G}^*$ *such that the classes of the* $x_i$ mod $\mathfrak{N}^2$ *generate* $\mathfrak{N}/\mathfrak{N}^2$ (i.e., *the smallest closed subspace of* $\mathfrak{N}/\mathfrak{N}^2$ *containing those classes is* $\mathfrak{N}/\mathfrak{N}^2$ *itself*). *Then the* $x^\alpha$ *for* $\alpha \in \mathbf{N}^{(\mathrm{I})}$ *generate* $\mathfrak{G}^*$ (*in the same sense*), *and the* $x^\alpha$ *with* $|\alpha| \geqslant m$ *generate* $\mathfrak{N}^m$.

As the filter base of the $\mathfrak{N}^m$ converges to 0 in $\mathfrak{G}^*$, to prove the first assertion it is enough to prove that for every integer $m \geqslant 2$, any element $\bar{y}$ in $\mathfrak{N}/\mathfrak{N}^m$ may be arbitrarily approximated by the class mod $\mathfrak{N}^m$ of a linear combination of the $x^\alpha$; we do this by induction on $m$, the result being a consequence of the definition for $m = 2$. If $\bar{y} \in \mathfrak{N}/\mathfrak{N}^{m+1}$, its class $\bar{\bar{y}}$ in $\mathfrak{N}/\mathfrak{N}^m$ can be approximated by linear combinations of classes of the $x^\alpha$ mod $\mathfrak{N}^m$; this means that for each open vector subspace V of $\mathfrak{N}/\mathfrak{N}^{m+1}$, there is a linear combination $\bar{z}$ of the classes of the $x^\alpha$ mod $\mathfrak{N}^{m+1}$ such that

$$\bar{y} - \bar{z} \in (\mathfrak{N}^m/\mathfrak{N}^{m+1}) + \mathrm{V}.$$

Hence we are reduced to proving the result for $\bar{y} \in \mathfrak{N}^m/\mathfrak{N}^{m+1}$. But then, by definition, if $y \in \mathfrak{N}^m$ is in the class $\bar{y}$, there exist polynomials in elements of $\mathfrak{R}$, *homogeneous* and of degree $m$, which arbitrarily approximate $y$. Replacing in these polynomials the elements of $\mathfrak{R}$ by elements in the same class mod $\mathfrak{R}^2$ does not change the classes of the polynomials mod $\mathfrak{R}^{m+1}$, hence each class of such a polynomial may be approximated arbitrarily by a linear combination of the classes of the $x^\alpha$ mod $\mathfrak{R}^{m+1}$, from the definition of the $x_i$; this proves the first assertion. This also shows that the classes mod $\mathfrak{R}^{m+1}$ of the $x^\alpha$ such that $|\alpha| = m$ generate $\mathfrak{R}^m/\mathfrak{R}^{m+1}$. A similar induction then shows that the classes mod $\mathfrak{R}^{m+n}$ of the $x^\alpha$ with $|\alpha| \geqslant m$ generate $\mathfrak{R}^m/\mathfrak{R}^{m+n}$, hence the second assertion.

These lemmas being proved, we use the natural duality between $\mathfrak{g}_0$ and $\mathfrak{R}/\mathfrak{R}^2$ (No. 2) to take in $\mathfrak{R}/\mathfrak{R}^2$ the pseudobasis $(\bar{x}_i)_{i\in I}$ dual to $(X_i)_{i\in I}$.

When I is infinite, if we choose *arbitrarily* an element $x_i$ in each class $\bar{x}_i$, these elements will *not* necessarily form a *pseudobasis* of a closed subspace in $\mathfrak{R}$; but since there always exists a closed subspace in $\mathfrak{R}$ which is a topological supplement to $\mathfrak{R}^2$ (I, §2, No. 2), it is always possible to choose the $x_i$ such that they form a pseudobasis of that subspace. Furthermore, when that condition is satisfied, it still holds if we modify arbitrarily a *finite* number of the $x_i$ in their classes (*loc. cit.*).

We will always suppose that the $x_i$ have been chosen satisfying the preceding condition.

**Theorem 1.** *Let $\mathfrak{G}$ be a bigebra of finite height $r$. It is possible to choose the $x_i$ such that in addition $x_i^{p^k} = 0$ for $i \in I_{k-1} - I_k$ and $1 \leqslant k \leqslant r$; for any such choice, the $x^\alpha$ for $\alpha \in L_r$ form a pseudo-basis of $\mathfrak{G}^*$, and the basis $(Z_\alpha)$ of $\mathfrak{G}$ dual to $(x^\alpha)$ is a privileged basis with respect to $(X_i)$.*

We first use induction on $r$: proposition 1 shows that the theorem is true for $r = 0$, and we assume it has been proved for bigebras of height $r - 1$, in particular for $\mathfrak{s}_{r-1}$. For a choice of the $x_i$ corresponding to the bigebra $\mathfrak{s}_{r-1}$, we therefore have, in $\mathfrak{G}^* = \mathfrak{s}_r^*$, since $\mathfrak{s}_{r-1}^* = \mathfrak{G}^*/\mathfrak{B}_r$,

$$(15) \qquad x_i^{p^k} \in \mathfrak{B}_r \oplus \mathfrak{R}^{p^r} \qquad \text{for} \quad i \in I_{k-1} - I_k, \quad 1 \leqslant k \leqslant r$$

(the case $k = r$ follows from the definition of the height). We are first going to show that it is possible to modify the choice of the $x_i$ in such a way that

$$(16) \qquad x_i^{p^k} \in \mathfrak{R}^{p^r+1} \qquad \text{for} \quad i \in I_{k-1} - I_k, \quad 1 \leqslant k \leqslant r + 1.$$

Let $E_k$ be the closed subspace generated by the $x_i$ such that $i \in I_{k-1} - I_k$, $1 \leqslant k \leqslant r + 1$; by assumption these $x_i$ form a pseudobasis of $E_k$. By definition, the image of $E_{r+1}$ in $\mathfrak{N}/\mathfrak{N}^2$ is supplementary to $N_r$; hence the restriction of $F^r$ to $E_{r+1}$ is a *diisomorphism* of $E_{r+1}$ onto a closed subspace $T_r \subset \mathfrak{B}_r \subset \mathfrak{N}^{p^r}$, which is supplementary to $\mathfrak{B}_r \cap \mathfrak{N}^{p^r+1}$ in $\mathfrak{B}_r$; let $f$ be the projection of $\mathfrak{B}_r$ onto $T_r$ corresponding to this direct sum decomposition; $z \mapsto f(z^{p^k})$ is then a continuous semilinear mapping of $E_k$ into $T_r$ for $1 \leqslant k \leqslant r$. On the other hand, $F^k$ is a diisomorphism of $F^{r-k}(E_{r+1})$ onto $T_r$; let $g$ be the inverse diisomorphism; then $z \mapsto g(f(z^{p^k}))$ is a continuous linear mapping of $E_k$ into $F^{r-k}(E_{r+1})$, and it follows at once from the definitions that

$$(z - g(f(z^{p^k})))^{p^k} \in \mathfrak{N}^{p^r+1}.$$

As $u : z \mapsto z - g(f(z^{p^k}))$ is a continuous linear mapping, the image $E'_k = u(E_k)$ is still a closed subspace of $\mathfrak{N}$ and as

$$g(f(z^{p^k})) \in \mathfrak{N}^2,$$

$u$ is an isomorphism of $E_k$ onto $E'_k$. We may therefore replace $x_i$ by $x'_i = u(x_i)$ for $i \in E_{k-1} - E_k$, $1 \leqslant k \leqslant r$ so that the $x'_i$ still form a pseudobasis of $E'_k$ and in addition verify (16) (for $k = r + 1$, one of course has $x_i^{p^r+1} = 0$ since $\mathfrak{G}$ has height $r$, so we have no need to modify the $x_i \in E_{r+1}$).

We may therefore assume that (16) is satisfied. The classes of the $x^\alpha$ mod $\mathfrak{N}_i^{p^r+1}$ for $\alpha \in L_{r-1}$ and $|\alpha| \leqslant p^r$ form by the inductive assumption a pseudobasis of $\mathfrak{G}^*/(\mathfrak{B}_r + \mathfrak{N}^{p^r+1})$, whose dual is a basis $(U_\alpha)$ of $M_{p^r} \cap \mathfrak{s}_{r-1}$, which is the subspace orthogonal to $\mathfrak{B}_r + \mathfrak{N}^{p^r+1}$. Let V be a closed supplementary to the direct sum $T_r + \mathfrak{N}^{p^r+1}$ in $\mathfrak{G}^*$, so that it is isomorphic to $\mathfrak{G}^*/(\mathfrak{B}_r + \mathfrak{N}^{p^r+1})$; $M_{p^r}$ is then the direct sum of $M_{p^r} \cap \mathfrak{s}_{r-1}$ and of the subspace orthogonal to $V + \mathfrak{N}^{p^r+1}$, which is isomorphic to the dual of the closed subspace $T_r$ of $\mathfrak{G}^*$. There is therefore a basis $(U_{p^r\varepsilon_i})$ (with $i \in I_r$) of $(V + \mathfrak{N}^{p^r+1})^0$, dual to the pseudobasis $(x_i^{p^r})_{i \in I_r}$ of the closed space $T_r$. It is then clear that we have

(17) $$\langle x^\lambda, U_\alpha \rangle = \delta_{\alpha\lambda} \qquad \text{(Kronecker delta)},$$

for all multiindices $\alpha$, $\lambda$ belonging to $L_r$ and of absolute values $|\alpha| \leqslant p^r$, $|\lambda| \leqslant p^r$. Furthermore, we have

(18) $$\langle x^\mu, U_\alpha \rangle = 0 \qquad \text{for} \quad \mu \notin L_r \quad \text{and} \quad \alpha \in L_r, \quad |\alpha| \leqslant p^r.$$

Indeed, if $\mu \notin L_r$, this means that for some $k$ with $1 \leqslant k \leqslant r + 1$ and some $i \in I_{k-1} - I_k$, we have $\mu(i) \geqslant p^k$: but as $x^\mu = x_i^{p^k} x^\nu$ with $\nu \geqslant 0$, it follows from (16) that $x^\mu \in \mathfrak{N}^{p^r+1}$, and this implies (18) by definition of $M_{p^r}$. The relations (17) and (18) imply

$$\text{(19)} \qquad \Delta(U_\alpha) = \sum_{\beta+\gamma=\alpha} U_\beta \otimes U_\gamma \qquad \text{for} \quad \alpha \in L_r, \quad |\alpha| \leqslant p^r.$$

Indeed, by (5) we may write $\Delta(U_\alpha) - 1 \otimes U_\alpha - U_\alpha \otimes 1$ as a linear combination of the $U_\lambda \otimes U_\mu$ with $|\lambda| < |\alpha|$, $|\mu| < |\alpha|$, $\lambda$ and $\mu$ belonging to $L_r$. To determine the coefficients of these basis elements, we have to compute

$$\langle \Delta(U_\alpha), x^\lambda \otimes x^\mu \rangle = \langle U_\alpha, x^{\lambda+\mu} \rangle,$$

and this gives 0 if

$$|\lambda| + |\mu| \geqslant p^r + 1,$$

or if

$$\lambda + \mu \notin L_r$$

by (18), or if

$$\lambda + \mu \in L_r \quad \text{and} \quad |\lambda| + |\mu| \leqslant p^r \quad \text{but} \quad \lambda + \mu \neq \alpha$$

by (17); hence (19).

We now propose to prove, by induction on $h \geqslant p^r$, that for any such integers $h$, there is a choice of the $x_i \in \bar{x}_i$ such that

$$\text{(20)} \qquad x_i^{p^k} \in \mathfrak{N}^{h+1} \qquad \text{for} \quad i \in I_{k-1} - I_k, \quad 1 \leqslant k \leqslant r+1$$

and the classes mod $\mathfrak{N}^{h+1}$ of the $x^\alpha$ such that $\alpha \in L_r$ and $|\alpha| \leqslant h$ form a *pseudobasis* of $\mathfrak{G}^*/\mathfrak{N}^{h+1}$.

We have already proved these assertions for $h = p^r$. We start with the inductive assumption for a value $h \geqslant p^r$ and we will prove it for $h + 1$. We at first keep the same choice of the $x_i$, and define, for each $\alpha \in L_r$ such that $|\alpha| = h + 1$, an element $Y_\alpha \in M_{h+1}$ such that, for all $\lambda \in L_r$ with $|\lambda| = h + 1$,

$$\text{(21)} \qquad \langle x^\lambda, Y_\alpha \rangle = \delta_{\alpha\lambda}.$$

From the definition of $L_r$ and the assumption $h + 1 > p^r$ it follows that $\alpha$ cannot be of the form $p^k \varepsilon_i$, hence it may be written $\beta + \gamma$ with $\beta \in \mathrm{L}_r$, $\gamma \in \mathrm{L}_r$, $\beta \neq 0$, $\gamma \neq 0$, and furthermore $\beta$ and $\gamma$ verifying condition (14) of Lemma 1. We denote by $(U_\mu)$ (for $\mu \in \mathrm{L}_r$ and $|\mu| \leqslant h$) the basis of $\mathrm{M}_h$ dual to the pseudobasis formed by the classes of the corresponding $x^\mu$ in $\mathfrak{G}^*/\mathfrak{N}^h$. The element

$$Y_\alpha = \frac{\beta!\gamma!}{\alpha!} U_\beta U_\gamma$$

is defined since $|\beta| \leqslant h$ and $|\gamma| \leqslant h$ and it belongs to $\mathrm{M}_{h+1}$ by (6) since by construction we have

$$U_\beta \in \mathrm{M}_{|\beta|} \qquad \text{and} \qquad U_\gamma \in \mathrm{M}_{|\gamma|}.$$

To prove (21), we first consider the case $\lambda = \alpha$, and write

$$\langle x^\alpha, Y_\alpha \rangle = \frac{\beta!\gamma!}{\alpha!} \langle x^\alpha, U_\beta U_\gamma \rangle = \frac{\beta!\gamma!}{\alpha!} \langle c(x^\alpha), U_\beta \otimes U_\gamma \rangle.$$

Now, for an element $z$ in any of the tensor products

$$\mathfrak{N}^j \mathbin{\hat{\otimes}} \mathfrak{N}^{h-j+2} \qquad (1 \leqslant j \leqslant h + 1),$$

we have either $|\beta| \leqslant j$ or $|\gamma| \leqslant h - j + 2$ since

$$|\beta| + |\gamma| = h + 1;$$

as $U_\beta$ is orthogonal to $\mathfrak{N}^{|\beta|+1}$ and $U_\gamma$ to $\mathfrak{N}^{|\gamma|+1}$, we have

$$\langle z, U_\beta \otimes U_\gamma \rangle = 0.$$

Lemma 1 and the inductive assumption then prove that $\langle x^\alpha, Y_\alpha \rangle = 1$. The same argument applies to prove that $\langle x^\lambda, Y_\alpha \rangle = 0$ for $|\lambda| = h + 1$, $\lambda \in \mathrm{L}_r$, and $\lambda \neq \alpha$.

We next modify the $Y_\alpha$ by adding to each an element of $\mathrm{M}_h$, so as to have

$$\langle x^\lambda, U_\alpha \rangle = \delta_{\alpha\lambda}$$

for the modified element $U_\alpha$ and for each $\lambda \in \mathrm{L}_r$ such that now $|\lambda| \leqslant h + 1$. Let

$$\langle x^\lambda, Y_\alpha \rangle = a_{\alpha\lambda}$$

for the multiindices $\lambda \in \mathrm{L}_r$ such that $|\lambda| \leqslant h$; as the classes of the $x^\lambda$ with $\lambda \in \mathrm{L}_r$ and $|\lambda| \leqslant h$ form a pseudobasis for $\mathfrak{G}^*/\mathfrak{N}^h$, we may have $a_{\alpha\lambda} \neq 0$ only for a finite number of multiindices $\lambda$, hence the element

$$U_\alpha = Y_\alpha - \sum_\lambda a_{\alpha\lambda} U_\lambda$$

is defined, belongs to $\mathrm{M}_{h+1}$, and satisfies our requirements, since

$$\langle x^\lambda, U_\alpha \rangle = \langle x^\lambda, Y_\alpha \rangle$$

by definition of $\mathrm{M}_h$ if $|\lambda| = h + 1$.

It is then obvious that the $U_\alpha$ for $\alpha \in \mathrm{L}_r$ and $|\alpha| \leqslant h + 1$ are linearly independent and therefore constitute a basis for a subspace $\mathrm{M}'_{h+1}$ of $\mathrm{M}_{h+1}$ containing $\mathrm{M}_h$; but it is *by no means obvious* that $\mathrm{M}'_{h+1} = \mathrm{M}_{h+1}$. In order to prove this we will first show how, by modifying the $x_i$, one may obtain the relations

$$(22) \qquad x_i'^{p^k} \in \mathrm{M}'^0_{h+1} \qquad \text{for} \quad i \in \mathrm{I}_{k-1} - \mathrm{I}_k, \quad 1 \leqslant k \leqslant r$$

for the modified $x'_i$. To do this, we first compute, for each $\alpha$ such that $\alpha \in \mathrm{L}_r$ and $|\alpha| = h + 1$, the elements

$$\frac{\alpha!}{\beta!\gamma!} \langle x_i^{p^k}, Y_\alpha \rangle = \langle x_i^{p^k}, U_\beta U_\gamma \rangle = \langle c(x_i^{p^k}), U_\beta \otimes U_\gamma \rangle,$$

with the above notations. Now

$$c(x_i^{p^k}) = (c(x_i))^{p^k}$$

can be written

$$1 \otimes x_i^{p^k} + x_i^{p^k} \otimes 1 + z,$$

where $z$ belongs to $\mathrm{F}^k(\mathfrak{N}) \hat{\otimes} \mathrm{F}^k(\mathfrak{N})$ (I, §2, No. 11, Eq. (8)). As $\mathrm{F}^k(\mathfrak{N}) \hat{\otimes} \mathrm{F}^k(\mathfrak{N})$ is the closed subspace generated by the elements $x^{p^k\lambda} \otimes x^{p^k\mu}$ for all multiindices $\lambda$, $\mu$, let us determine those elements which are not orthogonal to $U_\beta \otimes U_\gamma$. We must first exclude the cases where $p^k|\lambda| \geqslant h + 1$ or $p^k|\mu| \geqslant h + 1$, since $|\beta| \leqslant h$ and $|\gamma| \leqslant h$; similarly, we must exclude the cases in which $p^k\lambda \notin \mathrm{L}_r$ or $p^k\mu \notin \mathrm{L}_r$, for the same argument as in (18), starting from the assumption (20), shows that then

$$x^{p^k\lambda} \in \mathfrak{N}^{h+1} \qquad \text{or} \qquad x^{p^k\mu} \in \mathfrak{N}^{h+1}.$$

The only possibility is then

$$p^k\lambda = \beta, \qquad p^k\mu = \gamma,$$

which implies

$$h + 1 = |\beta| + |\gamma| = p^k m.$$

If we observe that, due to (20), we also have

$$\langle x_i^{p^k}, U_\alpha \rangle = 0 \qquad \text{for} \quad \alpha \in \mathrm{L}_r \quad \text{and} \quad |\alpha| \leqslant h,$$

we already see that if $h + 1$ is not divisible by $p^k$, we already have (22) for $i \in \mathrm{I}_{k-1} - \mathrm{I}_k$ without any modification of the $x_i$. If on the other hand $h + 1 = p^k m$, what we have proved is that

$$\mathrm{F}^k(\mathrm{E}_k) \subset \mathrm{F}^k(\mathrm{S}_m) + \mathrm{M}'^0_{h+1}$$

where $\mathrm{S}_m$ is the closed subspace generated by the $x^\lambda$ for $\lambda \in \mathrm{L}_r$ and $|\lambda| = m$. We may decompose the closed subspace $\mathrm{F}^k(\mathrm{S}_m)$ of $\mathfrak{G}^*$ into the direct sum of $\mathrm{F}^k(\mathrm{S}_m) \cap \mathrm{M}'^0_{h+1}$ and a closed subspace $\mathrm{Q}_m$; let $z \mapsto q(z)$ be the corresponding projection of $\mathrm{F}^k(\mathrm{S}_m)$ onto $\mathrm{Q}_m$. The inverse image of $\mathrm{Q}_m$ in $\mathrm{S}_m$ by $\mathrm{F}^k$ can similarly be decomposed into the direct sum of the kernel of $\mathrm{F}^k \mid \mathrm{S}_m$ and a closed subspace $\mathrm{P}_m$; the restriction of $\mathrm{F}^k$ to $\mathrm{P}_m$ is then a diisomorphism of $\mathrm{P}_m$ onto $\mathrm{Q}_m$; let $g$ be the inverse diisomorphism; then $z \mapsto g(q(z^{p^k}))$ is a continuous linear mapping of $\mathrm{E}_k$ into $\mathrm{P}_m$, and it follows from the definitions that

$$(z - g(q(z^{p^k})))^{p^k} \in \mathrm{M}'^0_{h+1}.$$

As

$$u\colon z \mapsto z - g(q(z^{p^k}))$$

is a continuous linear mapping, $\mathrm{E}'_k = u(\mathrm{E}_k)$ is still a closed subspace of $\mathfrak{N}$, and as

$$g(q(z^{p^k})) \in \mathfrak{R}^2,$$

$u$ is an isomorphism of $E_k$ onto $E'_k$. We may therefore replace $x_i$ by $x'_i = u(x_i)$ for $i \in I_{k-1} - I_k$, and $1 \leqslant k \leqslant r$, and (22) is then verified.

Taking now as representatives of the $\bar{x}_i$ the $x'_i$ instead of the $x_i$, we may repeat for them the arguments above leading to the definition of the $Y_\alpha$. We now obtain a new family $(U'_\alpha)$ of elements of $\mathfrak{G}$, for $\alpha \in L_r$ and $|\alpha| \leqslant h + 1$; they still form a basis of $M'_{h+1}$, and those corresponding to $|\alpha| \leqslant h$ form a basis of $M_h$. This is due to the fact that we have

$$x'_i - x_i \in \mathfrak{R}^{[(h+1)/p^r]}$$

by the preceding construction, hence

$$x'^\lambda - x^\lambda \in \mathfrak{R}^{|\lambda| + [(h+1)/p^r] - 1}$$

for any multiindex $\lambda$. This already shows that

$$U'_\alpha = U_\alpha \qquad \text{for} \quad |\alpha| < [(h + 1)/p^r];$$

for larger values of $|\alpha|$, the elements $\langle U_\alpha, x'^\lambda \rangle = a_{\alpha\lambda}$ are 0 for $|\lambda| \geqslant |\alpha|$ with the exception of the value $\lambda = \alpha$, for which $a_{\alpha\alpha} = 1$; for $|\lambda| < |\alpha|$, they are 0 except for a finite number of them (since the classes of the $x'^\lambda$ form a pseudobasis of $\mathfrak{G}^*/\mathfrak{R}^{h+2}$), and we have

$$U'_\alpha = U_\alpha - \sum_{|\lambda| < |\alpha|} a_{\alpha\lambda} U'_\lambda,$$

proving our assertion by induction on $|\alpha|$. We therefore still have

$$\langle x'^\lambda, U'_\alpha \rangle = \delta_{\alpha\lambda} \qquad \text{for} \quad \alpha, \lambda \text{ in } L_r, \quad |\alpha| \leqslant h + 1, \quad |\lambda| \leqslant h + 1$$

$$\langle x'^\mu, U'_\alpha \rangle = 0 \qquad \text{for} \quad \mu \notin L_r, \quad \alpha \in L_r \quad \text{and} \quad |\alpha| \leqslant h + 1.$$

From these relations one deduces

$$\Delta(U'_\alpha) = \sum_{\beta+\gamma=\alpha} U'_\beta \otimes U'_\gamma \qquad \text{for} \quad \alpha \in L_r, \quad |\alpha| \leqslant h + 1$$

by the same argument as in the case $h + 1 = p^r$ treated above.

It is now possible to prove that $M'_{h+1} = M_{h+1}$. Indeed, let $Z$ be an arbitrary element of $M_{h+1}$: using the fact that the $U'_\beta$ for $|\beta| \leqslant h$ form a basis of $M_h$, we consider the element

$$\Delta(Z) - 1 \otimes Z - Z \otimes 1,$$

and repeat the argument made in Proposition 1, which proves that there is an element $\sum_\beta b_\beta U'_\beta$ in $M'_{h+1}$ such that if

$$Y = Z - \sum_\beta b_\beta U'_\beta,$$

then

$$\Delta(Y) - 1 \otimes Y - Y \otimes 1 = 0;$$

as this implies

$$Y \in \mathfrak{g}_0 \subset M'_{h+1},$$

we have proved that $M'_{h+1} = M_{h+1}$, and completed the passage from $h$ to $h + 1$ in our inductive argument.

To conclude the proof of Theorem 1, observe that we have to make at each passage from $h$ to $h + 1$ a modification of representatives $x_i^{(h)}$ of the $\bar{x}_i$ to new representatives $x_i^{(h+1)}$ (at least when $h + 1$ is divisible by $p$) and this involves a corresponding passage from a basis $(U_\alpha^{(h)})$ of $M_h$ to a basis $(U_\alpha^{(h+1)})$ of $M_{h+1}$, which does not coincide with $(U_\alpha^{(h)})$ for $|\alpha| \leqslant h$. However, we have seen that we *do have*

$$U_\alpha^{(h+1)} = U_\alpha^{(h)} \qquad \text{for} \quad |\alpha| \leqslant [(h + 1)/p^r].$$

We may therefore, for *each* multiindex $\alpha \in L_r$, define $Z_\alpha$ as the common value of the $U_\alpha^{(h)}$ for all $h$ such that

$$[(h + 1)/p^r] > |\alpha|.$$

It is then clear that the $Z_\alpha$ form a privileged basis of $\mathfrak{G}$. Furthermore, the relation

$$x_i^{(h+1)} - x_i^{(h)} \in \mathfrak{N}^{[(h+1)/p^r]}$$

proves that for each $i \in I$ the sequence $(x_i^{(h)})_{h \geqslant p^r}$ tends to a limit $x_i$ in $\mathfrak{G}^*$ ($\mathfrak{G}^*$ being complete), and the corresponding monomials $x^\alpha$ for $\alpha \in L_r$

obviously form the pseudobasis of $\mathfrak{G}^*$ dual to $(Z_\alpha)$. Furthermore, for $i \in \mathrm{I}_{k-1} - \mathrm{I}_k$ and $1 \leqslant k \leqslant r$, we have shown that $(x_i^{(h)})^{pk}$ belongs to $\mathfrak{R}^{h+1}$; letting $h$ tend to $+\infty$, we obtain the relation $x_i^{pk} = 0$, and our proof is complete.

It may easily be deduced from this result that for an *arbitrary* choice of the $x_i$, there is still a basis $(U_\alpha)$ (for $\alpha \in \mathrm{L}_r$) of $\mathfrak{G}$ such that $\langle x^\lambda, U_\alpha \rangle = \delta_{\alpha\lambda}$ for any two indices in $\mathrm{L}_r$; but in general the basis $(U_\alpha)$ is not a privileged one, the $x^\lambda$ do not form its dual basis in $\mathfrak{G}^*$, and the $x_i^{pk}$ for $i \in \mathrm{I}_{k-1} - \mathrm{I}_k$, $1 \leqslant k \leqslant r$, are not necessarily 0.

**Corollary 1.** *Let*

$$\alpha = \alpha_0 + p\alpha_1 + \cdots + p^r\alpha_r, \qquad \beta = \beta_0 + p\beta_1 + \cdots + p^r\beta_r$$

*be two elements of* $\mathrm{L}_r$, *where the multiindices* $\alpha_j$, $\beta_j$ *have height* 0 *and are such that* $\alpha_j + \beta_j$ *also has height* 0 *for every* $j$; *then*

$$Z_\alpha Z_\beta - \frac{(\alpha + \beta)!}{\alpha!\,\beta!} Z_{\alpha+\beta} \in \mathrm{M}_{|\alpha| + |\beta| - 1}. \tag{23}$$

This is another form of (21).

Let us change the notations introduced at the beginning of this section by writing $X_{0i}$ instead of $X_i$ for the basis of $\mathfrak{g}_0$. For $s \leqslant r$ and $i \in \mathrm{I}_s$, let us write

$$X_{si} = Z_{p^s\varepsilon_i}. \tag{24}$$

Let us consider on the set of pairs $(s, i)$ such that $0 \leqslant s \leqslant r$ and $i \in \mathrm{I}_s$ an arbitrary total order. For each $\alpha \in \mathrm{L}_r$, write

$$\alpha = \alpha_0 + p\alpha_1 + \cdots + p^r\alpha_r,$$

where the $\alpha_h$ have height 0, and

$$X_\alpha = X_{h_1 i_1}^{\gamma_1} X_{h_2 i_2}^{\gamma_2} \cdots X_{h_m i_m}^{\gamma_m} \tag{25}$$

where

$$(h_1, i_1) < (h_2, i_2) < \cdots < (h_m, i_m)$$

is the strictly increasing sequence of the pairs $(h, i)$ such that $\alpha_h(i) \neq 0$ and where

$$\gamma_k = \alpha_{h_k}(i_k).$$

If we then write

$$\beta = \alpha - p^{h_1}\varepsilon_{i_1},$$

it follows from (23) that

$$X_{h_1 i_1} Z_\beta - \alpha_{h_1}(i_1) Z_\alpha \in \mathrm{M}_{|\alpha|-1}$$

hence, by an easy induction,

$$Z_\alpha - \frac{1}{\alpha!} X_\alpha \in \mathrm{M}_{|\alpha|-1}. \tag{26}$$

**Corollary 2.** *The $X_\alpha$ for $\alpha \in \mathrm{L}_r$ form a basis of $\mathfrak{G}$; more precisely, the $X_\alpha$ such that $|\alpha| = n$ form a basis of a supplementary to $\mathrm{M}_{n-1}$ in $\mathrm{M}_n$.*

As $X_\alpha \in \mathrm{M}_{|\alpha|}$ by (6), this follows from (26) and from the fact that the $Z_\alpha$ have the same properties.

**Corollary 3.** *If $\alpha$ and $\beta$ satisfy the conditions of Corollary 1, then*

$$X_{\alpha+\beta} - X_\alpha X_\beta \in \mathrm{M}_{|\alpha|+|\beta|-1} \tag{27}$$

*hence*

$$[X_\alpha, X_\beta] \in \mathrm{M}_{|\alpha|+|\beta|-1}. \tag{28}$$

This follows at once from (23) and (26).

*Example.* Let $\mathbf{G}_a$ (or $\mathbf{G}_a(k)$) be the formal group over $k$ associated to the additive group $\mathbf{G}_a$ (I, §2, No. 2). The bigebra $\mathfrak{G}^*$ is therefore the algebra $k[[x]]$ of power series in one indeterminate $x$, with the comultiplication defined by

$$c(x) = 1 \otimes x + x \otimes 1.$$

This implies, by the computation of Lemma 1, that

$$c(x^\alpha) = \sum_{(\beta,\gamma)} \frac{\alpha!}{\beta!\gamma!} x^\beta \otimes x^\gamma$$

where here $\alpha$, $\beta$, $\gamma$ run through the set $\mathbf{N}$ of integers $\geqslant 0$, but the factorials have to be defined as in No. 1, and the integers $\beta$, $\gamma$ must satisfy the conditions of lemma 1, when written in the base $p$. This formula gives dually the multiplication law of the privileged basis $(Z_\alpha)$ dual to the pseudobasis $(x^\alpha)$ of $\mathfrak{G}^*$:

$$Z_\beta Z_\gamma = \frac{\alpha!}{\beta!\gamma!} Z_\alpha \qquad \textit{if} \quad \beta + \gamma = \alpha,$$

*and* $\beta$ and $\gamma$ satisfy the conditions of Lemma 1;

$$Z_\beta Z_\gamma = 0 \qquad \text{otherwise.}$$

This implies that if we write here

$$X_0 = Z_1, \quad X_1 = Z_p, \quad \ldots, \quad X_h = Z_{p^h}, \quad \ldots,$$

the expression of $Z_\alpha$ as function of the $X_h$ is $(1/\alpha!)X_\alpha$; in other words, the expressions on the left-hand sides of (23) and (26) are here 0.

**7.** *The shift and the higher Lie algebras.* To the Frobenius homomorphism F of $\mathfrak{G}^*$ into $\mathfrak{G}^{*(1)}$ there corresponds by transposition (I, §2, No. 6) a *bigebra* homomorphism $\mathrm{V}: \mathfrak{G} \to \mathfrak{G}^{(-1)}$ which we call the *shift* in $\mathfrak{G}$, and which is defined by the identity

(29) $$\langle x, \mathrm{V}(X)\rangle^p = \langle x^p, X\rangle \qquad \text{for} \quad X \in \mathfrak{G} \quad \text{and} \quad x \in \mathfrak{G}^*.$$

It is possible to give for V a direct definition which does not use duality (P. Cartier [9]).

For any bigebra homomorphism $u: \mathfrak{G} \to \mathfrak{G}'$, ${}^t u: \mathfrak{G}'^* \to \mathfrak{G}^*$ may also be considered as a homomorphism of $\mathfrak{G}'^{*(1)}$ into $\mathfrak{G}^{*(1)}$, and as such, commutes with the Frobenius homomorphism; by transposition, we thus see that the diagram

$$\begin{array}{ccc} \mathfrak{G} & \xrightarrow{u} & \mathfrak{G}' \\ {\scriptstyle V}\downarrow & & \downarrow{\scriptstyle V} \\ \mathfrak{G}^{(-1)} & \xrightarrow[u]{} & \mathfrak{G}'^{(-1)} \end{array}$$

is commutative. In particular, every subbigebra and every biideal in $\mathfrak{G}$ is *stable* by V.

As, for each $k \leqslant r$, the Frobenius subbigebra $\mathfrak{s}_k$ is orthogonal to the closed biideal $\mathfrak{B}_{k+1}$ generated by $F^{k+1}(\mathfrak{N})$, it may also be defined as *the largest subbigebra having its augmentation ideal contained in the biideal* $\mathrm{Ker}(V^{k+1})$ (I, §2, No. 3).

If $(Z_\alpha)$ is a privileged basis of $\mathfrak{G}$, then

$$V(Z_\alpha) = Z_{\alpha/p} \tag{30}$$

where $\alpha/p$ is the multiindex $i \mapsto \alpha(i)/p$ when all the integers $\alpha(i)$ are divisible by $p$, and by convention $Z_{\alpha/p} = 0$ if that condition is not satisfied. Indeed, by (29), we have, for any two multiindices $\alpha$, $\beta$ in $L_r$,

$$\langle x^\beta, V(Z_\alpha)\rangle^p = \langle x^{p\beta}, Z_\alpha\rangle$$

from which (30) immediately follows.

**Proposition 2.** *For each $k \leqslant r$,*

$$\mathfrak{g}_k = (M_{p^k} + \mathfrak{s}_{k-1}) \cap \mathfrak{G}^+$$

*is a p-Lie algebra (also written $\mathfrak{g}_k(\mathfrak{G})$), in which $\mathfrak{s}_{k-1} \cap \mathfrak{G}^+$ is a p-ideal, and $\mathfrak{s}_k$ is the subalgebra of $\mathfrak{G}$ generated by $\mathfrak{g}_k$. If $(Z_\alpha)$ is a privileged basis of $\mathfrak{G}$, the $Z_\alpha$ such that $\alpha \neq 0$ and either $\mathrm{ht}(\alpha) \leqslant k$ or $\alpha = p^k\varepsilon_i$ with $i \in I_k$ form a basis of $\mathfrak{g}_k$.*

The fact that $\mathfrak{g}_k$ generates the algebra $\mathfrak{s}_k$ follows at once from the other assertions and from (26), by induction on $|\alpha|$.

Let us first show that

$$[X_{ki}, Y] \in \mathfrak{s}_{k-1} \qquad \text{for} \quad i \in I_k, \quad Y \in \mathfrak{s}_{k-1}.$$

By definition of the ideal $\mathfrak{B}_k$ in $\mathfrak{G}^*$, it is enough to prove that for $i \in I_k$, $j \in I_k$, $\lambda \in \mathbf{N}^{(I)}$, and $\alpha \in L_{k-1}$, one has

$$(31) \qquad \langle x_j^{p^k} x^\lambda, [X_{ki}, Z_\alpha] \rangle = 0.$$

Use induction on $|\alpha|$, the result being trivial if $|\alpha| = 0$. The left-hand side of (31) is equal to

$$\langle x_j^{p^k} \otimes x^\lambda, [\Delta(X_{ki}), \Delta(Z_\alpha)] \rangle.$$

But by (12),

$$\Delta(X_{ki}) - 1 \otimes X_{ki} - X_{ki} \otimes 1 \in \mathfrak{s}_{k-1} \otimes \mathfrak{s}_{k-1} \quad \text{and} \quad \Delta(Z_\alpha) \in \mathfrak{s}_{k-1} \otimes \mathfrak{s}_{k-1};$$

as $x_j^{p^k}$ is orthogonal to $\mathfrak{s}_{k-1}$, the only terms in $[\Delta(X_{ki}), \Delta(Z_\alpha)]$ which might give a term $\neq 0$ are those of type

$$[X_{ki}, Z_\beta] \otimes Z_\gamma \qquad \text{and} \qquad Z_\beta \otimes [X_{ki}, Z_\gamma]$$

with $\beta + \gamma = \alpha$. If both $\beta$ and $\gamma$ are $\neq 0$, these terms belong to $\mathfrak{s}_{k-1} \otimes \mathfrak{s}_{k-1}$ owing to the inductive assumption, hence give again 0 in (31). We thus are reduced to considering the cases $\beta = 0$ and $\gamma = 0$; the case $\beta = 0$ gives again 0, since $x_j^{p^k}$ is orthogonal to $\mathfrak{s}_{k-1}$. For $\gamma = 0$, we have only to consider the case $\lambda = 0$, since the unit 1 in $\mathfrak{G}^*$ is the only monomial $x^\lambda$ not orthogonal to $Z_0$. We thus have only to prove that

$$\langle x_j^{p^k}, [X_{ki}, Z_\alpha] \rangle = 0;$$

in fact, we will show that we already have

$$\langle x_j^{p^k}, X_{ki} Z_\alpha \rangle = 0 \qquad \text{and} \qquad \langle x_j^{p^k}, Z_\alpha X_{ki} \rangle = 0.$$

Indeed, we have

$$\langle x_j^{p^k}, X_{ki} Z_\alpha \rangle = \langle c(x_j^{p^k}), X_{ki} \otimes Z_\alpha \rangle = \langle (c(x_j))^{p^k}, X_{ki} \otimes Z_\alpha \rangle.$$

But $c(x_j)^{p^k}$ belongs to $\mathrm{F}^k(\mathfrak{N}) \hat{\otimes} \mathrm{F}^k(\mathfrak{N})$, and as $\alpha \in \mathrm{L}_{k-1}$, $Z_\alpha$ is orthogonal to $\mathrm{F}^k(\mathfrak{N})$, hence our assertion.

Next we show that

$$[X_{ki}, X_{kj}] \in \mathfrak{g}_k \qquad \text{for} \quad i, j \text{ in } \mathrm{I}_k;$$

this will result from the fact that, for $h \in \mathrm{I}_k$, $\lambda \in \mathbf{N}^{(\mathrm{I})}$, *and* $\lambda \neq 0$,

(32) $$\langle x_h^{p^k} x^{\lambda}, [X_{ki}, X_{kj}] \rangle = 0$$

since by definition of $\mathfrak{g}_k$, it is orthogonal to the intersection of $\mathfrak{B}_k$ and $\mathfrak{N}^{p^k+1}$. We use the same method as above, observing that, for $\alpha$ and $\beta$ in $\mathrm{L}_{k-1}$, we have

$$[X_{ki} \otimes Z_0, Z_\alpha \otimes Z_\beta] = [X_{ki}, Z_\alpha] \otimes Z_\beta \in \mathfrak{s}_{k-1} \otimes \mathfrak{s}_{k-1},$$

due to what has been proved above. We are therefore reduced to proving that

$$\langle x_h^{p^k} \otimes x^{\lambda}, [X_{ki}, X_{kj}] \otimes Z_0 \rangle = 0,$$

which follows at once from the fact that $\lambda \neq 0$.

Finally, we prove that $X_{ki}^p \in \mathfrak{g}_k$ for $i \in \mathrm{I}_k$; under the same assumptions as in (32), we must show that

$$\langle x_h^{p^k} x^{\lambda}, X_{ki}^p \rangle = 0,$$

which, as before, is equivalent to

(33) $$\langle x_h^{p^k} \otimes x^{\lambda}, (\Delta(X_{ki}))^p \rangle = 0.$$

But from (12) and the Jacobson formula giving the $p$-th power of a sum of noncommuting elements [32], it follows that $(\Delta(X_{ki}))^p$ is the sum of

$$Z_0 \otimes X_{ki}^p + X_{ki}^p \otimes Z_0$$

(which give 0 in (33) since $\lambda \neq 0$), and of "alternants" formed from

$$Z_0 \otimes X_{ki} + X_{ki} \otimes Z_0$$

and terms of the form $Z_\alpha \otimes Z_\beta$ with $\alpha$, $\beta$ in $\mathrm{L}_{k-1}$, by repeated bracketing. However, the first part of the proof shows that such terms all belong to $\mathfrak{s}_{k-1} \otimes \mathfrak{s}_{k-1}$, and this proves (33) and the proposition.

**Corollary.** *The Lie algebra $\mathfrak{g}_k$ may be defined as the set of elements $Y \in \mathfrak{G}^+$ such that*

$$\Delta(Y) - Z_0 \otimes Y - Y \otimes Z_0$$

*belongs to $\mathfrak{s}_{k-1} \otimes \mathfrak{s}_{k-1}$.*

If we express $Y$ as a linear combination of the $Z_\alpha$ and use the description of the basis of $\mathfrak{g}_k$ given in Proposition 2, this follows at once from (12).

It follows also from Proposition 2 and (30) that

(34) $$\mathrm{V}^k(\mathfrak{g}_k) \subset \mathfrak{g}_{k-h}$$

for any $h > 0$, $\mathfrak{g}_{k-h}$ being replaced by 0 if $h > k$. In particular,

(35) $$\mathrm{V}^h(\mathfrak{g}_k) = \mathfrak{g}_0 \cap \mathrm{V}^k(\mathfrak{G}) = \mathfrak{g}'_k$$

which shows that the subspaces $\mathfrak{g}'_k$ of $\mathfrak{g}_0$, defined in (11), are *sub-p-Lie algebras* of $\mathfrak{g}_0$.

We will say that the $\mathfrak{g}_k$ $(k \leqslant r)$ are the *higher Lie algebras* of $\mathfrak{G}$. The fact that

$$\mathrm{V}^k(X_{ki}) = X_{0i} \qquad \text{for} \quad i \in \mathrm{I}_k$$

shows that the structure of the quotient $p$-Lie algebras $\mathfrak{g}_k/(\mathfrak{s}_{k-1} \cap \mathfrak{G}^+)$ is already determined by the structure of the sub-Lie algebras $\mathfrak{g}'_k$ of $\mathfrak{g}_0$.

**8.** *The existence of antipodisms.* An antipodism $\alpha$ in the bigebra $\mathfrak{G}$ is characterized by the fact that it makes the two diagrams in (I, §2, No. 13) commutative. This first implies that, for any primitive element $X \in \mathfrak{g}_0 = \mathrm{P}_1$, we must have

(36) $$\alpha(X) = -X;$$

indeed, as $\eta(X) = 0$, the commutativity of the diagrams implies

$$(\mu \circ (1 \otimes \alpha) \circ \Delta)\,(X) = 0,$$

and as

$$\Delta(X) = Z_0 \otimes X + X \otimes Z_0 \qquad \text{and} \qquad \alpha(Z_0) = Z_0,$$

we get $\alpha(X) + X = 0$ since $Z_0$ is the unit in $\mathfrak{G}$.

By induction on $n$, we deduce that, for any cogebra $\mathrm{M}_n$ of the fundamental filtration of $\mathfrak{G}$ (No. 2),

(37) $$\alpha(\mathrm{M}_n) \subset \mathrm{M}_n,$$

for if $X \in \mathrm{M}_n$, we have, by (2),

$$(1 \otimes \alpha)(\Delta(X) - Z_0 \otimes X - X \otimes Z_0) \subset \mathrm{M}_1 \otimes \alpha(\mathrm{M}_{n-1}) + \mathrm{M}_2 \otimes \alpha(\mathrm{M}_{n-2}) + \cdots + \mathrm{M}_{n-1} \otimes \alpha(\mathrm{M}_1)$$

or, from the inductive assumption,

$$(1 \otimes \alpha)(\Delta(X)) - (Z_0 \otimes \alpha(X) + X \otimes Z_0) \subset \mathrm{M}_1 \otimes \mathrm{M}_{n-1} + \cdots + \mathrm{M}_{n-1} \otimes \mathrm{M}_1.$$

The commutativity of the diagrams in (I, §2, No. 13) then yields

$$\alpha(X) + X \in \mathrm{M}_n$$

in view of (6), which proves (37).

**Proposition 3.** *Under the assumptions of No. 2, there exists in the bigebra $\mathfrak{G}$ a unique antipodism.*

As an antipodism $\alpha$ must satisfy (37), it will be enough to prove that, if a cogebra homomorphism of $\mathrm{M}_n$ into itself makes the diagrams in (I, §2, No. 13) commutative, then it may be uniquely extended to $\mathrm{M}_{n+1}$ in order to have the same properties in that cogebra. As any $\mathrm{M}_n$ is contained in a Frobenius subbigebra $\mathfrak{s}_r$ of finite height, the proposition will be true for any bigebra $\mathfrak{G}$ of the type considered in No. 2 if it is true for such bigebras of *finite height*, and we will therefore assume $\mathfrak{G}$ has finite height. We may then take in $\mathfrak{G}$ a privileged basis $(Z_\lambda)_{\lambda \in \mathrm{L}_r}$, and the commutativity of the diagrams in (I, §2, No. 13) is then expressed by the following systems of equations:

$$\alpha(Z_\lambda) + Z_\lambda + \sum_{0<\mu<\lambda} \alpha(Z_\mu) Z_{\lambda-\mu} = 0, \tag{38}$$

$$Z_\lambda + \alpha(Z_\lambda) + \sum_{0<\mu<\lambda} Z_\mu \alpha(Z_{\lambda-\mu}) = 0. \tag{39}$$

Clearly, the system (38) already entirely determines $\alpha$, by induction on $|\lambda|$. We have to show that the solution of that system is also a solution of (39) and is a *cogebra* homomorphism, i.e.,

$$\Delta(\alpha(Z_\lambda)) = (\alpha \otimes \alpha)(\Delta(Z_\lambda)) \tag{40}$$

Let us first prove (40) by induction on $|\lambda|$. For $|\lambda| = 1$, $\alpha(Z_\lambda) = -Z_\lambda$, and as $Z_\lambda$ is a primitive element, Eq. (40) is trivial. In general, we deduce from (38) that

$$(\alpha \otimes \alpha)(\Delta(Z_\lambda)) = -Z_0 \otimes Z_\lambda - Z_\lambda \otimes Z_0 - Z_0 \otimes \Big(\sum_{0<\mu<\lambda} \alpha(Z_\mu)Z_{\lambda-\mu}\Big)$$
$$- \Big(\sum_{0<\mu<\lambda} \alpha(Z_\mu)Z_{\lambda-\mu}\Big) \otimes Z_0 + \sum_{0<\mu<\lambda} \alpha(Z_\mu) \otimes \alpha(Z_{\lambda-\mu}).$$

On the other hand, the inductive assumption gives

$$\Delta(\alpha(Z_\lambda)) = -\Delta(Z_\lambda) - \sum_{0<\mu<\lambda} \Delta(\alpha(Z_\lambda))\Delta(Z_{\lambda-\mu})$$
$$= -Z_0 \otimes Z_\lambda - Z_\lambda \otimes Z_0 - \sum_{0<\mu<\lambda} Z_\mu \otimes Z_{\lambda-\mu}$$
$$- \sum_{0<\mu<\lambda} ((\alpha \otimes \alpha)(\Delta(Z_\mu)))(\Delta(Z_{\lambda-\mu})).$$

The last sum can be written

$$\sum_{0<\mu<\lambda} \Big(\sum_{0\leqslant\nu\leqslant\mu} \alpha(Z_\nu) \otimes \alpha(Z_{\mu-\nu})\Big)\Big(\sum_{0\leqslant\rho\leqslant\lambda-\mu} Z_\rho \otimes Z_{\lambda-\mu-\rho}\Big)$$

or equivalently

(41) $$\Big(\sum_{\gamma,\pi} \alpha(Z_\nu)Z_\pi\Big) \otimes \Big(\sum_{\rho,\sigma} \alpha(Z_\rho)Z_\sigma\Big)$$

where the multiindices have to satisfy the conditions

$$\nu + \pi + \rho + \sigma = \lambda, \quad \nu + \rho > 0, \quad \pi + \sigma > 0.$$

Let us put together in that sum the terms for which $\gamma = \nu + \pi$ is fixed; for $\gamma = 0$ and $\gamma = \lambda$, this gives the terms

$$Z_0 \otimes \Big(\sum_{0<\mu<\lambda} \alpha(Z_\mu)Z_{\lambda-\mu}\Big) + \Big(\sum_{0<\mu<\lambda} \alpha(Z_\mu)Z_{\lambda-\mu}\Big) \otimes Z_0.$$

Next, for each $\gamma$ such that $0 < \gamma < \lambda$, we note that by (38)

$$\sum_{0\leqslant\nu\leqslant\gamma} \alpha(Z_\nu)Z_{\gamma-\nu} = 0, \qquad \sum_{0\leqslant\rho\leqslant\lambda-\gamma} \alpha(Z_\rho)Z_{\lambda-\gamma-\rho} = 0.$$

But the tensor product of these two sums gives the terms in (41) with given $\gamma = \nu + \pi$, plus the terms for which either $\nu = \rho = 0$ or $\nu = \gamma$ and $\rho = \lambda - \gamma$; hence the terms in (41) with given $\nu + \pi = \gamma$ add up to

$$-Z_\gamma \otimes Z_{\lambda-\gamma} - \alpha(Z_\gamma) \otimes \alpha(Z_{\lambda-\gamma})$$

and this ends the verification.

The proof that $\alpha$ satisfies (39) is quite similar: the inductive assumption gives, for $0 < \mu < \lambda$,

$$\alpha(Z_\mu) = -Z_\mu - \sum_{0<\nu<\mu} Z_\nu \alpha(Z_{\mu-\nu})$$

hence

$$\sum_{0<\mu<\lambda} \alpha(Z_\mu) Z_{\lambda-\mu} = -\sum_{0<\mu<\lambda} Z_\mu Z_{\lambda-\mu} - \sum_{\rho,\sigma,\tau} Z_\rho \alpha(Z_\sigma) Z_\tau$$

where in the last sum none of the multiindices $\rho$, $\sigma$, $\tau$ may be 0 and

$$\rho + \sigma + \tau = \lambda.$$

But for each $\rho$ such that $0 < \rho < \lambda$, the inductive assumption gives

$$\sum_{0<\sigma<\rho} \alpha(Z_\sigma) Z_{\lambda-\rho-\sigma} = -Z_{\lambda-\rho} - \alpha(Z_{\lambda-\rho}).$$

Using this relation in the previous equation, we prove (39) for every $\lambda$.

In fact, this last verification is really not needed: it is well known that to define a group it is enough to have an associative law with two-sided unit element and *one-sided* inverses. If we go back to the definition of a $\boldsymbol{C}$-group as an object representing a functor $\boldsymbol{C}^0 \to \boldsymbol{Gr}$ (I, §1, No. 3), we conclude that, in the definition of a $\boldsymbol{C}$-group given in (I, §1, No. 1), only one of the diagrams concerning $i$ is needed, the commutativity of the second being a consequence of the other axioms.

The uniqueness of the antipodism implies that every subbigebra of $\mathfrak{G}$ is stable under the antipodism of $\mathfrak{G}$. The same is true for every biideal J of $\mathfrak{G}$, for $\mathrm{J}^0$ is a closed subbigebra of $\mathfrak{G}^*$, hence stable under the antipodism of $\mathfrak{G}^*$, hence our assertion by transposition. There is no need therefore to talk about covariant subbigebras or biideals of $\mathfrak{G}$, or contravariant subbigebras and biideals of $\mathfrak{G}^*$ (I, §3, No. 3).

Having thus proved that the bigebras $\mathfrak{G}$ and $\mathfrak{G}^*$ considered in this paragraph have unique antipodisms, we may now speak of them as *bigebras of an infinitesimal formal group*.

**9.** *Stable bigebras and reduced formal groups.* When $\mathfrak{G}$ has infinite height, it is still true that, for each $r$,

$$\mathfrak{g}_r = (\mathrm{M}_{p^r} + \mathfrak{s}_{r-1}) \cap \mathfrak{G}^+$$

is a $p$-Lie algebra, in which $\mathfrak{s}_{r-1} \cap \mathfrak{G}^+$ is a $p$-ideal, that $\mathfrak{g}_r$ generates the subalgebra $\mathfrak{s}_r$, and that $\mathrm{V}^r(\mathfrak{g}_r) = \mathfrak{g}'_r$, so that $\mathfrak{g}'_r$ is a $p$-Lie subalgebra of $\mathfrak{g}_0$ for each $r$, for these results are established in a Frobenius subbigebra $\mathfrak{s}_r$ of $\mathfrak{G}$, of finite height.

We will say that $\mathfrak{G}$ is *stable* if the sequence $(\mathfrak{g}'_r)$ is *stationary*, i.e., there is an $r_0$ such that $\mathfrak{g}'_r = \mathfrak{g}'_{r_0}$ for $r \geqslant r_0$; we will say that $\mathfrak{G}$ is *reduced* if $\mathfrak{g}'_r = \mathfrak{g}_0$ for every $r$. We will use the same qualifications for the dual bigebra $\mathfrak{G}^*$ and the formal group **G** corresponding to $\mathfrak{G}$.

When $\mathfrak{G}$ is stable, we may still choose a basis $(X_i)_{i\in \mathrm{I}}$ of $\mathfrak{g}_0$ as in No. 6. We then define the set $\mathrm{L} \subset \mathbf{N}^{(\mathrm{I})}$ of multiindices by the conditions $\alpha(i) < p^{k+1}$ for $i \in \mathrm{I}_k - \mathrm{I}_{k+1}$, *but only for* $0 \leqslant k \leqslant r_0 - 1$; the values of the $\alpha(i)$ for $i \in \mathrm{I}_{r_0}$ ($= \mathrm{I}_r$ for all $r \geqslant r_0$) remain *arbitrary*. With the replacement of $\mathrm{L}_r$ by L, the definition of a *privileged basis* $(Z_\alpha)_{\alpha\in \mathrm{L}}$ of $\mathfrak{G}$, corresponding to $(X_i)$, remains otherwise unchanged, and *Theorem 1 is still valid.* The proof runs exactly in the same way, except that, for each $r > r_0$, in the inductive process, we have to repeat the first part of the argument (corresponding to the construction of the basis of $\mathrm{M}_{p^r}$), and we have, for $1 \leqslant k \leqslant r_0$ and $i \in \mathrm{I}_{k-1} - \mathrm{I}_k$,

$$x_i^{(h+1)} - x_i^{(h)} \in \mathfrak{R}^{[(h+1)/p^k]} \qquad \text{for all} \quad h > p^{r_0};$$

this enables one to conclude the proof as in Theorem 1.

For such a privileged basis, we still have relation (30) for the shift; in particular, $\mathrm{V}^r(Z_\alpha) = 0$ for $r \geqslant r_0$ and for all multiindices $\alpha$ for which $\alpha(i) \neq 0$ for at least one index $i \notin \mathrm{I}_{r_0}$. The other multiindices may be identified with the elements of $\mathbf{N}^{(\mathrm{I}_{r_0})}$; for each such multiindex $\alpha$, $p\alpha$ is still in L, hence $Z_\alpha$ belongs to $\mathrm{V}(\mathfrak{G})$. In other words, the subbigebras $\mathrm{V}^r(\mathfrak{G})$ for $r \geqslant r_0$ are all *the same*, and it can obviously be defined as *the largest reduced subbigebra of* $\mathfrak{G}$.

When $\mathfrak{G}$ is stable and $\mathfrak{g}'_{r_0}$ ($= \mathfrak{g}'_r$ for $r \geqslant r_0$) has finite dimension, it is called the *reduced dimension* of $\mathfrak{G}$. When $\mathfrak{g}_0$ itself has finite dimension, of course $\mathfrak{G}$ is stable and the dimension of $\mathfrak{g}_0$ is called the *dimension* of $\mathfrak{G}$ (or *algebraic dimension* if any confusion should arise with the dimension of $\mathfrak{G}$ as a vector space) and written dim $\mathfrak{G}$; these definitions and notations are of course transferred to infinitesimal groups.

As a final remark, observe that the definition of the basis $(X_i)$ cannot be extended in general when the bigebra $\mathfrak{G}$ is not stable; this is due to the well-known pathology in infinite-dimensional vector spaces, according to which the intersection of the subspaces $\mathfrak{g}'_r$ of $\mathfrak{g}_0$ may well be reduced to 0, whereas if $\mathfrak{h}_r$ is a supplementary subspace to $\mathfrak{g}'_{r+1}$ in $\mathfrak{g}'_r$, the direct sum of the $\mathfrak{h}_r$ may well be distinct from the whole space $\mathfrak{g}_0$ (an example consists of a Hilbert space and the decreasing sequence of the closed subspaces orthogonal to the first $n$ vectors of a Hilbert basis). Hence an extension of Theorem 1 to that case remains an open problem.

## §3. Reduced infinitesimal formal groups

**1.** *Canonical bases.* Let $\mathfrak{G}$ be a *reduced* infinitesimal group bigebra; for any basis $(X_i)_{i \in \mathrm{I}}$ of the Lie algebra $\mathfrak{g}_0$, it is clear that *any* choice of representatives $x_i \in \bar{x}_i$, provided they constitute a pseudobasis of a closed subspace supplementary to $\mathfrak{R}^2$ in $\mathfrak{R}$, gives rise to a privileged basis $(Z_\alpha)$ (with $\alpha \in \mathbf{N}^{(\mathrm{I})}$) of $\mathfrak{G}$, the dual pseudobasis of $\mathfrak{G}^*$ being $(x^\alpha)$. As a linearly compact commutative algebra, $\mathfrak{G}^*$ is therefore isomorphic to the algebra of formal power series $k[[\mathrm{T}_i]]_{i \in \mathrm{I}}$ by an isomorphism sending every indeterminate $\mathrm{T}_i$ onto $x_i$ ($i \in \mathrm{I}$); for finite I, we completely recover the "naive formal Lie groups" (I, §2, No. 12). For any set of indices I, we observe that $\mathfrak{G}^*$ is a local *integral domain.*

The arbitrariness of the choice of the $x_i \in \bar{x}_i$ allows one to impose additional conditions on the privileged basis $(Z_\alpha)$ corresponding to a given basis $(X_i)_{i \in \mathrm{I}}$. For any integer $n > 1$, consider the elements

$$x'_i = x_i - (b_{\alpha i} x^\alpha)$$

where in the element $(b_{\alpha i} x^\alpha)$ of $\mathfrak{G}^*$ considered as a product, $b_{\alpha i}$ are scalars which are 0 except for $|\alpha| = n$, and for any $\alpha$ such that $|\alpha| = n$, the number of indices $i \in \mathrm{I}$ such that $b_{\alpha i} \neq 0$ is *finite.* Then, for any multiindex $\lambda \in \mathbf{N}^{(\mathrm{I})}$, we may write

$$x'^\lambda = x^\lambda - (b_{\mu\lambda} x^\mu)$$

and for each given $\mu$, there are again only a *finite* number of $\lambda \in \mathbf{N}^{(\mathrm{I})}$ such that $b_{\mu\lambda} \neq 0$: indeed, we have

$$x'^{\lambda} = \prod_i (x_i - (b_{\alpha i} x^{\alpha}))^{\lambda(i)};$$

a term $b_{\mu\lambda} x^{\mu}$ with $b_{\mu\lambda} \neq 0$ can only come from products of terms $x_i$, but then $\varepsilon_i \leqslant \mu$, or from terms $b_{\alpha i} x^{\alpha}$, but then $\alpha \leqslant \mu$; as the multiindices $\nu \leqslant \mu$ are in finite number, it follows from the assumption on $(b_{\alpha i})$ that the $\lambda$ such that $b_{\mu\lambda} \neq 0$ must have $\lambda(i) \neq 0$ only for a finite number of indices $i \in \mathrm{I}$; furthermore one must have $\lambda(i) \leqslant |\mu|$, and this shows that the possible $\lambda$ are in finite number. Now if we put

$$Z'_{\mu} = Z_{\mu} - \sum_{\lambda} b_{\mu\lambda} Z_{\lambda} \tag{1}$$

it is clear (as $b_{\mu\lambda} \neq 0$ only if $|\lambda| < |\mu|$) that the $Z'_{\mu}$ form again a basis of $\mathfrak{G}$, and that we have $\langle x'^{\lambda}, Z'_{\mu} \rangle = \delta_{\lambda\mu}$ (Kronecker delta), hence the $x'^{\lambda}$ form the pseudobasis of $\mathfrak{G}^*$ dual to $(Z'_{\lambda})$ and our choice of the new representatives $x'_i$ is such that they again constitute a pseudobasis of a supplementary of $\mathfrak{R}^2$ in $\mathfrak{R}$.

It is clear from the definition that, for $|\alpha| < n$, we have $b_{\alpha\beta} = 0$ for every $\beta$, and therefore $Z'_{\alpha} = Z_{\alpha}$; furthermore, for $|\alpha| = n$, relation (1) boils down to

$$Z'_{\alpha} = Z_{\alpha} - \sum_{i \in \mathrm{I}} b_{\alpha i} X_{0i}. \tag{2}$$

Now, from Eq. (26) and Corollary 2 of §2, No. 6, it follows that, for any $\alpha \in \mathbf{N}^{(\mathrm{I})}$, we may write

$$Z_{\alpha} = \frac{1}{\alpha!} X_{\alpha} + \sum_{|\beta| < |\alpha|} e_{\alpha\beta} X_{\beta}. \tag{3}$$

**Proposition 1.** *Let $(Z_{\alpha})$, $(Z'_{\alpha})$ be two privileged bases of $\mathfrak{G}$ such that:*

$1^0$ $X_{hi} = X'_{hi}$ *for any pair $(h, i)$ $(h \geqslant 0, i \in \mathrm{I})$;*

$2^0$ *in the expressions*

$$Z_{\alpha} = \frac{1}{\alpha!} X_{\alpha} + \sum_{|\beta| < |\alpha|} e_{\alpha\beta} X_{\beta}, \qquad Z'_{\alpha} = \frac{1}{\alpha!} X'_{\alpha} + \sum_{|\beta| < |\alpha|} e'_{\alpha\beta} X'_{\beta}$$

*(where the same total order has been chosen on $\mathbf{N} \times \mathrm{I}$ to define the $X_{\alpha}$ and $X'_{\alpha}$), we have $e'_{\alpha\varepsilon_i} = e_{\alpha\varepsilon_i}$ for every $\alpha \in \mathrm{N}^{(\mathrm{I})}$ and every $i \in \mathrm{I}$.*

*Then $Z'_{\alpha} = Z_{\alpha}$ for every $\alpha$.*

As by definition $X'_\alpha = X_\alpha$ for every $\alpha$, it is enough to show that $e'_{\alpha\beta} = e_{\alpha\beta}$ for every pair of indices such that $|\beta| < |\alpha|$; we use induction on $|\alpha|$, the statement being trivial for $|\alpha| = 1$. We may write by §2, No. 6, Eq. (26)

$$(4) \qquad Z_\alpha = \frac{1}{\alpha!} X_\alpha + \sum_{|\beta|<|\alpha|} c_{\alpha\beta} Z_\beta$$

and we have similar relations for the $Z'_\alpha$, denoting the coefficients by $c'_{\alpha\beta}$; it is clear from (3) and (4) that, for $|\beta| < |\alpha|$,

$$(5) \qquad e_{\alpha\beta} = \frac{1}{\beta!} c_{\alpha\beta} + \sum_{|\beta|<|\gamma|<|\alpha|} c_{\alpha\gamma} e_{\gamma\beta},$$

hence the $e_{\alpha\beta}$ are entirely determined by the $c_{\alpha\gamma}$ for $|\gamma| < |\alpha|$; using the change of basis inverse to (3), we see likewise that the $c_{\alpha\beta}$ are entirely determined by the $e_{\alpha\gamma}$ for $|\gamma| < |\alpha|$. We assume that, for $|\alpha| = n$, we have

$$e'_{\beta\gamma} = e_{\beta\gamma} \qquad \text{for} \quad |\beta| < n, \quad |\gamma| < |\beta|,$$

hence also

$$c'_{\beta\gamma} = c_{\beta\gamma} \qquad \text{for} \quad |\beta| < n, \quad |\gamma| < |\beta|.$$

We may write

$$(6) \qquad \Delta(X_\alpha) = 1 \otimes X_\alpha + X_\alpha \otimes 1 + \sum_{\beta>0,\gamma>0,|\beta|+|\gamma|\leqslant|\alpha|} d_{\alpha\beta\gamma} X_\beta \otimes X_\gamma$$

where the assumption $X_{hi} = X'_{hi}$ implies that the $d_{\alpha\beta\gamma}$ are the same for the two bases $(Z_\alpha)$, $(Z'_\alpha)$. Now from (4) we deduce, by computing $\Delta(Z_\alpha)$,

$$1 \otimes Z_\alpha + Z_\alpha \otimes 1 + \sum_{0<\gamma<\alpha} Z_\gamma \otimes Z_{\alpha-\gamma}$$

$$= \frac{1}{\alpha!} \Delta(X_\alpha) + \sum_{|\beta|<|\alpha|} c_{\alpha\beta}(1 \otimes Z_\beta + Z_\beta \otimes 1 + \sum_{0<\gamma<\alpha} Z_\gamma \otimes Z_{\beta-\gamma}).$$

If in this equation we replace $\Delta(X_\alpha)$ by its value on the right-hand side of (6), where each $X_\beta$ is replaced by its value taken from (4), we must have equality for the coefficients of the $Z_\lambda \otimes Z_\mu$ on both sides. For each $\beta$ such that $|\beta| > 1$, we have on the right-hand side one term $c_{\alpha\beta} Z_\gamma \otimes Z_{\beta-\gamma}$ with $\gamma \neq 0$, which cannot be equal to any term on the left-hand side, so it must cancel with one of the terms coming from $\Delta(X_\alpha)$, which means that it is a polynomial in the $c_{\lambda\mu}$ with $|\lambda| < n$, the coefficients being the $d_{\alpha\beta\gamma}$; hence the inductive assumption shows that in that case $c'_{\alpha\beta} = c_{\alpha\beta}$. There remains the case $\beta = \varepsilon_i$; but by what has been proved and (5), we have

$$c_{\alpha\varepsilon_i} - e_{\alpha\varepsilon_i} = c'_{\alpha\varepsilon_i} - e'_{\alpha\varepsilon_i},$$

and the assumption $e_{\alpha\varepsilon_i} = e'_{\alpha\varepsilon_i}$ yields $c'_{\alpha\varepsilon_i} = c_{\alpha\varepsilon_i}$, thus ending the proof by induction.

**Proposition 2.** *For a given basis $(X_{0i})$ of $\mathfrak{g}_0$ and a family $(a_{\alpha i})$ of elements of $k$ with indices in $\mathbf{N}^{(\mathrm{I})} \times \mathrm{I}$ and $|\alpha| > 1$, such that for every $\alpha \in \mathbf{N}^{(\mathrm{I})}$ with $|\alpha| > 1$ the number of indices $i \in \mathrm{I}$ for which $\alpha_{\alpha i} \neq 0$ is finite, there exists a privileged basis $(Z_\alpha)$ such that, for any $\alpha \in \mathbf{N}^{(\mathrm{I})}$ with $|\alpha| > 1$ and any $i \in \mathrm{I}$, the element $e_{\alpha\varepsilon_i} = a_{\alpha i}$ in* (3).

Assume that, for an integer $n$, we have found elements $x_i^{(n)}$ in $\bar{x}_i$ for $i \in \mathrm{I}$, such that, when we build up the corresponding privileged basis, we have $e_{\alpha\varepsilon_i} = a_{\alpha i}$ for $i \in \mathrm{I}$ and *for all $\alpha$ such that* $1 < |\alpha| < n$. Replacing $x_i^{(n)}$ by

$$x_i^{(n+1)} = x_i^{(n)} + \sum_{|\alpha|=n} b_{\alpha i}(x^{(n)})^\alpha$$

with a suitable choice of the $b_{\alpha i}$, we will then have for the new privileged basis corresponding to that new choice, $e_{\alpha\varepsilon_i} = a_{\alpha i}$ for $1 < |\alpha| < n + 1$ by (2), and for each $\alpha$, only finitely many $b_{\alpha i}$ will be $\neq 0$, since this is true for the $e_{\alpha\varepsilon_i}$. Furthermore, we have

$$x_i^{(n+1)} - x_i^{(n)} \in \mathfrak{R}^n,$$

and as $\mathfrak{G}^*$ is complete, the sequence $(x_i^{(n)})_{n>1}$ converges to an $x_i \in \mathfrak{R}$ for each $i \in \mathrm{I}$. It is then clear that when we take the $x_i$ as representatives of the $\bar{x}_i$, the corresponding privileged basis $(Z_\alpha)$ satisfies the condition $e_{\alpha\varepsilon_i} = a_{\alpha i}$ for each $i \in \mathrm{I}$ and each $\alpha$ with $|\alpha| > 1$.

In particular, we always have privileged bases for which $e_{\alpha\varepsilon_i} = 0$ for all $i \in \mathrm{I}$ and all $\alpha$ of degree $> 1$; we will say that such a privileged basis is *canonical*. It follows from Proposition 1 that any two canonical privileged bases for which the $X_{hi}$ are the same, not only for $h = 0$ but for all $h > 0$, are necessarily identical.

**2.** *Formal subgroups of reduced infinitesimal groups.* Let $\mathfrak{G}$ be a reduced infinitesimal group bigebra, and let $\mathfrak{H}$ be an arbitrary *stable* subbigebra of $\mathfrak{G}$. We first observe that the cogebras $M_n \cap \mathfrak{H}$ form the fundamental filtration of $\mathfrak{H}$, as follows from relations (4) and (5) of §2, No. 2 which define that filtration; in particular, $\mathfrak{g}_0 \cap \mathfrak{H} = \mathfrak{h}_0$ is the Lie algebra of $\mathfrak{H}$. On the other hand, the functoriality of the shift (§2, No. 7) shows that the restriction of V to $\mathfrak{H}$ is the shift on $\mathfrak{H}$. We have therefore in $\mathfrak{h}_0$ a decreasing sequence of Lie algebras

$$\mathfrak{h}'_k = V^k(\mathfrak{H}) \cap \mathfrak{h}_0 = V^k(\mathfrak{H}) \cap \mathfrak{g}_0,$$

which by assumption is stationary; furthermore the Frobenius subbigebras of $\mathfrak{H}$ are the intersections $\mathfrak{s}_r \cap \mathfrak{H}$ of those of $\mathfrak{G}$ by $\mathfrak{H}$: indeed, the image of $\mathfrak{N}/\mathfrak{H}^0$ under the $r$th iterate of the Frobenius homomorphism in $\mathfrak{G}^*/\mathfrak{H}^0$ is $(F^r(\mathfrak{N}) + \mathfrak{H}^0)/\mathfrak{H}^0$; the ideal generated by that image is therefore $(\mathfrak{B}_r + \mathfrak{H}^0)/\mathfrak{H}^0$, which proves our assertion, since $(\mathfrak{s}_r \cap \mathfrak{H})^0 = \mathfrak{B}_r + \mathfrak{H}^0$. If we take into account the definition of the higher Lie algebras of $\mathfrak{G}$ (§2, No. 7, Corollary), we see that the $\mathfrak{g}_k \cap \mathfrak{H}$ are the higher Lie algebras of $\mathfrak{H}$, and that we may also write $\mathfrak{h}'_k = V^k(\mathfrak{g}_k \cap \mathfrak{H})$.

We take now in $\mathfrak{g}_0$ a basis $(X_{0i})_{i \in I}$ having the property that there exists a decreasing stationary sequence $(I_k)$ of subsets of I such that $I_0 = I$ and the $X_{0i}$ for $i \in I_k$ form a basis of $\mathfrak{h}'_{k-1}$ for $k \geqslant 1$.

The dual bigebra $\mathfrak{H}^*$ is identified to the quotient $\mathfrak{G}^*/\mathfrak{H}^0$, $\mathfrak{H}^0$ being a closed biideal of $\mathfrak{G}^*$. By the structure theorem applied to $\mathfrak{H}$ (§2, No. 6, Theorem 1) we may find in the augmentation ideal $(\mathfrak{N} + \mathfrak{H}^0)/\mathfrak{H}^0$ of $\mathfrak{H}^*$ a family $(y_i)_{i \in I_1}$ of elements forming a pseudobasis of a closed subspace supplementary to $(\mathfrak{N}^2 + \mathfrak{H}^0)/\mathfrak{H}^0$ and such that $\langle y_i, X_{0j} \rangle = \delta_{ij}$, and $y_i^{p^k} = 0$ for $k \geqslant 1$, $i \in I_k - I_{k+1}$, $j \in I$. We may then lift back the $y_i$ to elements $x_i$ of $\mathfrak{G}^*$ by taking in $\mathfrak{N}$ a closed subspace $S_1$ supplementary to $\mathfrak{N}^2 + \mathfrak{H}^0$ $(i \in I_1)$; the $x_i$ for $i \in I - I_1$ are taken to form the pseudobasis of a closed subspace in $\mathfrak{H}^0$ supplementary to $S_1 + \mathfrak{N}^2$, and such that $\langle x_i, X_{0j} \rangle = \delta_{ij}$ for $j \in I$. This choice of the $x_i$ in the classes $\bar{x}_i$ in $\mathfrak{N}/\mathfrak{N}^2$ is such therefore that we have

(7) $$x_i^{p^k} \in \mathfrak{H}^0 \quad \text{for} \quad k \geqslant 0, \quad i \in I_k - I_{k+1}.$$

Now, starting from this choice of $x_i$, we may form the corresponding privileged basis $(Z_\alpha)$ of $\mathfrak{G}$ for $\alpha \in \mathbf{N}^{(I)}$. Let us denote by L the set of indices $\alpha \in \mathbf{N}^{(I)}$ such that $\alpha(i) < p^k$ for $k \geqslant 0$ and $i \in I_k - I_{k+1}$.

**Proposition 3.** *With the preceding notations, the $Z_\alpha$ such that $\alpha \in L$ form a privileged basis of $\mathfrak{H}$.*

Indeed, the structure theorem applied to $\mathfrak{H}$ shows that the $y^\alpha$ for $\alpha \in L$ form a pseudobasis of $\mathfrak{G}^*/\mathfrak{H}^0$; on the other hand from (7) and the fact that $\mathfrak{H}^0$ is an ideal it follows that the $x^\alpha$ for $\alpha \notin L$ are all in $\mathfrak{H}^0$. Now the $x^\alpha$ for $\alpha \in L$ form a pseudobasis of a closed subspace $E_1$ of $\mathfrak{G}^*$ and the $x^\alpha$ for $\alpha \notin L$ a pseudobasis of a closed subspace $E_2$ of $\mathfrak{G}^*$, supplementary to $E_1$; together with the preceding remarks and the fact that $y^\alpha$ is the class mod $\mathfrak{H}^0$ of $x^\alpha$ for $\alpha \in L$, this proves that $E_2 = \mathfrak{H}^0$, hence the $Z_\alpha$ for $\alpha \in L$ form a basis of $\mathfrak{H}$.

When the privileged basis $(Z_\alpha)$ has been chosen in the preceding way, we will say it is *adapted* to the subbigebra $\mathfrak{H}$.

It should be observed that the sub-Lie algebras $\mathfrak{h}'_r = V^r(\mathfrak{g}_r \cap \mathfrak{H})$ *do not determine* the subbigebra $\mathfrak{H}$. For instance, consider the formal group which is the direct product of two additive groups $\mathbf{G}_a$; $\mathfrak{G}^*$ is therefore the algebra $k[[x, y]]$ of power series in two indeterminates $x$, $y$, with the comultiplication defined by

$$c(x) = 1 \otimes x + x \otimes 1, \qquad c(y) = 1 \otimes y + y \otimes 1.$$

The subbigebras $\mathfrak{H}_1$, $\mathfrak{H}_2$ of $\mathfrak{G}$ are defined by taking for $\mathfrak{H}_1^0$ the ideal generated by $y$, and for $\mathfrak{H}_2^0$ the ideal generated by $y - x^p$; both are easily seen to be biideals, and $\mathfrak{H}_1$, $\mathfrak{H}_2$ are *reduced* subbigebras of $\mathfrak{G}$, which are distinct although their intersection with $\mathfrak{g}_0$ is the same (hence the sub-Lie algebras $V^r(\mathfrak{g}_r \cap \mathfrak{H}_1)$ and $V^r(\mathfrak{g}_r \cap \mathfrak{H}_2)$ are equal to that intersection). This also gives an example of two reduced subbigebras of $\mathfrak{G}$ whose intersection $\mathfrak{H}_1 \cap \mathfrak{H}_2$ *is not reduced*; it is the subbigebra of $\mathfrak{s}_0$ generated by

$$\mathfrak{g}_0 \cap \mathfrak{H}_1 = \mathfrak{g}_0 \cap \mathfrak{H}_2.$$

However, we have the following corollary:

**Corollary 1.** *If two subbigebras $\mathfrak{H}_1$, $\mathfrak{H}_2$ of a reduced bigebra $\mathfrak{G}$ are such that $\mathfrak{H}_1 \subset \mathfrak{H}_2$, and*

$$V^r(\mathfrak{g}_r \cap \mathfrak{H}_1) = V^r(\mathfrak{g}_r \cap \mathfrak{H}_2)$$

*for every $r \geqslant 0$, then $\mathfrak{H}_1 = \mathfrak{H}_2$.*

Indeed, it is enough to prove that

$$\mathfrak{s}_r \cap \mathfrak{H}_1 = \mathfrak{s}_r \cap \mathfrak{H}_2$$

for every $r$, hence one may assume $\mathfrak{H}_1$ and $\mathfrak{H}_2$ are of finite height. We then construct a privileged basis $(Z_\alpha)$ of $\mathfrak{G}$ adapted to the subbigebra $\mathfrak{H}_2$ (Proposition 3) and prove by induction or $r$ that

$$\mathfrak{s}_r \cap \mathfrak{H}_1 = \mathfrak{s}_r \cap \mathfrak{H}_2.$$

Assuming that

$$\mathfrak{s}_{r-1} \cap \mathfrak{H}_1 = \mathfrak{s}_{r-1} \cap \mathfrak{H}_2,$$

it is then enough, by §2, No. 7, Proposition 2, to prove that

$$\mathfrak{g}_r \cap \mathfrak{H}_1 = \mathfrak{g}_r \cap \mathfrak{H}_2.$$

Now, for $i \in \mathrm{I}_r$,

$$\mathrm{V}^r(X_{ri}) = X_{0i} \in \mathrm{V}^r(\mathfrak{g}_r \cap \mathfrak{H}_1),$$

and therefore there is an $X'_{ri} \in \mathfrak{g}_r \cap \mathfrak{H}_1$ such that

$$\mathrm{V}^r(X'_{ri} - X_{ri}) = 0,$$

which implies that $X'_{ri} - X_{ri}$, which belongs to $\mathfrak{g}_r \cap \mathfrak{H}_2$, is in fact in

$$\mathfrak{s}_{r-1} \cap \mathfrak{H}_2 = \mathfrak{s}_{r-1} \cap \mathfrak{H}_1;$$

hence

$$X_{ri} \in \mathfrak{H}_1 \qquad \text{for} \quad i \in \mathrm{I}_r,$$

which proves that $\mathfrak{g}_r \cap \mathfrak{H}_2 = \mathfrak{g}_r \cap \mathfrak{H}_1$.

**Corollary 2.** *Let $\mathfrak{H}_1 \subset \mathfrak{H}_2 \subset \cdots \subset \mathfrak{H}_m$ a strictly increasing finite sequence of* reduced *subbigebras of a reduced bigebra $\mathfrak{G}$. Then there exists a privileged basis $(Z_\alpha)$ of $\mathfrak{G}$ corresponding to a basis $(X_{0i})_{i \in \mathrm{I}}$ of $\mathfrak{g}_0$, and an increasing sequence $(\mathrm{I}_s)_{1 \leqslant s \leqslant m}$ of subsets of* I *such that the $Z_\alpha$ for $\alpha \in \mathbf{N}^{(\mathrm{I}_s)}$ form a privileged basis of $\mathfrak{H}_s$ for $1 \leqslant s \leqslant m$.*

We may take the $I_s$ such that for $i \in I_s$, the $X_{0i}$ form a basis of $\mathfrak{g}_0 \cap \mathfrak{H}_s$. The pseudobasis $(\bar{x}_i)$ of $\mathfrak{N}/\mathfrak{N}^2$ dual to $(X_{0i})_{i\in I}$ is then such that for each $s$, the $\bar{x}_i$ such that $i \in I_s$ form a pseudobasis of $(\mathfrak{H}_s^0 + \mathfrak{N}^2)/\mathfrak{N}^2$. As

$$\mathfrak{H}_1^0 \supset \mathfrak{H}_2^0 \supset \cdots \supset \mathfrak{H}_m^0,$$

we may determine in succession the $x_i$ such that the $x_i$ for $i \in I_1$ form a pseudobasis of a closed subspace of $\mathfrak{N}$ supplementary to $\mathfrak{H}_1^0 + \mathfrak{N}^2$, the $x_i$ such that $i \in I_{s+1} - I_s$ a pseudobasis of a closed subspace of $\mathfrak{H}_s^0 + \mathfrak{N}^2$ supplementary to $\mathfrak{H}_{s+1}^0 + \mathfrak{N}^2$, for $1 \leqslant s \leqslant m-1$, and finally the $x_i$ such that $i \in I - I_m$ a pseudobasis of a closed subspace of $\mathfrak{H}_m^0 + \mathfrak{N}^2$ supplementary to $\mathfrak{N}^2$. Then the privileged basis $(Z_\alpha)$ of $\mathfrak{G}$ dual to the pseudobasis $(x^\alpha)$ satisfies the required conditions, since here the relations (7) disappear for all $\mathfrak{H}_s$.

*Remark.* Suppose only that $\mathfrak{H}_1, \mathfrak{H}_2, \ldots, \mathfrak{H}_{m-1}$ are reduced and $\mathfrak{H}_m$ is an arbitrary subbigebra of $\mathfrak{G}$ containing $\mathfrak{H}_{m-1}$. Then we still may, for every $r \geqslant 0$, determine a basis $(X_{0i})_{i\in I}$ of $\mathfrak{g}_0$ and an increasing sequence $(I_s)_{1\leqslant s\leqslant m}$ of subsets of I, together with a decreasing sequence $(I_{mh})_{0\leqslant h\leqslant r+1}$ such that the $X_{0i}$ form a basis of $\mathfrak{g}_0 \cap \mathfrak{H}_s$ for $i \in I_s$ and $1 \leqslant s \leqslant m$, $I_{m0} = I_m$, $I_{m,r+1} = I_{m-1}$, and the $X_{0i}$ for $i \in I_{mh}$ form a basis for $V^h(\mathfrak{H}_m) \cap \mathfrak{g}_0$ and $0 \leqslant h \leqslant r$. We then pick the $x_i$ in the classes $\bar{x}_i$ mod $\mathfrak{N}^2$ as before, and in addition such that we have $x_i^{p^h} \in \mathfrak{H}_m^0$ for $i \in I_{mh} - I_{m,h+1}$ for $0 \leqslant h \leqslant r-1$. If we form the privileged basis $(Z_\alpha)$ of $\mathfrak{G}$ corresponding to those choices, the $Z_\alpha$ for $\alpha \in \mathbf{N}^{(I_s)}$ still form a privileged basis for $\mathfrak{H}_s$ and $s \leqslant m-1$; furthermore the $Z_\alpha$ such that $\alpha(i) < p^{h+1}$ for $i \in I_{mh} - I_{m,h+1}$ and $0 \leqslant h \leqslant r-1$, and $\alpha(i) < p^r$ for $i \in I_{mr}$, form a privileged basis of $\mathfrak{s}_r \cap \mathfrak{H}_m$.

**Proposition 4.** *In a reduced bigebra $\mathfrak{G}$ of finite algebraic dimension* (§2, No. 9), *the subbigebras satisfy the minimal condition* (i.e., every decreasing sequence of subbigebras is stationary).

Let $(\mathfrak{H}_i)_{i\geqslant 0}$ be a decreasing sequence of subbigebras of $\mathfrak{G}$, and for each $i$, let $\mathfrak{H}_i'$ be the largest reduced subbigebra of $\mathfrak{H}_i$ (§2, No. 9); it is clear that the sequence $(\mathfrak{H}_i')$ is also decreasing, and so is the sequence of Lie algebras $\mathfrak{g}_0 \cap \mathfrak{H}_i'$; as $\mathfrak{g}_0$ is finite dimensional, this last sequence is stationary, hence also the sequence $(\mathfrak{H}_i')$ by the corollary 1 to Proposition 3. We may therefore suppose that all the $\mathfrak{H}_i'$ are identical. Let $r$ be the smallest integer such that

$V^r(\mathfrak{H}_0) = \mathfrak{g}_0 \cap \mathfrak{H}'_0$; for each integer $k \leqslant r$, in finite number, the sequence $(\mathfrak{g}_k \cap \mathfrak{H}_i)$ is decreasing and consists of finite-dimensional vector spaces, hence is stationary. There is therefore an integer $m$ such that, for $i \geqslant m$,

$$\mathfrak{g}_k \cap \mathfrak{H}_i = \mathfrak{g}_k \cap \mathfrak{H}_m \qquad \text{for} \quad 1 \leqslant k \leqslant r;$$

for $k > r$, we have

$$\mathfrak{g}_k \cap \mathfrak{H}_i = (\mathfrak{g}_r \cap \mathfrak{H}_i) + (\mathfrak{g}_k \cap \mathfrak{H}'_i),$$

hence

$$\mathfrak{g}_k \cap \mathfrak{H}_i = \mathfrak{g}_k \cap \mathfrak{H}_m \qquad \text{for} \quad \text{all} \quad k,$$

and this implies $\mathfrak{H}_i = \mathfrak{H}_m$.

The following counterexample, due to P. Gabriel, shows that Proposition 3 cannot be extended to *nonreduced* bigebras. Take $\mathfrak{G}^*$ to be the quotient of $k[x, y]$ ($x$, $y$ indeterminates) by the ideal generated by $x^p$ and $y^{p^3}$; the comultiplication is defined by

$$c(x) = 1 \otimes x + x \otimes 1, \qquad c(y) = 1 \otimes y + y \otimes 1.$$

Take for $\mathfrak{H}^0$ the ideal generated by $x - y^p$, which is a biideal in $\mathfrak{G}^*$. The only privileged bases of $\mathfrak{G}$ correspond to a choice of the $x_i$ consisting (up to constant factors) of

$$x(1 + P(x, y)) + y^{p^2}(a + Q(x, y))$$

$$\text{and} \qquad y(1 + R(x, y)) + x(b + S(x, y)),$$

where P, Q, R, S are polynomials without constant terms and $a$ and $b$ scalars; however, none of these elements may belong to $\mathfrak{H}^0$.

**3.** *Invariant formal subgroups of reduced infinitesimal groups.* Let J be a biideal in a reduced bigebra $\mathfrak{G}$. As the dual of $\mathfrak{G}/J$ is the subbigebra $J^0$ of $\mathfrak{G}^*$ orthogonal to J, $J^0$ contains no nilpotent elements, hence $\mathfrak{G}/J$ is also a reduced bigebra; we have already seen (§2, No. 8) that J is stable under the unique antipodism of $\mathfrak{G}$; it is also stable under the shift, since it is the kernel of a bigebra homomorphism $\mathfrak{G} \to \mathfrak{G}/J$ (§2, No. 7).

**Proposition 5.**     *Let J be a biideal in a reduced bigebra $\mathfrak{G}$, contained in $\mathfrak{G}^+$; if $\mathfrak{H}$ is the largest subbigebra of $\mathfrak{G}$ contained in $k \cdot 1 + J$ (I, §3, No. 5), we have, for every integer $r \geqslant 0$,*

$$(8) \qquad\qquad V^r(\mathfrak{g}_r \cap J) = V^r(\mathfrak{g}_r \cap \mathfrak{H})$$

*and J is the left ideal (and the right ideal) of $\mathfrak{G}$ generated by $\mathfrak{H}^+ = \mathfrak{H} \cap \mathfrak{G}^+$.*

It is clear that for each $r \geqslant 0$, $\mathfrak{s}_r \cap J$ is a biideal *in* $\mathfrak{s}_r$, and that $\mathfrak{s}_r \cap \mathfrak{H}$ is the largest subbigebra contained in $k \cdot 1 + (\mathfrak{s}_r \cap J)$. It will therefore be enough to prove that

$$V^s(\mathfrak{g}_s \cap J) = V^s(\mathfrak{g}_s \cap \mathfrak{H}) \qquad \text{for} \quad s < r$$

and that $\mathfrak{s}_r \cap J$ is the left (or right) ideal of $\mathfrak{s}_r$ generated by $\mathfrak{H}^+ \cap \mathfrak{s}_r$.
Define

$$\mathfrak{h}'_s = V^s(\mathfrak{g}_s \cap J) \subset \mathfrak{g}_0 \qquad \text{for} \quad s \leqslant r;$$

we take in $\mathfrak{g}_0$ a basis $(X_{0i})_{i \in I}$ for which there is a decreasing sequence $(I_s)_{0 \leqslant s \leqslant r+1}$ of subsets of $I$ such that $I_0 = I$, $I_{r+1} = \varnothing$, and the $X_{0i}$ for $i \in I_s$ form a basis of $\mathfrak{h}'_{s-1}$ for $1 \leqslant s \leqslant r$. By definition (§2, No. 7, formula (29)), the orthogonal $\mathfrak{h}'^0_s$ of $\mathfrak{h}_s$ in $\mathfrak{G}^*$ is

$$F^{-s}(\mathfrak{g}^0_s + J^0) = F^{-s}(\mathfrak{g}^0_s) + F^{-s}(J^0)$$

since $F^s$ is injective in $\mathfrak{G}^*$. As $\mathfrak{g}^0_s = \mathfrak{R}^{p^s+1}$, we have $F^{-s}(\mathfrak{g}^0_s) \subset \mathfrak{R}^2$ for $s \geqslant 1$, hence the images of $\mathfrak{h}'^0_s$ and of $F^{-s}(J^0)$ in $\mathfrak{R}/\mathfrak{R}^2$ are the same. Define now in $\mathfrak{R}$ a closed subspace $T_0$ contained in $J^0$ and supplementary to $J^0 \cap \mathfrak{R}^2$, so that it is also supplementary to $\mathfrak{R}^2$ in $J^0 + \mathfrak{R}^2$; similarly, for $1 \leqslant s \leqslant r - 1$, define a closed subspace $T_s$ contained in $F^{-s}(J^0)$ supplementary to

$$F^{-(s-1)}(J^0) + (F^{-s}(J^0) \cap \mathfrak{R}^2),$$

so that it is also supplementary to

$$F^{-(s-1)}(J^0) + \mathfrak{R}^2 \qquad \text{in} \quad F^{-s}(J^0) + \mathfrak{R}^2;$$

finally take for $T_r$ a closed subspace supplementary to

$$\mathrm{F}^{-(r-1)}(\mathrm{J}^0) + \mathfrak{N}^2 \quad \text{in} \quad \mathfrak{N}.$$

The space $\mathrm{T}_s$ is in natural duality with the subspace of $\mathfrak{g}_0$ generated by the $X_{0i}$ for $i \in \mathrm{I}_s - \mathrm{I}_{s+1}$, and we choose the $x_i$ for $i \in \mathrm{I}_s - \mathrm{I}_{s+1}$ as forming the pseudobasis of $\mathrm{T}_s$ dual to the basis of the $X_{0i}$ for the same indices. We then have by definition

(9) $$x_i^{p^s} \in \mathrm{J}^0 \quad \text{for} \quad i \in \mathrm{I}_s - \mathrm{I}_{s+1}, \quad 0 \leqslant s \leqslant r-1.$$

With this choice of the $x_i$, we form the corresponding privileged basis $(Z_\alpha)$ for $\mathfrak{G}$ (with $\alpha \in \mathbf{N}^{(\mathrm{I})}$). Let $\mathrm{L}_r$ be the set of multiindices $\alpha$ of height $\leqslant r$, so that the $Z_\alpha$ for $\alpha \in \mathrm{L}_r$ form a privileged basis for $\mathfrak{s}_r$. We define $\mathrm{L}'_r$ to be the subset of $\mathrm{L}_r$ consisting of the $\alpha$ *such that* $\alpha(i) < p^s$ *for* $i \in \mathrm{I}_s - \mathrm{I}_{s+1}$ $(0 \leqslant s \leqslant r)$, and $\mathrm{L}''_r$ to be the subset of $\mathrm{L}_r$ consisting of the $\alpha$ *such that* $p^s$ *divides* $\alpha(i)$ *for* $i \in \mathrm{I}_s$, $0 \leqslant s \leqslant r$. It is clear that $\mathrm{L}'_r \subset \mathrm{L}_r - \mathrm{L}''_r$; furthermore, from (9) and the fact that $\mathrm{J}^0$ is a subalgebra of $\mathfrak{G}^*$, it follows that $x^\alpha \in \mathrm{J}^0$ *for* $\alpha \in \mathrm{L}''_r$. We will prove that the $Z_\alpha$ *for* $\alpha \in \mathrm{L}_r - \mathrm{L}''_r$ *belong to* $\mathfrak{s}_r \cap \mathrm{J}$; this and the previous result will at once prove that they form a *basis* of $\mathfrak{s}_r \cap \mathrm{J}$ and the classes of the $x^\alpha$ with $\alpha \in \mathrm{L}''_r$ a *pseudobasis* of $(\mathrm{J}^0 + \mathfrak{B}_{r+1})/\mathfrak{B}_{r+1}$ (notations of §2, No. 6).

We assume the result has been proved for $\mathfrak{s}_{r-1} \cap \mathrm{J}$ (for $r = 0$, the assumption is vacuous), and we prove that for $\alpha \in \mathrm{L}_r - \mathrm{L}''_r$ and $|\alpha| \geqslant p^r$, $Z_\alpha$ belongs to J, by induction on $|\alpha|$ (for $|\alpha| < p^r$, to say that $\alpha \in \mathrm{L}_r - \mathrm{L}''_r$ is equivalent to saying that it is in $\mathrm{L}_{r-1} - \mathrm{L}''_{r-1}$).

Suppose $|\alpha| = p^r$, and consider first the case in which $\alpha = p^r \varepsilon_i$; as $\alpha \notin \mathrm{L}''_r$, this is only possible if $i \in \mathrm{I} - \mathrm{I}_r$. By definition $Z_\alpha = X_{ri}$ and $\mathrm{V}^r(X_{ri}) = X_{0i}$, and there exists an $Y \in \mathfrak{g}_r \cap \mathrm{J}$ such that $\mathrm{V}^r(X_{ri} - Y) = 0$, hence

$$X_{ri} - Y = \sum_\lambda c_\lambda Z_\lambda, \qquad \text{where } Z_\lambda \text{ runs through a basis of } \mathfrak{s}_{r-1}.$$

Now $\langle X_{ri}, x^\lambda \rangle = 0$ for $\mathrm{ht}(\lambda) \leqslant r-1$, by definition; on the other hand, as $Y \in \mathrm{J}$, we have $\langle Y, x^\lambda \rangle = 0$ for all $\lambda \in \mathrm{L}''_r$. This proves that the only coefficients $c_\lambda \neq 0$ for $\mathrm{ht}(\lambda) \leqslant r-1$ are those for which $\lambda \in \mathrm{L}_r - \mathrm{L}''_r$, hence in $\mathrm{L}_{r-1} - \mathrm{L}''_{r-1}$. But then the inductive assumption implies that $\sum_\lambda c_\lambda Z_\lambda \in \mathrm{J}$, hence $X_{ri} \in \mathrm{J}$.

If on the other hand $|\alpha| = p^r$ and $\alpha$ is not of the form $p^r \varepsilon_i$, then $\mathrm{ht}(\alpha) \leqslant r-1$, and again $Z_\alpha \in \mathrm{J}$ by the inductive assumption.

Suppose now that we have proved $Z_\alpha \in \mathrm{J}$ for $\alpha \in \mathrm{L}_r - \mathrm{L}''_r$ and $p^r \leqslant |\alpha| \leqslant h$, and consider the indices $\alpha \in L_r - L_r''$ with $|\alpha| = h + 1$. Observe that *any* multiindex $\alpha \in L_r$ can be written uniquely $\alpha = \beta + \gamma$ where $\beta \in L_r'$ and $\gamma \in L_r''$, for if $i \in I_s - I_{s+1}$ we may write uniquely

$$\alpha(i) = \beta(i) + p^s\lambda(i) \qquad \text{with} \quad 0 \leqslant \beta(i) < p^s,$$

and our assertion follows from the definitions. Now when we take $\alpha \in L_r - L_r''$, we have $\beta \neq 0$; if $\gamma \neq 0$, we have $\beta < \alpha$ and $\beta \in L_r - L_r''$; the inductive hypothesis and the fact that J is a two-sided ideal of $\mathfrak{G}$ imply that $Z_\beta Z_\gamma \in J$. By §2, No. 6, Corollary 1, the element

$$Y_\alpha = Z_\alpha - \frac{\alpha!}{\beta!\gamma!} Z_\beta Z_\gamma$$

therefore belongs to $\mathfrak{s}_r \cap M_h$, and as $\alpha \notin L_r''$ and $Z_\beta Z_\gamma \in J$, $Y_\alpha$ is orthogonal to all $x^\lambda$ with $\lambda \in L_r''$. However, the inductive hypothesis implies that the $Z_\lambda$ with $|\lambda| \leqslant h$ and $\lambda \in L_r - L_r''$ form a basis of $\mathfrak{s}_r \cap J \cap M_h$, hence $Y_\alpha \in J$ and $Z_\alpha \in J$.

If on the other hand $\gamma = 0$, hence $\alpha \in L_r'$, as $\alpha$ has not the form $p^r\varepsilon_i$, it can be written $\lambda + \mu$ with $\lambda$, $\mu$ in $L_r'$ satisfying the conditions of §2, No. 6, Lemma 1, and both $\neq 0$, hence in $M_h \cap \mathfrak{s}_r$. However, as $L_r' \subset L_r - L_r''$, the inductive assumption shows that $Z_\lambda Z_\mu \in J$; the conclusion follows as above by considering the element

$$Y_\alpha = Z_\alpha - \frac{\alpha!}{\lambda!\mu!} Z_\lambda Z_\mu.$$

We now have proved that the $Z_\alpha$ for $\alpha \in L_r - L_r''$ form a basis of $\mathfrak{s}_r \cap J$, and in addition, for any $s \leqslant r$,

$$V^s(\mathfrak{s}_r \cap J) \cap \mathfrak{g}_0 = \mathfrak{h}_s'. \tag{10}$$

Now it follows from the definition that the multiindices $\alpha \in L_r - L_r'$ are exactly the elements of $L_r$ which *majorize an element* $\neq 0$ of $L_r''$, in other words, those which may be written $\alpha = \beta + \gamma$, with $\beta \in L_r''$, $\beta \neq 0$ and $\gamma \geqslant 0$. As we have proved that the classes mod $\mathfrak{B}_{r+1}$ of the $x^\beta$ with $\beta \in L_r''$ form a pseudobasis of $(J^0 + \mathfrak{B}_{r+1})/\mathfrak{B}_{r+1}$ and on the other hand the classes mod $\mathfrak{B}_{r+1}$ of the $x^\alpha$ for all $\alpha \in L_r$ form a pseudobasis of $\mathfrak{G}^*/\mathfrak{B}_{r+1}$, it is clear that the classes mod $\mathfrak{B}_{r+1}$ of the $x^\alpha$ with $\alpha \in L_r - L_r'$ form a pseudobasis of the *closed ideal* in $\mathfrak{G}^*/\mathfrak{B}_{r+1}$ generated by the augmentation ideal of the subbigebra $(\mathrm{J}^0 + \mathfrak{B}_{r+1})/\mathfrak{B}_{r+1}$. But by definition this closed ideal is the orthogonal of the largest subbigebra $\mathfrak{H} \cap \mathfrak{s}_r$ in $k \cdot 1 + (\mathrm{J} \cap \mathfrak{s}_r)$, hence the $Z_\alpha$ with $\alpha \in \mathrm{L}'_r$ form a *privileged basis* of $\mathfrak{s}_r \cap \mathfrak{H}$, and both sides of (8) are equal to the sub-Lie algebra of $\mathfrak{g}_0$ spanned by the $X_{0i}$ with $i \in \mathrm{I}_r$.

Finally, to show that for instance J is the right ideal in $\mathfrak{G}$ generated by $\mathfrak{H}^+$, it is enough to prove that, for each $r$, $\mathfrak{s}_r \cap \mathrm{J}$ is the right ideal of the algebra $\mathfrak{s}_r$ generated by $\mathfrak{s}_r \cap \mathfrak{H}^+$. To do this, we use again induction on $r$, assuming therefore that $\mathfrak{s}_{r-1} \cap \mathrm{J}$ is the right ideal of $\mathfrak{s}_{r-1}$ generated by $\mathfrak{s}_{r-1} \cap \mathfrak{H}^+$. We then prove by induction on $h$ that the intersection of $\mathfrak{s}_r \cap \mathrm{J}$ with $\mathrm{M}_h$, for $h \geqslant p^r$, is in the right ideal generated by $\mathfrak{s}_r \cap \mathfrak{H}^+$. From what we have proved above it follows that

$$\mathfrak{g}_r \cap \mathrm{J} = (\mathfrak{s}_{r-1} \cap \mathrm{J}) + (\mathfrak{g}_r \cap \mathfrak{H}),$$

hence our contention is already proved for $h = p^r$. For $h \geqslant p^r$ and $|\alpha| = h + 1$, we have seen above that if $\alpha \in \mathrm{L}_r - \mathrm{L}''_r$, then either $\alpha \in \mathrm{L}'_r$, and then $Z_\alpha \in \mathfrak{s}_r \cap \mathrm{H}$, or we can write

$$Z_\alpha = \frac{\alpha!}{\beta!\gamma!} Z_\beta Z_\gamma + Y_\alpha, \qquad \text{where} \quad Z_\beta \in \mathfrak{s}_r \cap \mathrm{H} \quad \text{and} \quad Y_\alpha \in \mathrm{M}_h \cap \mathfrak{s}_r \cap \mathrm{J}.$$

This clearly ends the proof.

We may now characterize the *stable invariant subbigebras* of a reduced bigebra $\mathfrak{G}$ and answer in that case a question left open in I, §3, No. 6.

**Proposition 6.** *Let $\mathfrak{H}$ be a stable subbigebra of a reduced bigebra $\mathfrak{G}$. The following properties are equivalent:*

a) *$\mathfrak{H}$ is an invariant subbigebra.*

b) *The left and right ideals of the algebra $\mathfrak{G}$ generated by $\mathfrak{H}^+$ are identical.*

c) *If $\mathrm{J}_r$ is the two-sided ideal of $\mathfrak{s}_r$ generated by $\mathfrak{s}_r \cap \mathfrak{H}^+$, then $(\mathfrak{g}_r \cap \mathfrak{H}) + \mathrm{J}_{r-1}$ is an ideal of the Lie algebra $\mathfrak{g}_r$ for every $r \geqslant 0$ (for $r = 0$, this simply means that $\mathfrak{g}_0 \cap \mathfrak{H}$ is an ideal in $\mathfrak{g}_0$).*

We have seen in Proposition 5 that a) implies b); it also implies c), for it is enough to prove that, for $i \in \mathrm{I}$ and every $Y \in \mathfrak{g}_r \cap \mathfrak{H}$, we have

$$[X_{ri}, Y] \in (\mathfrak{g}_r \cap \mathfrak{H}) + (\mathfrak{s}_{r-1} \cap \mathrm{J}).$$

However, we have $[X_{ri}, Y] \in \mathfrak{g}_r$, and the fact that J is a two-sided ideal implies that $[X_{ki}, Y] \in \mathrm{J}$; but the relation

$$\mathfrak{g}_r \cap \mathrm{J} = (\mathfrak{g}_r \cap \mathfrak{H}) + (\mathfrak{s}_{r-1} \cap \mathrm{J})$$

results from the properties of the basis $(Z_\alpha)$ constructed in Proposition 5.

To show that c) implies b), it is enough to prove that the left and right ideals generated by $\mathfrak{s}_r \cap \mathfrak{H}$ in the algebra $\mathfrak{s}_r$ coincide; as $\mathfrak{s}_r \cap \mathfrak{H}$ is the subalgebra of $\mathfrak{s}_r$ generated by $\mathfrak{g}_r \cap \mathfrak{H}$ (§2, No. 7, Proposition 2), this will be proved if we show that, for $Y \in \mathfrak{g}_r \cap \mathfrak{H}$, $X_{ri}Y - YX_{ri}$ belongs to $\mathrm{J}_{r-1} + (\mathfrak{g}_r \cap \mathfrak{H})$, which is exactly assumption c).

Finally, to prove that b) implies a), we apply Proposition 3 of No. 2 to the stable subbigebra $\mathfrak{H}$ of $\mathfrak{G}$, and keep the notations of that proposition. Then $(X_\alpha)_{\alpha \in \mathrm{L}}$ is also a basis of $\mathfrak{H}$ (§2, No. 6, Corollary 2); the $X_\alpha X_\beta$ for $\alpha \in \mathbf{N}^{(\mathrm{I})}$ and $\beta \in \mathrm{L}$, $\beta \neq 0$ therefore generate (as a vector subspace of $\mathfrak{G}$) the left ideal J (which by assumption is two-sided) generated by $\mathfrak{H}^+$. Denote by L″ the set of multiindices $\alpha \in \mathbf{N}^{(\mathrm{I})}$ *such that* $p^s$ *divides* $\alpha(i)$ *for* $i \in \mathrm{I}_s$, $s \geqslant 0$ arbitrary (so that if $\mathrm{I}_r = \mathrm{I}_{r_0}$ for $r \geqslant r_0$, we must have $\alpha(i) = 0$ for $i \in \mathrm{I}_{r_0}$); it is then immediately verified that any $\alpha \in \mathbf{N}^{(\mathrm{I})}$ may be uniquely written $\alpha = \beta + \gamma$ with $\beta \in \mathrm{L}$ and $\gamma \in \mathrm{L}''$. Using §2, No. 6, Corollary 3, it is then easily verified that the $X_\gamma X_\beta$ for $\beta \in \mathrm{L}$ and $\gamma \in \mathrm{L}''$ form a *basis of* $\mathfrak{G}$. We will see that those elements for which $\beta \neq 0$ form a *basis of* J. As they are in J and linearly independent, all we have to prove is that for *any* $\alpha \in \mathbf{N}^{(\mathrm{I})}$ and any $\beta \in \mathrm{L}$ with $\beta \neq 0$, $X_\alpha X_\beta$ is a linear combination of these elements. But $X_\alpha$ can be written as a linear combination of elements $X_\mu X_\lambda$ with $\lambda \in \mathrm{L}$ and $\mu \in \mathrm{L}''$, and $X_\lambda X_\beta$ as a linear combination of elements $X_\nu$ with $\nu \in \mathrm{L}$ and $\nu \neq 0$, since $\mathfrak{H}^+$ is an ideal in $\mathfrak{H}$; this proves our assertion.

Now, for each integer $h$, the $X_\gamma X_\beta$ with $\beta \in \mathrm{L}$, $\beta \neq 0$, $\gamma \in \mathrm{L}''$, and $|\beta| + |\gamma| = h$ form the basis of a supplementary subspace to $\mathrm{M}_{h-1} \cap \mathrm{J}$ in $\mathrm{M}_h \cap \mathrm{J}$. From this it follows at once that

$$\mathrm{V}^r(\mathfrak{g}_r \cap \mathrm{J}) = \mathrm{V}^r(\mathfrak{g}_r \cap \mathfrak{H});$$

if $\mathfrak{H}'$ is the largest subbigebra whose augmentation ideal is contained in the ideal J (which is a biideal, being the two-sided ideal generated by the augmentation ideal of a subbigebra), we therefore have, by Proposition 5,

$$\mathrm{V}^r(\mathfrak{g}_r \cap \mathfrak{H}) = \mathrm{V}^r(\mathfrak{g}_r \cap \mathfrak{H}') \qquad \text{for every } r \geqslant 0,$$

and Corollary 1 to Proposition 3 of No. 2 implies that $\mathfrak{H}' = \mathfrak{H}$.

**Corollary.** *If $\mathfrak{H}$ is an invariant stable subbigebra of a reduced bigebra $\mathfrak{G}$, then, for every $r \geqslant 0$, $V^r(\mathfrak{g}_r \cap \mathfrak{H})$ is a p-ideal in the p-Lie algebra $\mathfrak{g}_0$.*

Since

$$V^r(\mathfrak{g}_r \cap \mathfrak{H}) = \mathfrak{g}_0 \cap V^r(\mathfrak{H}),$$

this follows from the fact that $V^r(\mathfrak{H})$ is an invariant subbigebra of $\mathfrak{G}^{(r)}$, $V^r$ being surjective.

**Proposition 7.** *If $(\mathfrak{H}_\alpha)$ is a family of invariant stable subbigebras of a reduced bigebra $\mathfrak{G}$, and if the smallest subbigebra $\mathfrak{H}$ containing the $\mathfrak{H}_\alpha$ is stable, then it is invariant and the biideal generated by $\mathfrak{H}^+$ is the sum of the biideals generated by the $\mathfrak{H}^+$.*

The subbigebra $\mathfrak{H}$ is the subalgebra of $\mathfrak{G}$ generated by the sum $\sum_\alpha \mathfrak{H}_\alpha$ of the vector spaces $\mathfrak{H}_\alpha$; hence it is immediate that $\mathfrak{H}$ satisfies condition b) of Proposition 6; furthermore, if $J_\alpha$ is the two-sided ideal generated by $\mathfrak{H}_\alpha^+$, $J = \sum_\alpha J_\alpha$ is a biideal and as it contains the sum $\sum_\alpha \mathfrak{H}_\alpha^+$, it is *a fortiori* generated by $\mathfrak{H}^+$.

**4.** *Quotient groups and inverse images of subgroups.* Let $\mathfrak{G}$ be a reduced bigebra and $\mathfrak{H}$ an invariant stable subbigebra of $\mathfrak{G}$; let J be the biideal of $\mathfrak{G}$ generated by $\mathfrak{H}^+$. The construction of the privileged basis $(Z_\alpha)$ in the proof of Proposition 5 yields immediately a privileged basis of the reduced quotient bigebra $\mathfrak{G}/J$ (which is the bigebra of the formal group **G**/**H**). Indeed, with the notations of Proposition 5, we have seen that if L″ is the set of all multiindices $\alpha \in \mathbf{N}^{(I)}$ such that $p^s$ divides $\alpha(i)$ for $i \in I_s$, $s \geqslant 0$, then the $Z_\alpha$ such that $\alpha \in L''$ form a basis of a subspace of $\mathfrak{G}$ *supplementary* to J, hence the images of these elements by the natural homomorphism $\varphi: \mathfrak{G} \to \mathfrak{G}/J$ form a basis of $\mathfrak{G}/J$. We *re-index* this basis in the following way: for each $\alpha \in L''$, let $\rho(\alpha)$ be the index in $\mathbf{N}^{(I)}$ such that

$$\rho(\alpha)(i) = \alpha(i)/p^s \quad \text{if} \quad i \in I_s - I_{s+1}, \quad s \geqslant 0,$$

and

$$\rho(\alpha)\,(i) = 0 \qquad \text{if } i \text{ belongs to the intersection of all the } I_s.$$

We then define in $\mathfrak{G}/J$,

$$\bar{Z}_{\rho(\alpha)} = \varphi(Z_\alpha) \qquad \text{for each} \quad \alpha \in L''; $$

if $\bar{\Delta}$ is the comultiplication in $\mathfrak{G}/J$, we have

$$\bar{\Delta}(\bar{Z}_{\rho(\alpha)}) = \bar{\Delta}(\varphi(Z_\alpha)) = (\varphi \otimes \varphi)\,(\Delta(Z_\alpha)) = \sum_{\beta + \gamma = \alpha} \varphi(Z_\beta) \otimes \varphi(Z_\gamma).$$

But in the last term, all products $\varphi(Z_\beta) \otimes \varphi(Z_\gamma)$ are 0 except if *both* $\beta \in L''$ and $\gamma \in L''$, and then $\varphi(Z_\beta) = \bar{Z}_{\rho(\beta)}$ and $\varphi(Z_\gamma) = \bar{Z}_{\rho(\gamma)}$, and we have

$$\rho(\beta) + \rho(\gamma) = \rho(\alpha).$$

It is clear that the $\rho(\alpha)$ are the multiindices in $\mathbf{N}^{(I)}$ such that $\rho(\alpha)(i) = 0$ for $i \in I'$, where $I'$ is the intersection of all the $I_s$; they are naturally identified with the multiindices in $\mathbf{N}^{(I-I')}$, and the preceding argument shows that they form a privileged basis for the bigebra $\mathfrak{G}/J$.

Another interpretation of this result is that the subbigebra $J^0$ of $\mathfrak{G}^*$, dual to $\mathfrak{G}/J$, has as pseudobasis the $x^\gamma$ with $\gamma \in L''$.

We may now complete for reduced infinitesimal groups the results of I, §3, No. 4:

**Proposition 8.**   *Let $\mathfrak{G}$, $\mathfrak{G}'$ be two reduced bigebras of finite dimension, $u: \mathfrak{G} \to \mathfrak{G}'$ a surjective homomorphism, $J$ its kernel, $\mathfrak{H}$ the largest subbigebra of $\mathfrak{G}$ contained in $k \cdot 1 + J$. In order that, for any algebra $A \in Alc_k$, the group homomorphism $\mathbf{u}_A: \mathbf{G}_A \to \mathbf{G}'_A$ be surjective, it is necessary and sufficient that $\mathfrak{H}$ be reduced.*

We may identify $\mathfrak{G}'$ with $\mathfrak{G}/J$, hence $\mathfrak{G}'^*$ with the subbigebra $J^0$ of $\mathfrak{G}$. All subbigebras of $\mathfrak{G}$ being stable, we may apply the description of privileged bases of $\mathfrak{G}$ and $\mathfrak{G}/J$ given above. The surjectivity of $\mathbf{u}_A$ for every $A \in Alc_k$ is equivalent to the possibility of extending a continuous homomorphism of $J^0$ into A, to a continuous homomorphism of $\mathfrak{G}^*$ into A. Let us in particular take

$$A = J^0/(J^0 \cap \mathfrak{R}^{h+1}),$$

and consider the natural homomorphism of $J^0$ onto A; suppose that it can be extended to a continuous homomorphism of $\mathfrak{G}^*$ onto A for all $h$. Now, if $\mathfrak{H}$ is not reduced, there would exist an index $i$ such that $x_i^m \notin J^0$ for $m < p^r$ but $x_i^{p^r} \in J^0$ with $r \geqslant 1$; let $\varphi$ be the extension to $\mathfrak{G}^*$ of the natural homomorphism of $J^0$ onto $J^0/(J^0 \cap \mathfrak{N}^{p^r+1})$. As the $x^\beta$ with $\beta \in L''$ form a pseudobasis of $J^0$,

$$\varphi(x_i^{p^r}) = (\varphi(x_i))^{p^r}$$

would be the class mod $J^0 \cap \mathfrak{N}^{p^r+1}$ of an element in the closed subspace E of $J^0$ having the $x^{p^r\beta}$ with $\beta \in L''$ as pseudobasis. But by definition, this element is also in the class of $x_i^{p^r}$ mod $J^0 \cap \mathfrak{N}^{p^r+1}$; this would mean that $x_i^{p^r}$ belongs to the subspace $E + \mathfrak{N}^{p^r+1}$ having as pseudobasis the $x^\alpha$ such that either $|\alpha| > p^r$ or $\alpha = p^r\beta$ with $\beta \in L''$; as by the choice of $i$, $\varepsilon_i$ is not in $L''$, the conclusion is absurd.

Let us now suppose that $\mathfrak{H}$ is reduced. Then $L''$ is simply the set of multi-indices $\alpha$ such that $\alpha(i) = 0$ for $i \in I_1$, since $I_r = I_1$ for every $r \geqslant 1$; it follows at once that the $x^\alpha$ for $\alpha \in L - L''$ form the pseudobasis of a *closed ideal* $\mathfrak{N}$ of $\mathfrak{G}^*$, supplementary to the subalgebra $J^0$; the projection of $\mathfrak{G}^*$ onto $J^0$ corresponding to the direct sum decomposition $\mathfrak{G}^* = J^0 \oplus \mathfrak{N}$ is therefore a continuous *algebra* homomorphism, and its existence obviously implies the possibility of extending to $\mathfrak{G}^*$ any continuous homomorphism of $J^0$ into an algebra $A \in \boldsymbol{Alc}_k$.

We now complete the results of I, §3, No. 9 on formal subgroups of quotient groups:

**Proposition 9.** *Let $u: \mathfrak{G} \to \mathfrak{G}'$ be a surjective homomorphism of reduced bigebras, J its kernel and $\mathfrak{N}$ the largest subbigebra of $\mathfrak{G}$ contained in $k \cdot 1 + J$; suppose $\mathfrak{N}$ is stable. Then, for any subbigebra $\mathfrak{H}'$ of $\mathfrak{G}'$, the largest subbigebra $\mathfrak{H}$ of $\mathfrak{G}$ contained in the subalgebra $u^{-1}(\mathfrak{H}')$ is such that $u(\mathfrak{H}) = \mathfrak{H}'$. If $\mathfrak{H}'$ is an invariant subbigebra of $\mathfrak{G}'$, $\mathfrak{H}$ is an invariant subbigebra of $\mathfrak{G}$; if K′ is the two-sided ideal of $\mathfrak{G}'$ generated by $\mathfrak{H}'^+$ and K the two-sided ideal of $\mathfrak{G}$ generated by $\mathfrak{H}^+$, we have $K = u^{-1}(K')$, and $\mathfrak{G}/K$ is naturally isomorphic to $\mathfrak{G}'/K'$.*

We may identify $\mathfrak{G}'$ with $\mathfrak{G}/J$. We first have to prove that if $\mathfrak{H}'^0$ is a closed biideal in $J^0$, then the closed biideal $\mathfrak{H}^0$ of $\mathfrak{G}^*$ which it generates is such that $\mathfrak{H}^0 \cap J^0 = \mathfrak{H}'^0$. Now we have seen that with the privileged basis $(Z_\alpha)$ chosen at the beginning of this section, $J^0$ is identified with the subbigebra of $\mathfrak{G}^*$ having as pseudobasis the $x^\gamma$ with $\gamma \in L''$; if $L'$ is the set of multiindices $\beta \in N^{(I)}$ such that $\beta(i) < p^s$ for each $i \in I_s - I_{s+1}$ and each $s \geqslant 0$, we have seen in the proof of Proposition 5 that each $\alpha \in N^{(I)}$ is uniquely written $\alpha = \beta + \gamma$ with $\beta \in L'$ and $\gamma \in L''$, so that we may identify $\mathfrak{G}^*$, as a linearly compact vector space, to the product of the closed subspaces $x^\beta J^0$ where $\beta \in L'$, the mapping $y \mapsto x^\beta y$ of $J^0$ onto $x^\beta J^0$ being a continuous linear bijection. This shows at once that $\mathfrak{H}^0$ is the *product* of the vector subspaces $x^\beta \mathfrak{H}'^0$ for $\beta \in L'$, and it is then obvious that $\mathfrak{H}^0 \cap J^0 = \mathfrak{H}'^0$.

If $\mathfrak{H}'$ is an invariant subbigebra of $\mathfrak{G}'$, it is clear that $K_1 = u^{-1}(K')$ is a biideal in $\mathfrak{G}$, kernel of the surjective bigebra homomorphism

$$\mathfrak{G} \xrightarrow{u} \mathfrak{G}' \longrightarrow \mathfrak{G}'/K'.$$

Let $\mathfrak{H}_1$ be the largest subbigebra contained in $k \cdot 1 + K_1$; we have

$$u(\mathfrak{H}_1) \subset k \cdot 1 + K',$$

hence $u(\mathfrak{H}_1) \subset \mathfrak{H}'$, or equivalently $\mathfrak{H}_1 \subset u^{-1}(\mathfrak{H}')$, hence $\mathfrak{H}_1 \subset \mathfrak{H}$; on the other hand, as $u(\mathfrak{H}) = \mathfrak{H}'$, one has $u(\mathfrak{H}^+) = \mathfrak{H}'^+$, hence $u(K) \subset K'$, and therefore

$$\mathfrak{H}^+ \subset K_1 \qquad \text{and} \qquad \mathfrak{H} \subset k \cdot 1 + K_1$$

which by definition implies $\mathfrak{H} \subset \mathfrak{H}_1$. We have thus proved that $\mathfrak{H}_1 = \mathfrak{H}$ and $K_1 = K$, taking into account Proposition 5.

*Remark.* I do not know if, in general, there may exist a subbigebra $\mathfrak{L}$ of $\mathfrak{H}$, different from $\mathfrak{H}$, containing $\mathfrak{N}$ and such that $u(\mathfrak{L}) = \mathfrak{H}'$. This is impossible, however, when $\mathfrak{N}$ is *reduced.* To see that, we first prove that necessarily

$$\mathfrak{g}_0 \cap \mathfrak{H} = \mathfrak{g}_0 \cap \mathfrak{L};$$

by assumption,

$$\mathfrak{g}_0 \cap \mathfrak{L} \supset \mathfrak{g}_0 \cap \mathfrak{N};$$

suppose $Y \in \mathfrak{g}_0 \cap \mathfrak{H}$ is such that $Y \notin \mathfrak{g}_0 \cap \mathfrak{N}$. We then have $u(Y) \in \mathfrak{g}'_0$, Lie algebra of $\mathfrak{G}'$; as by assumption $u(\mathfrak{H}) = u(\mathfrak{L})$, $u(Y)$ is the image of an element $Y'$ of $\mathfrak{L}$ in some $\mathfrak{s}_r \cap \mathfrak{L}$. Now we may apply the Remark of No. 2 to the subbigebras $\mathfrak{s}_r \cap \mathfrak{N}$ and $\mathfrak{s}_r \cap \mathfrak{L}$, since $\mathfrak{N}$ is reduced; with such a choice of the privileged basis $(Z_\alpha)$ of $\mathfrak{G}$, it follows that the $u(\mathfrak{s}_r \cap \mathfrak{L}) \cap \mathfrak{g}'_0$ is equal to the image $u(\mathfrak{g}_0 \cap \mathfrak{L})$, hence we may assume that $Y' \in \mathfrak{g}_0 \cap \mathfrak{L}$;

this implies that

$$Y' - Y \in \mathfrak{g}_0 \cap \mathrm{J} = \mathfrak{g}_0 \cap \mathfrak{N},$$

hence

$$Y' - Y \in \mathfrak{g}_0 \cap \mathfrak{L}$$

and this proves that

$$\mathfrak{g}_0 \cap \mathfrak{L} = \mathfrak{g}_0 \cap \mathfrak{H}.$$

Now observe that, for any $r \geqslant 0$, $\mathrm{V}^r(\mathfrak{L})$ and $\mathrm{V}^r(\mathfrak{H})$ have the same image $\mathrm{V}^r(\mathfrak{H}')$, and $\mathrm{V}^r(\mathfrak{H})$ is the largest subbigebra of $u^{-1}(\mathrm{V}^r(\mathfrak{H}'))$: indeed, by duality, this amounts to saying (with the notations of Proposition 9) that the ideal in $\mathfrak{G}^*$ generated by the inverse image $(\mathrm{F}^r)^{-1}(\mathfrak{H}'^0)$ is equal to $(\mathrm{F}^r)^{-1}(\mathfrak{H}^0)$; however, here all the subsets $\mathrm{I}_s$ are the same for $s \geqslant 1$, $\mathrm{L}''$ is the set of multiindices $\alpha$ such that $\alpha(i) = 0$ for $i \in \mathrm{I}_1$ and $\mathrm{L}'$ the set of multiindices $\alpha$ such that $\alpha(i) = 0$ for $i \in \mathrm{I} - \mathrm{I}_1$. Then an element $z$ of $\mathfrak{H}^0$ belonging to $x^\beta \mathfrak{H}'^0$ for a $\beta \in \mathrm{L}'$ can only belong to $\mathrm{F}^r(\mathfrak{G}^*)$ if $\beta$ has the form $p^r\lambda$ with $\lambda \in \mathrm{L}'$, and if $z = x^\beta y$ with $y \in \mathfrak{H}'^0$, one must in addition have $y \in \mathrm{F}^r(\mathfrak{G}^*)$, hence

$$y = y'^{p^r} \qquad \text{with} \quad y' \in (\mathrm{F}^r)^{-1}(\mathfrak{H}'^0).$$

This proves our assertion.

From the first result applied to $\mathrm{V}^r(\mathfrak{N})$ and $\mathrm{V}^r(\mathfrak{H})$, it then follows that

$$\mathfrak{g}_0 \cap \mathrm{V}^r(\mathfrak{H}) = \mathfrak{g}_0 \cap \mathrm{V}^r(\mathfrak{L});$$

but as $\mathfrak{L} \subset \mathfrak{H}$, this implies that $\mathfrak{L} = \mathfrak{H}$ (Corollary 1 to Proposition 3).

For *invariant* subbigebras $\mathfrak{H}$, the question raised above always has a negative answer. More precisely, we have:

**Proposition 10.** *With the general assumptions of Proposition 9, suppose $\mathfrak{H}'$ is an invariant subbigebra of $\mathfrak{G}'$; the largest subbigebra $\mathfrak{H}$ of $\mathfrak{G}$ contajned in $u^{-1}(\mathfrak{H}')$ is then the only invariant subbigebra of $\mathfrak{G}$ such that $u(\mathfrak{H}) = \mathfrak{H}'$.*

Indeed, let $\mathfrak{L} \supset \mathfrak{N}$ be an invariant subbigebra of $\mathfrak{G}$ such that $u(\mathfrak{L}) = \mathfrak{H}'$.

This means that $\mathfrak{L}^0 \cap J^0 = \mathfrak{H}'^0$. If I is the two-sided ideal of $\mathfrak{G}$ generated by $\mathfrak{L}^+$, we know that $\mathfrak{L}^0$ is the closed ideal of $\mathfrak{G}^*$ generated by the angmentation ideal $I^{0+}$ (Proposition 5) and that $I^0$ is the largest subbigebra of $\mathfrak{G}^*$ contained in $k.1 + \mathfrak{L}^0$. As $\mathfrak{L}^0 \subset \mathfrak{N}^0$ and $J^0$ is the largest subbigebra of $\mathfrak{G}^*$ contained in $k.1 + \mathfrak{N}^0$ (Proposition 5), we necessarily have

$$I^0 \subset J^0 \cap \mathfrak{L}^0 = \mathfrak{H}'^0,$$

and as $\mathfrak{H}^0$ is the ideal in $\mathfrak{G}^*$ generated by $\mathfrak{H}'^0$, we have finally $\mathfrak{L}^0 \subset \mathfrak{H}^0$ or $\mathfrak{L} \supset \mathfrak{H}$, hence $\mathfrak{L} = \mathfrak{H}$.

These results can of course be translated, as in I, §3, into the language of infinitesimal groups. If $\mathbf{G}$ is a *reduced* infinitesimal group over $k$, and $\mathbf{N}$ a *stable* invariant subgroup of $\mathbf{G}$, then, for every subgroup $\mathbf{H}'$ of $\mathbf{G}/\mathbf{N}$, there is a *largest* subgroup $\mathbf{H} \supset \mathbf{N}$ in $\mathbf{G}$ such that $\mathbf{H}'$ is equal to $\mathbf{H}/\mathbf{N}$; if in addition $\mathbf{N}$ is *reduced*, then $\mathbf{H}'_A = \mathbf{H}_A/\mathbf{N}_A$ for every $A \in \boldsymbol{Alc}_k$. We say that $\mathbf{H}$ is the *inverse image* of $\mathbf{H}'$ by the natural homomorphism $\mathbf{u}: \mathbf{G} \to \mathbf{G}/\mathbf{N}$, and we write $\mathbf{H} = \mathbf{u}^{-1}(\mathbf{H}')$. The mapping $\mathbf{H} \mapsto \mathbf{u}(\mathbf{H})$ of the set of all formal subgroups of $\mathbf{G}$ containing $\mathbf{N}$ into the set of all formal subgroups of $\mathbf{G}/\mathbf{N}$ is *surjective*, and it is *bijective* if $\mathbf{N}$ is *reduced*, the inverse mapping being then $\mathbf{H}' \mapsto \mathbf{u}^{-1}(\mathbf{H}')$. If $\mathbf{H} \supset \mathbf{N}$ is invariant in $\mathbf{G}$, $\mathbf{u}(\mathbf{H}) = \mathbf{H}/\mathbf{N}$ is invariant in $\mathbf{G}/\mathbf{N}$ and $(\mathbf{G}/\mathbf{N})/(\mathbf{H}/\mathbf{N})$ is isomorphic to $\mathbf{G}/\mathbf{H}$ ("second isomorphism theorem"); furthermore, $\mathbf{H} \mapsto \mathbf{u}(\mathbf{H})$ is bijective in the set of invariant subgroups containing $\mathbf{H}$, without supposing that $\mathbf{N}$ is reduced.

## 5. *Normalizers and centralizers.*

**Proposition 11.** *Let* $\mathbf{G}$ *be a reduced infinitesimal group,* $\mathbf{H}$ *a reduced subgroup of* $\mathbf{G}$. *Then there exists a largest reduced subgroup* $\mathscr{N}(\mathbf{H})$ *of* $\mathbf{G}$ *such that* $\mathbf{H}$ *is an invariant subgroup of* $\mathscr{N}(\mathbf{H})$.

We work with the corresponding covariant bigebras. We prove that the set $\mathscr{F}$ of reduced subbigebras $\mathfrak{L}$ of $\mathfrak{G}$ containing $\mathfrak{H}$ and such that $\mathfrak{H}$ is invariant in $\mathfrak{L}$ is a *directed* set for the inclusion relation. Indeed, let $\mathfrak{L}_1$, $\mathfrak{L}_2$ be two elements of $\mathscr{F}$ and let $\mathfrak{L}$ be the subbigebra of $\mathfrak{G}$ generated by their union, which is the *subalgebra* of $\mathfrak{G}$ generated by the subcogebra $\mathfrak{L}_1 + \mathfrak{L}_2$ (I, §3, No. 3), and is obviously reduced. One has only to apply criterion (b) of Proposition 6; every element of $\mathfrak{L}$ is sum of products $Y_1 Y_2 \cdots Y_n$, where each $Y_j$ belongs to $\mathfrak{L}_1$ or to $\mathfrak{L}_2$; we have to show that for every $Z \in \mathfrak{H}$, $(Y_1 Y_2 \cdots Y_n)Z$ can be written as sum of terms of the form $Z'(Y_1' \cdots Y_m')$ with $Z' \in \mathfrak{H}$ and the $Y_j'$ belong to $\mathfrak{L}_1$ or $\mathfrak{L}_2$. The proof is done by induction on $n$; for $n = 1$, it follows from the assumption that $\mathfrak{H}$ is invariant both in $\mathfrak{L}_1$ and $\mathfrak{L}_2$ and from criterion b) of Proposition 6. The passage from $n - 1$ to $n$ is immediate, since $Y_n Z$ is a sum of terms $Z'Y'$ with $Z' \in \mathfrak{H}$ and $Y'$ in $\mathfrak{L}_1$ or $\mathfrak{L}_2$, and the inductive assumption shows that $(Y_1 \cdots Y_{n-1})Z'$ is sum of terms $Z''(Y_1'' \cdots Y_q'')$ with $Z'' \in \mathfrak{H}$ and the $Y_k''$ in $\mathfrak{L}_1$ or $\mathfrak{L}_2$.

Since $\mathfrak{H}$ is reduced, it belongs to $\mathscr{F}$, which therefore is not empty. The union $\mathscr{N}(\mathfrak{H})$ of the subbigebras of the set $\mathscr{F}$ is then obviously a reduced subbigebra of $\mathfrak{G}$, in which $\mathfrak{H}$ is invariant by criterion b) of Proposition 6, and it is obviously the largest element of $\mathscr{F}$.

We will say that the formal group $\mathscr{N}(\mathbf{H})$ having as subbigebra $\mathscr{N}(\mathfrak{H})$ is the *normalizer* of the reduced subgroup $\mathbf{H}$.

It may well happen that the *centralizer* of a reduced subgroup of a reduced group is *not reduced*; in particular the *center* of a reduced group is not necessarily reduced. A simple example is obtained by taking the two-dimensional reduced group $\mathbf{G}$, for which, in $\mathfrak{G}^* = k[[x_1, x_2]]$, the co-multiplication is given by

$$c(x_1) = 1 \otimes x_1 + x_1 \otimes 1, \qquad c(x_2) = 1 \otimes x_2 + x_2 \otimes 1 + x_1^p \otimes x_2^{p^2}$$

(it corresponds to the algebraic group law

$$(x, y)(x', y') = (x + x', y + y' + x^p x'^{p^2})).$$

One easily verifies that

$$\langle X_{h1} \otimes X_{k2}, c(x_1^\alpha x_2^\beta)\rangle = \langle X_{k2} \otimes X_{h1}, c(x_1^\alpha x_2^\beta)\rangle$$

and has the value 0 except for $\alpha = p^h$ and $\beta = p^k$, where it has the value 1; furthermore the $X_{k2}$ commute for all values of $k$, hence the center of $\mathbf{G}$ has a subbigebra which contains the $X_{k2}$; but in addition

$$X_{h1} X_{01} = X_{01} X_{h1} \qquad \text{for every} \quad h,$$

because there is no term in $x_1 \otimes x_1^{p^h}$ nor in $x_1^{p^h} \otimes x_1$ in $c(x_1^\alpha x_2^\beta)$ if $\beta > 0$. Therefore the subbigebra of $\mathscr{Z}(\mathbf{G})$ also contains $X_{01}$, hence is not reduced (it cannot of course be $\mathfrak{G}$ itself).

**Proposition 12.** *Let* **G** *be a reduced infinitesimal group of finite dimension, and* **H** *a reduced subgroup of* **G**. *Then the largest reduced subgroup in* $\mathscr{Z}(\mathbf{H})$ *is an invariant subgroup of* $\mathscr{N}(\mathbf{H})$, *written* ${}^r\mathscr{Z}(\mathbf{H})$ *and called the reduced centralizer of* **H**. (${}^r\mathscr{Z}(\mathbf{G})$ *is called the reduced center of* **G**).

We work again with the corresponding bigebras, and note that all subbigebras of $\mathfrak{G}$ are stable due to the assumption of finite dimension; hence there exists a largest reduced subbigebra $\mathfrak{L}$ in the subbigebra $\mathscr{Z}(\mathfrak{H})$, and it follows from the definitions that $\mathfrak{L} \subset \mathscr{N}(\mathfrak{H})$. We will prove slightly more than stated, by showing that the subbigebra

$$\mathfrak{L}' = \inf(\mathscr{Z}(\mathfrak{H}), \mathscr{N}(\mathfrak{H}))$$

is invariant in $\mathscr{N}(\mathfrak{H})$. As there is an integer $r$ such that $\mathrm{V}^r(\mathfrak{L}') = \mathfrak{L}$, and $\mathrm{V}^r(\mathscr{N}(\mathfrak{H})) = \mathscr{N}(\mathfrak{H})$, this will prove that $\mathfrak{L}$ is invariant in $\mathscr{N}(\mathfrak{H})$. Consider the algebra

$$\mathrm{A} = \mathfrak{H}^* \hat{\otimes} \mathscr{N}(\mathfrak{H})^* \hat{\otimes} \mathfrak{L}'^* \in \boldsymbol{Alc}_k,$$

and let $x$, $y$, $z$ be the canonical points of **H**, $\mathscr{N}(\mathbf{H})$, and $\mathbf{L}'$ in A (I, §3, No. 2); let $\mathbf{j}: \mathbf{H} \to \mathscr{N}(\mathbf{H})$ and $\mathbf{j}': \mathbf{L}' \to \mathscr{N}(\mathbf{H})$ be the natural formal homomorphisms. By definition, $y^{-1}\mathbf{j}(x)y$ is a point of $\mathbf{j}_{\mathrm{A}}(\mathbf{H}_{\mathrm{A}}) \subset \mathscr{N}(\mathbf{H})_{\mathrm{A}}$, and $\mathbf{j}'(z)$ commutes with it (I, §3, No. 10); hence we have in $\mathscr{N}(\mathbf{H})_{\mathrm{A}}$

$$(y\mathbf{j}'(z)y^{-1})\mathbf{j}(x)(y\mathbf{j}'(z)y^{-1})^{-1} = \mathbf{j}(x)$$

and this shows that $y\mathbf{j}'(z)y^{-1}$ is a point of $\mathscr{Z}(\mathbf{H})_{\mathrm{A}}$ (*loc. cit.*), hence a point of $\mathbf{j}'_{\mathrm{A}}(\mathbf{L}'_{\mathrm{A}})$ (cf. I, §3, No. 7, formula (1)). This shows that $\mathbf{L}'$ is invariant in $\mathscr{N}(\mathbf{H})$ (I, §3, No. 6).

**6.** *Isogenies.* If $\mathfrak{G}$, $\mathfrak{G}'$ are two reduced bigebras and $u: \mathfrak{G} \to \mathfrak{G}'$ a homomorphism, the largest subbigebra $\mathfrak{N}$ contained in $k \cdot 1 + \mathrm{J}$, where J is the kernel of $u$, is not necessarily reduced: for instance, if $\mathfrak{G}' = \mathfrak{G}^{(-r)}$ and $u = \mathrm{V}^r$, $\mathfrak{N}$ is equal to the Frobenius subbigebra $\mathfrak{s}_{r-1}$. We will say that $u$ is an *isogeny* of reduced bigebras if $u(\mathfrak{G}) = \mathfrak{G}'$ and if $\mathfrak{N}$ has *finite height* $t - 1$; the number $t$ is called the *height* of the isogeny.

**Proposition 13.** *If* $u: \mathfrak{G} \to \mathfrak{G}'$ *is an isogeny of height* $t$, *there exists a unique isogeny* $v: \mathfrak{G}' \to \mathfrak{G}^{(-t)}$ *such that* $v \circ u = \mathrm{V}^t$.

We may assume that $\mathfrak{G}' = \mathfrak{G}/\mathrm{J}$, where J is the ideal generated by the augmentation ideal of the invariant subbigebra $\mathfrak{N}$, and $u$ is the natural homomorphism. We use the privileged bases $(Z_\alpha)$ of $\mathfrak{G}$ and $(\overline{Z}_{\rho(\alpha)})$ of $\mathfrak{G}/\mathrm{J}$ constructed in Nos. 4 and 5, which are such that $u(Z_\alpha) = \overline{Z}_{\rho(\alpha)}$. If the multiplication table of $(Z_\alpha)$ is

$$Z_\alpha Z_\beta = \sum_\gamma d_{\alpha\beta\gamma} Z_\gamma \tag{11}$$

the multiplication table of $(\overline{Z}_{\rho(\alpha)})$ is therefore

$$\overline{Z}_{\rho(\alpha)} \overline{Z}_{\rho(\beta)} = \sum_\gamma d_{\alpha\beta\gamma} \overline{Z}_{\rho(\gamma)} \tag{12}$$

where $\alpha$ and $\beta$ are in L″, and on the right-hand side $\gamma$ must only run through L″. From the assumption on $\mathfrak{N}$, we have $\mathrm{I}_{t+1} = \varnothing$. We may write any multiindex $\alpha \in \mathbf{N}^{(\mathrm{I})}$ as

$$\alpha_0 + \alpha_1 + \cdots + \alpha_t, \qquad \text{where} \quad \alpha_s(i) = 0 \quad \text{for} \quad i \in \mathrm{I}_s - \mathrm{I}_{s+1}.$$

The set L″ can be described as consisting of the $\alpha$ for which $\alpha_s$ is divisible by $p^s$ for $0 \leqslant s \leqslant t$, and

$$\rho(\alpha) = \alpha_0 + p^{-1}\alpha_1 + p^{-2}\alpha_2 + \cdots + p^{-t}\alpha_t \qquad \text{for all} \quad \alpha \in \mathrm{L}''.$$

Now define $\sigma(\alpha)$ as equal to

$$\alpha_t + p^{-1}\alpha_{t-1} + \cdots + p^{-t}\alpha_0$$

for the multiindices $\alpha$ such that $\alpha_{t-s}$ is divisible by $p^s$ for $0 \leqslant s \leqslant t$; this implies that $\sigma(\rho(\alpha))$ is defined for the multiindices $\alpha$ which are *divisible by* $p^t$, and then

$$\sigma(\rho(\alpha)) = p^{-t}\alpha. \tag{13}$$

Now define a linear mapping $v\colon \mathfrak{G}' \to \mathfrak{G}^{(-t)}$ by the conditions

$$\begin{aligned} v(\overline{Z}_{\rho(\alpha)}) &= Z_{\sigma(\rho(\alpha))} && \text{when } p^t \text{ divides } \alpha,\\ &= 0 && \text{otherwise.} \end{aligned}$$

Using the fact that $(Z_\alpha)$ and $(\overline{Z}_{\rho(\alpha)})$ are privileged bases, it can be immediately verified that $v$ is a cogebra homomorphism. On the other hand, we have, by definition of $\mathfrak{G}^{(-t)}$ and formula (12),

$$v(\overline{Z}_{\rho(\alpha)}\overline{Z}_{\rho(\beta)}) = \sum_{\gamma} d_{\alpha\beta\gamma}^{p^{-t}}\overline{Z}_{\sigma(\rho(\gamma))} \tag{14}$$

where $\alpha$ and $\beta$ are divisible by $p^t$, and on the right-hand side $\gamma$ runs through the indices divisible by $p^t$. From (11) and the definition of multiplication in $\mathfrak{G}^{(-t)}$, it follows that the right-hand side of (14) is equal to $v(\overline{Z}_{\rho(\alpha)})v(\overline{Z}_{\rho(\beta)})$ in that case. On the other hand, if one of the multiindices $\alpha$, $\beta$ is not divisible by $p^t$, then, on the right-hand side of (12), there is no coefficient $d_{\alpha\beta\gamma} \neq 0$ such that $\gamma$ is divisible by $p^t$; for it would then follow from (11) that $\mathrm{V}^t(Z_\alpha Z_\beta) \neq 0$, which is absurd since

$$\mathrm{V}^t(Z_\alpha)\mathrm{V}^t(Z_\beta) = 0.$$

This implies that when one of the multiindices $\alpha$, $\beta$ is not divisible by $p^t$, the terms $\overline{Z}_{\rho(\gamma)}$ on the right-hand side of (12) have images by $v$ which are 0, and ends the proof that $v$ is a bigebra homomorphism; furthermore it follows from (13) that $v \circ u = \mathrm{V}^t$. The same construction shows that

$$u \circ v = \mathrm{V}^t \tag{15}$$

where $u$ is considered as a bigebra homomorphism of $\mathfrak{G}^{(-t)}$ into $\mathfrak{G}'^{(-t)}$.

Conversely, if

$$u\colon \mathfrak{G} \to \mathfrak{G}' \qquad \text{and} \qquad v\colon \mathfrak{G}' \to \mathfrak{G}^{(-r)}$$

are two bigebra homomorphisms (for reduced bigebras $\mathfrak{G}$, $\mathfrak{G}'$) such that

$$v \circ u = \mathrm{V}^r, \qquad u \circ v = \mathrm{V}^r \tag{16}$$

(where $u$ is considered as a homomorphism of $\mathfrak{G}^{(-r)}$ into $\mathfrak{G}'^{(-r)}$), then $u$ and $v$ are isogenies, since $u$ and $v$ are surjective ($\mathrm{V}^r$ being surjective), and their kernels are contained in that of $\mathrm{V}^r$.

We will say that two reduced bigebras $\mathfrak{G}$, $\mathfrak{G}'$ are *isogenous* if there exists a (positive or negative) integer $s$ and an isogeny $\mathfrak{G} \to \mathfrak{G}'^{(s)}$. This is in fact an *equivalence relation* between reduced bigebras: indeed, it is obviously reflexive; it is symmetric, for if $u\colon \mathfrak{G} \to \mathfrak{G}'^{(s)}$ is an isogeny of height $t$, there exists an isogeny

$$v : \mathfrak{G}^{(s)} \to \mathfrak{G}^{(-t)}$$

by Proposition 13, and $v$ is also an isogeny of $\mathfrak{G}'$ onto $\mathfrak{G}^{(-s-t)}$. Finally the relation is transitive, for if

$$u' : \mathfrak{G}' \to \mathfrak{G}''^{(r)}$$

is an isogeny, it is also an isogeny of $\mathfrak{G}'^{(s)}$ onto $\mathfrak{G}''^{(r+s)}$, and $u'' \circ u'$ is an isogeny by definition.

Another way of expressing the isogeny relation is the following one: $\mathfrak{G}_1$ and $\mathfrak{G}_2$ are isogenous if and only if there exists a reduced bigebra $\mathfrak{G}_3$ and two isogenies $\mathfrak{G}_3 \to \mathfrak{G}_1$, $\mathfrak{G}_3 \to \mathfrak{G}_2$. Indeed, if $u\colon \mathfrak{G}_1 \to \mathfrak{G}_2^{(s)}$ is an isogeny, we may assume $s < 0$ (otherwise $V^s \circ u$ is already an isogeny $\mathfrak{G}_1 \to \mathfrak{G}_2$); if we take $\mathfrak{G}_3 = \mathfrak{G}_1^{(-s)}$, we have two isogenies $V^{-s}\colon \mathfrak{G}_3 \to \mathfrak{G}_1$ and $u\colon \mathfrak{G}_3 \to \mathfrak{G}_2$. Conversely, suppose $v\colon \mathfrak{G}_3 \to \mathfrak{G}_1$ and $w\colon \mathfrak{G}_3 \to \mathfrak{G}_2$ are isogenies; Proposition 13 then shows that there exists an isogeny $u\colon \mathfrak{G}_1 \to \mathfrak{G}_3^{(t)}$, and then $w \circ u\colon \mathfrak{G}_1 \to \mathfrak{G}_2^{(t)}$ is an isogeny.

**Proposition 14.** *If $u\colon \mathfrak{G} \to \mathfrak{G}'$ is an isogeny of reduced bigebras, the mapping $\mathfrak{H} \mapsto u(\mathfrak{H})$ is a strictly increasing bijection of the set of reduced subbigebras* (resp. *reduced invariant subbigebras*) *of $\mathfrak{G}$ onto the set of reduced subbigebras* (resp. *reduced invariant subbigebras*) *of $\mathfrak{G}'$.*

Taking account of criterion b) of Proposition 5, it is clear that, if $\mathfrak{H}$ is a reduced subbigebra (resp. an invariant reduced subbigebra) of $\mathfrak{G}$, $\mathfrak{H}^{(s)}$ is a reduced subbigebra (resp. an invariant reduced subbigebra) of $\mathfrak{G}^{(s)}$; the proposition is therefore obvious when $\mathfrak{G}' = \mathfrak{G}^{(-t)}$ and $u = V^t$, since then

$$V^t(\mathfrak{H}) = \mathfrak{H}^{(-t)}$$

for every reduced subbigebra $\mathfrak{H}$. In general, we use Proposition 13: as there exists an isogeny

$$v\colon \mathfrak{G}' \to \mathfrak{G}^{(-t)}$$

such that $v \circ u = V^t$, $\mathfrak{H} \mapsto u(\mathfrak{H})$ is injective in the set of reduced subbigebras of $\mathfrak{G}$; and as $u \circ v = V^t$ ($u$ being considered as an isogeny of $\mathfrak{G}^{(-t)}$ onto $\mathfrak{G}'^{(-t)}$), this also shows that $\mathfrak{H} \mapsto u(\mathfrak{H})$ is surjective.

Of course the definitions given for reduced bigebras are immediately transferred to reduced infinitesimal groups.

Let $\mathfrak{G}$ be a reduced bigebra of *finite* dimension, and let $u: \mathfrak{G} \to \mathfrak{G}'$ be any homomorphism of $\mathfrak{G}$ into an infinitesimal bigebra $\mathfrak{G}'$; let J be the kernel of $u$, and $\mathfrak{N}$ the largest subbigebra of $\mathfrak{G}$ contained in $k \cdot 1 + \mathrm{J}$, which is stable since $\mathfrak{G}$ has finite dimension. If $\mathfrak{N}'$ is the largest *reduced* subbigebra contained in $\mathfrak{N}$ (§2, No. 9), it is equal to some $\mathrm{V}^r(\mathfrak{N})$, and as $\mathrm{V}^r(\mathfrak{G}) = \mathfrak{G}$, $\mathfrak{N}'$ is also an *invariant* subbigebra of $\mathfrak{G}$ by the criterion b) of Proposition 6. Let J′ be the two-sided ideal generated by $\mathfrak{N}'^+$ in the algebra $\mathfrak{G}$, which is also a biideal; we may then factorize $u$ as

$$u : \mathfrak{G} \xrightarrow{p} \mathfrak{G}/\mathrm{J}' \xrightarrow{v} \mathfrak{G}/\mathrm{J} \xrightarrow{j} \mathfrak{G}' \tag{17}$$

where $p$ is the natural surjection, $j$ the natural injection, and $v$ is an *isogeny*, since the kernel $v^{-1}(0)$ is obviously generated by the image of $\mathfrak{N}^+$ in $\mathfrak{G}/\mathrm{J}'$, which has finite height, as results from the construction of the privileged basis in $(Z_\alpha)$ of $\mathfrak{G}$ in Proposition 5. We will say (17) is the *natural decomposition* of $u$; it follows from Proposition 5 that

$$\dim \mathfrak{G} - \dim \mathfrak{N}' = \dim(\mathfrak{G}/\mathrm{J}'). \tag{18}$$

The homomorphism $u$ will be said to be *separable* if, in the decomposition (17), $v$ is the identity, in other words if $\mathfrak{N}$ is reduced. From Proposition 5 we get the immediate criterion:

**Proposition 15.** *Let $\mathfrak{G}$ be a reduced finite-dimensional bigebra, $u: \mathfrak{G} \to \mathfrak{G}'$ a homomorphism; $u$ is separable if and only if*

$$\dim(u(\mathfrak{G})) = \dim(u(\mathfrak{g}_0)).$$

**Corollary.** *Let $\mathfrak{G}$, $\mathfrak{G}'$, $\mathfrak{G}''$ be three bigebras, $\mathfrak{G}$ and $\mathfrak{G}'$ being reduced and finite dimensional; let*

$$u: \mathfrak{G} \to \mathfrak{G}', \qquad v: \mathfrak{G}' \to \mathfrak{G}''$$

*be two separable homomorphisms. If both $u$ and $v$ are injective, or if both are surjective, then $v \circ u$ is separable.*

The first statement follows from the fact that *all* injective homomorphisms are separable; the second follows from the fact that if $u$ is surjective and separable, then $u(\mathfrak{g}_0) = \mathfrak{g}_0'$, the Lie algebra of $\mathfrak{G}'$.

An example in which $u$ and $v$ are separable, $u$ injective, $v$ surjective and $v \circ u$ not separable, is given by the example after Proposition 3 of No. 2: if $J_1$ is the biideal generated by $\mathfrak{H}_1^+$, take for $u$ the natural injection $\mathfrak{H}_2 = \mathfrak{G}$ and for $v$ the natural surjection $\mathfrak{G} \to \mathfrak{G}/J_1$.

**7.** *Theory of reduced groups "up to isogeny".* The elementary theory of reduced infinitesimal groups cannot be developed as simply as the classical theory of groups, due to two facts which have been stressed above: in a formal reduced group $\mathbf{G}$, two formal reduced subgroups $\mathbf{H}_1$, $\mathbf{H}_2$ may be such that $\inf(\mathbf{H}_1, \mathbf{H}_2)$ is not reduced, and the kernel of a formal homomorphism $\mathbf{u}: \mathbf{G} \to \mathbf{G}'$ of reduced groups needs not be reduced.

This leads us to introduce new notions for reduced groups of *finite* dimension: for such a group $\mathbf{G}$, all subgroups are stable, and therefore any subgroup $\mathbf{H}$ contains a largest reduced subgroup, which is invariant in $\mathbf{G}$ if $\mathbf{H}$ is. We will write $\mathbf{H}_1 \wedge \mathbf{H}_2$ the largest reduced subgroup in $\inf(\mathbf{H}_1, \mathbf{H}_2)$; for the sake of symmetry, we will also write $\mathbf{H}_1 \vee \mathbf{H}_2$ instead of $\sup(\mathbf{H}_1, \mathbf{H}_2)$ (which is always reduced). For a formal homomorphism $\mathbf{u}: \mathbf{G} \to \mathbf{G}'$, we will say that the largest reduced subgroup contained in the kernel of $\mathbf{u}$ is the *reduced kernel* of $\mathbf{u}$, and we will write it ${}^r\mathbf{u}^{-1}(\mathbf{e})$; from the decomposition (17) it follows that there is a *natural isogeny* $\mathbf{G}/{}^r\mathbf{u}^{-1}(\mathbf{e}) \to \mathbf{u}(\mathbf{G})$, and we have

$$\dim \mathbf{u}(\mathbf{G}) + \dim {}^r\mathbf{u}^{-1}(\mathbf{e}) = \dim \mathbf{G}. \tag{19}$$

More generally:

**Proposition 16.** *Let* $\mathbf{u}: \mathbf{G} \to \mathbf{G}'$ *be a formal homomorphism of infinitesimal groups, where* $\mathbf{G}$ *is reduced and finite dimensional. For every infinitesimal subgroup* $\mathbf{H}'$ *of* $\mathbf{G}'$, *there is a largest reduced subgroup of* $\mathbf{G}$ *contained in* $\mathbf{u}^{-1}(\mathbf{H}')$; *it is called the reduced inverse image of* $\mathbf{H}'$ *and written* ${}^r\mathbf{u}^{-1}(\mathbf{H}')$. *The subgroup* ${}^r\mathbf{u}^{-1}(\mathbf{H}')$ *is the only reduced subgroup of* $\mathbf{G}$ *containing* ${}^r\mathbf{u}^{-1}(\mathbf{e})$ *and whose image by* $\mathbf{u}$ *is the same as that of* ${}^r\mathbf{u}^{-1}(\mathbf{H}')$; *that image* $\mathbf{u}({}^r\mathbf{u}^{-1}(\mathbf{H}'))$ *is the largest reduced subgroup contained in* $\inf(\mathbf{H}', \mathbf{u}(\mathbf{G}))$.

We have only to prove the corresponding statements for the covariant bigebras; we may obviously suppose that $u(\mathfrak{G}) = \mathfrak{G}'$ and that $\mathfrak{H}'$ is reduced (since the image of a reduced subgroup of $\mathbf{G}$ is reduced). If $\mathfrak{H}$ is the largest subbigebra contained in $u^{-1}(\mathfrak{H}')$, we have $u(\mathfrak{H}) = \mathfrak{H}'$ (Proposition 9); as for each $t > 0$, we have

$$u(\mathrm{V}^t(\mathfrak{H})) = \mathrm{V}^t(u(\mathfrak{H})) = \mathrm{V}^t(\mathfrak{H}') = \mathfrak{H}',$$

and $\mathrm{V}^t(\mathfrak{H})$ is the largest reduced subbigebra contained in $\mathfrak{H}$ for $t$ large enough, this proves the second assertion. To prove the first one, we are reduced to showing that if $\mathfrak{N}$ is the largest subbigebra such that $\mathfrak{N}^+$ is contained in the kernel J of $u$, $\mathfrak{G}$ is the unique reduced subbigebra of $\mathfrak{G}$ containing the largest reduced subbigebra $\mathfrak{N}'$ of $\mathfrak{N}$, and whose image is $u(\mathfrak{G}) = \mathfrak{G}'$; but if J′ is the biideal generated by $\mathfrak{N}'^+$, we have seen that $\mathfrak{G}/\mathrm{J}$ and $\mathfrak{G}/\mathrm{J}'$ are isogenous, and Proposition 14 allows us to replace $u$ by the natural homomorphism $\mathfrak{G} \to \mathfrak{G}/\mathrm{J}'$; but then the proposition is a consequence of the remark following Proposition 9.

**Corollary.** *Let* $\mathbf{G}$, $\mathbf{G}'$, $\mathbf{G}''$ *be three infinitesimal groups,* $\mathbf{G}$ *and* $\mathbf{G}'$ *being reduced and finite dimensional. Let*

$$\mathbf{u} : \mathbf{G} \to \mathbf{G}' \qquad \mathbf{v} : \mathbf{G}' \to \mathbf{G}''$$

*be two formal homomorphisms and let* $\mathbf{w} = \mathbf{v} \circ \mathbf{u}$; *then, for any infinitesimal subgroup* $\mathbf{H}''$ *of* $\mathbf{G}''$,

$$^r\mathbf{w}^{-1}(\mathbf{H}'') = {}^r\mathbf{u}^{-1}({}^r\mathbf{v}^{-1}(\mathbf{H}'')).$$

We may assume that $\mathbf{H}''$ is reduced and contained in $\mathbf{w}(\mathbf{G})$. Let

$$\mathbf{H} = {}^r\mathbf{u}^{-1}({}^r\mathbf{v}^{-1}(\mathbf{H}''));$$

then $\mathbf{w}(\mathbf{H}) = \mathbf{H}''$ by Proposition 16 applied to $\mathbf{u}$ and $\mathbf{v}$. On the other hand, if

$$\mathbf{N}' = {}^r\mathbf{v}^{-1}(\mathbf{e})$$

we have

$$^r\mathbf{u}^{-1}(\mathbf{N}') = {}^r\mathbf{w}^{-1}(\mathbf{e}):$$

indeed, if $\mathbf{M}$ is a reduced subgroup of $\mathbf{G}$ such that

$$\mathbf{v}(\mathbf{u}(\mathbf{M})) = \mathbf{e},$$

this implies $\mathbf{u}(\mathbf{M}) \subset \mathbf{N}'$, hence

$$\mathbf{M} \subset {}^r\mathbf{u}^{-1}(\mathbf{N}').$$

On the other hand, we have

$$\mathbf{v}(\mathbf{u}({}^r\mathbf{u}^{-1}(\mathbf{N}'))) = \mathbf{v}(\mathbf{N}') = \mathbf{e},$$

which proves that

$${}^r\mathbf{u}^{-1}(\mathbf{N}') = {}^r\mathbf{w}^{-1}(\mathbf{e}).$$

Now as $\mathbf{N}' \subset {}^r\mathbf{v}^{-1}(\mathbf{H}'')$, we have

$${}^r\mathbf{u}^{-1}(\mathbf{N}') \subset {}^r\mathbf{u}^{-1}({}^r\mathbf{v}^{-1}(\mathbf{H}'')),$$

and from Proposition 16 and the relation $\mathbf{w}(\mathbf{H}) = \mathbf{H}''$, we deduce finally that $\mathbf{H} = {}^r\mathbf{w}^{-1}(\mathbf{H}'')$.

**Proposition 17.** *Let* $\mathbf{G}$ *be a reduced finite-dimensional infinitesimal group,* $\mathbf{u}: \mathbf{G} \to \mathbf{G}'$ *the natural homomorphism of* $\mathbf{G}$ *onto a quotient group* $\mathbf{G}'$, $\mathbf{N} = {}^r\mathbf{u}^{-1}(\mathbf{e})$ *its reduced kernel.*

a) *The mapping* $\mathbf{H} \mapsto \mathbf{u}(\mathbf{H})$ *is a bijection of the set of reduced subgroups of* $\mathbf{G}$ *containing* $\mathbf{N}$, *on the set of all reduced subgroups of* $\mathbf{G}'$, *the inverse bijection being* $\mathbf{H}' \mapsto {}^r\mathbf{u}^{-1}(\mathbf{H}')$.

b) *The preceding bijection is strictly increasing and induces a bijection of the set of reduced invariant subgroups containing* $\mathbf{N}$ *on the set of all reduced invariant subgroups of* $\mathbf{G}'$. *Furthermore, if* $\mathbf{H} \supset \mathbf{N}$ *is a reduced invariant subgroup of* $\mathbf{G}$, *the formal homomorphism* $\mathbf{G}/\mathbf{H} \to \mathbf{G}'/\mathbf{u}(\mathbf{H})$ *deduced from* $\mathbf{u}$ *is an isogeny; it is an isomorphism if* $\mathbf{u}$ *is separable.*

c) *If* $\mathbf{L}$ *is any reduced subgroup of* $\mathbf{G}$, $\mathbf{L} \wedge \mathbf{N}$ *is the reduced kernel of the restriction of* $\mathbf{u}$ *to* $\mathbf{L}$, *and*

$$\mathbf{L} \vee \mathbf{N} = {}^r\mathbf{u}^{-1}(\mathbf{u}(\mathbf{L}));$$

*the formal homomorphisms*

$$\mathbf{L}/(\mathbf{L} \wedge \mathbf{N}) \to (\mathbf{L} \vee \mathbf{N})/\mathbf{N} \qquad \textit{and} \qquad (\mathbf{L} \vee \mathbf{N})/\mathbf{N} \to \mathbf{u}(\mathbf{L})$$

*are isogenies; the second is an isomorphism if* $\mathbf{u}$ *is separable.*

The assertion a) is Proposition 16; furthermore, $^r\mathbf{u}^{-1}(\mathbf{H}')$ being the largest reduced subgroup contained in $\mathbf{u}^{-1}(\mathbf{H}')$, it is invariant in $\mathbf{G}$ if $\mathbf{H}'$ is invariant in $\mathbf{G}'$, due to Proposition 9. Under the assumptions of b), if $\mathbf{M} = \mathbf{u}^{-1}(\mathbf{e})$, $\sup(\mathbf{H}, \mathbf{M})$ is an invariant subgroup of $\mathbf{G}$ containing $\mathbf{M}$ (I, §3, No. 9) and such that

$$\mathbf{u}(\sup(\mathbf{H}, \mathbf{M})) = \mathbf{u}(\mathbf{H}),$$

hence

$$\sup(\mathbf{H}, \mathbf{M}) = \mathbf{u}^{-1}(\mathbf{u}(\mathbf{H}))$$

by Proposition 10, and the homomorphism $\mathbf{G}/\mathbf{H} \to \mathbf{G}'/\mathbf{u}(\mathbf{H})$ factorizes into

$$\mathbf{G}/\mathbf{H} \xrightarrow{\mathbf{v}} \mathbf{G}/\sup(\mathbf{H}, \mathbf{M}) \xrightarrow{\mathbf{w}} \mathbf{G}'/\mathbf{u}(\mathbf{H}),$$

where $\mathbf{v}$ is an isogeny; on the other hand, $\mathbf{w}$ is an isomorphism by Proposition 9, and this ends the proof of b).

By definition, $\mathbf{L} \wedge \mathbf{N}$ is the largest reduced subgroup contained in $\inf(\mathbf{L}, \mathbf{N})$, hence (I, §3, No. 9) the reduced kernel of the restriction of $\mathbf{u}$ to $\mathbf{L}$; as by definition

$$\mathbf{L} \vee \mathbf{N} = \sup(\mathbf{L}, \mathbf{N}) \quad \text{and} \quad \mathbf{u}(\mathbf{L} \vee \mathbf{N}) = \mathbf{u}(\mathbf{L}),$$

it follows first from Proposition 16 that

$$\mathbf{L} \vee \mathbf{N} = {}^r\mathbf{u}^{-1}(\mathbf{u}(\mathbf{L})).$$

On the other hand, let $\mathbf{f}\colon \mathbf{G} \to \mathbf{G}/\mathbf{N}$ the natural homomorphism; there is a natural isomorphism

$$(\mathbf{L} \vee \mathbf{N})/\mathbf{N} \to \mathbf{f}(\mathbf{L})$$

(I, §3, No. 9), and on the other hand there is a natural isogeny $\mathbf{f}(\mathbf{L}) \to \mathbf{u}(\mathbf{L})$ (formula (17)), hence by composition an isogeny

$$(\mathbf{L} \vee \mathbf{N})/\mathbf{N} \to \mathbf{u}(\mathbf{L}).$$

There is also (I, §3, No. 9) a natural isomorphism

$$\mathbf{L}/\inf(\mathbf{L}, \mathbf{N}) \to (\mathbf{L} \vee \mathbf{N})/\mathbf{N},$$

and as $\mathbf{L} \wedge \mathbf{N}$ is the largest reduced subgroup in $\inf(\mathbf{L}, \mathbf{N})$, we have a natural isogeny

$$\mathbf{L}/(\mathbf{L} \wedge \mathbf{N}) \to \mathbf{L}/\inf(\mathbf{L}, \mathbf{N}) \qquad \text{(formula (17))},$$

hence by composition the isogeny

$$\mathbf{L}/(\mathbf{L} \wedge \mathbf{N}) \to (\mathbf{L} \vee \mathbf{N})/\mathbf{N}.$$

**Corollary 1.** *Under the assumptions of Proposition 17c), we have*

$$\dim \mathbf{L} + \dim \mathbf{N} = \dim(\mathbf{L} \wedge \mathbf{N}) + \dim(\mathbf{L} \vee \mathbf{N}). \tag{20}$$

This follows from Proposition 17c) and formula (18).

**Corollary 2.** *Let* $\mathbf{H}_1$, $\mathbf{H}_2$ *be two reduced subgroups of* $\mathbf{G}$, $\mathbf{H}'_1$, $\mathbf{H}'_2$ *two reduced subgroups of* $\mathbf{G}'$; *then*

$$^r\mathbf{u}^{-1}(\mathbf{H}'_1 \vee \mathbf{H}'_2) = {}^r\mathbf{u}^{-1}(\mathbf{H}'_1) \vee {}^r\mathbf{u}^{-1}(\mathbf{H}'_2) \tag{21}$$

$$^r\mathbf{u}^{-1}(\mathbf{H}'_1 \wedge \mathbf{H}'_2) = {}^r\mathbf{u}^{-1}(\mathbf{H}'_1) \wedge {}^r\mathbf{u}^{-1}(\mathbf{H}'_2) \tag{22}$$

$$\mathbf{u}(\mathbf{H}_1 \vee \mathbf{H}_2) = \mathbf{u}(\mathbf{H}_1) \vee \mathbf{u}(\mathbf{H}_2). \tag{23}$$

Formulas (21) and (22) are direct consequences of the existence of the strictly increasing bijection defined in Proposition 17a); so is (23) if we observe that

$$\mathbf{H}_1 \vee \mathbf{H}_2 \vee \mathbf{N} = (\mathbf{H}_1 \vee \mathbf{N}) \vee (\mathbf{H}_2 \vee \mathbf{N}) \qquad \text{and} \quad \mathbf{u}(\mathbf{L} \vee \mathbf{N}) = \mathbf{u}(\mathbf{L})$$

for any reduced subgroup $\mathbf{L}$ of $\mathbf{G}$.

The results of Proposition 17 are, for infinitesimal reduced groups of finite dimension, the analogs of the "isomorphism theorems" of classical group theory. By the usual arguments, one can therefore deduce from them theorems which correspond to the Zassenhaus, Schreier, and Jordan-Hölder theorems of classical group theory. Of course, in the statement of these theorems (which we leave to the reader) one must always replace "isomorphic" by "isogenous"; furthermore, a *simple* reduced infinitesimal group of finite dimension must be defined as one having no invariant *reduced* subgroup distinct from $e$ and itself (remember that any reduced bigebra $\mathfrak{G}$ of dimension $\geqslant 1$ always has infinitely many invariant subbigebras, namely the Frobenius subbigebras $\mathfrak{s}_r$). An elementary criterion for simple reduced groups is the following one:

**Proposition 18.** *If the Lie algebra $\mathfrak{g}_0$ of a reduced finite-dimensional group* $\mathbf{G}$ *is simple, then* $\mathbf{G}$ *is simple; more precisely, the Frobenius subbigebras $\mathfrak{s}_r$ ($r \geqslant 0$) are the only nontrivial invariant subbigebras of $\mathfrak{G}$.*

Indeed, if $\mathfrak{H}$ is an invariant subbigebra of $\mathfrak{G}$, $\mathfrak{g}_0 \cap V^r(\mathfrak{H})$ is a $p$-ideal in $\mathfrak{g}_0$ (corollary to Proposition 6), hence equal to 0 or to $\mathfrak{g}_0$ by assumption, and the result follows from the structure theorem (Proposition 3).

It should be emphasized that the sufficient condition of Proposition 18 is by no means necessary; in Chapter III, we will show that there are simple *commutative* reduced groups of arbitrary dimension, for which of course the Lie algebra is not at all simple.

We will say that an infinitesimal reduced group $\mathbf{G}$ is *almost direct product* of two of its reduced subgroups $\mathbf{H}_1$, $\mathbf{H}_2$ if there exist two reduced groups $\mathbf{L}_1$, $\mathbf{L}_2$, and an *isogeny* $\mathbf{u}: \mathbf{L}_1 \times \mathbf{L}_2 \to \mathbf{G}$ such that

$$\mathbf{u}(\mathbf{L}_1) = \mathbf{H}_1 \quad \text{and} \quad \mathbf{u}(\mathbf{L}_2) = \mathbf{H}_2.$$

**Proposition 19.** *Let* $\mathbf{G}$ *be an infinitesimal reduced group of finite dimension,* $\mathbf{H}_1$, $\mathbf{H}_2$ *two reduced subgroups of* $\mathbf{G}$. *The following conditions are equivalent:*

a) $\mathbf{G}$ *is almost direct product of* $\mathbf{H}_1$ *and* $\mathbf{H}_2$.

b) $\mathbf{H}_1 \wedge \mathbf{H}_2 = \mathbf{e}$, $\mathbf{H}_1 \vee \mathbf{H}_2 = \mathbf{G}$, *and* $\mathbf{H}_1$ *and* $\mathbf{H}_2$ *are invariant subgroups.*

c) $\mathbf{H}_1 \wedge \mathbf{H}_2 = \mathbf{e}$, $\mathbf{H}_1 \vee \mathbf{H}_2 = \mathbf{G}$, *and* $\mathbf{H}_1$ *and* $\mathbf{H}_2$ *centralize each other.*

It follows from Proposition 14 that a) implies b) and c). Let us prove that c) implies b). As the subbigebras $\mathfrak{H}_1$, $\mathfrak{H}_2$ centralize each other (I, §3, No. 10) and $\mathfrak{G} = \mathfrak{H}_1\mathfrak{H}_2$ (I, §3, No. 10), $\mathfrak{H}_1$ and $\mathfrak{H}_2$ satisfy the criterion b) of Proposition 6, hence are invariant. Finally, we show that b) implies a): it follows from Proposition 17c) that there is a surjective formal homomorphism $\mathbf{v}_i$ of $\mathbf{G}$ onto a reduced group $\mathbf{L}_i$ isogenous to $\mathbf{H}_i$ ($i = 1,2$), such that $\mathbf{H}_2$ is the reduced kernel of $\mathbf{v}_1$ and $\mathbf{H}_1$ the reduced kernel of $\mathbf{v}_2$. Let us consider the homomorphism $\mathbf{v} = (\mathbf{v}_1, \mathbf{v}_2)$ of $\mathbf{G}$ into $\mathbf{L}_1 \times \mathbf{L}_2$; as

$$\mathbf{v}(\mathbf{H}_1) = \mathbf{L}_1 \quad \text{and} \quad \mathbf{v}(\mathbf{H}_2) = \mathbf{L}_2,$$

we have $\mathbf{v}(\mathbf{G}) = \mathbf{L}_1 \times \mathbf{L}_2$ by (23) and the reduced kernel of $\mathbf{v}$ is contained in $\inf(\mathbf{H}_1, \mathbf{H}_2)$, hence reduced to $\mathbf{e}$. This shows that $\mathbf{v}$ is an isogeny, and therefore there exists an isogeny $u$ of $\mathfrak{L}_1^{(t)} \times \mathfrak{L}_2^{(t)}$ onto $\mathfrak{G}$ for a sufficiently large $t$ such that $u \circ v = \mathrm{V}^t$ (Proposition 13), and the definition of $v$ shows that

$$u(\mathfrak{L}_1^{(t)}) = \mathfrak{H}_1 \quad \text{and} \quad u(\mathfrak{L}_2^{(t)}) = \mathfrak{H}_2.$$

A simple example of a reduced group $\mathbf{G}$ which is almost direct product of two reduced subgroups $\mathbf{H}_1$, $\mathbf{H}_2$, but is not their direct product, is the one given after Proposition 3.

**8.** *A new criterion for invariant subgroups.* Let $\mathrm{K} \supset k$ be a perfect field. For any reduced infinitesimal bigebra $\mathfrak{G}$ over $k$, it is obvious that the extension

$$\mathfrak{G}_{(\mathrm{K})} = \mathfrak{G} \otimes_k \mathrm{K}$$

is a reduced infinitesimal bigebra over K, a privileged basis $(Z_\alpha)$ of $\mathfrak{G}$ over $k$ being also (after identification of $Z_\alpha$ with $Z_\alpha \otimes 1$) a privileged basis of $\mathfrak{G}_{(\mathrm{K})}$ over K. As the contravariant bigebra $\mathfrak{G}^*$ is identified to the algebra of formal power series $k[[x]]_{i \in \mathrm{I}}$, with $(x^\alpha)$ as the pseudobasis dual to $(Z_\alpha)$, the contravariant bigebra $(\mathfrak{G}_{(\mathrm{K})})^*$ is identified to $\mathrm{K}[[x_i]]_{i \in \mathrm{I}}$, and the coproduct of that bigebra is the unique continuous K-algebra homomorphism of $(\mathfrak{G}_{(\mathrm{K})})^*$ into $(\mathfrak{G}_{(\mathrm{K})})^* \hat{\otimes} (\mathfrak{G}_{(\mathrm{K})})^*$ which extends the coproduct of $\mathfrak{G}^*$, and is therefore entirely determined by its values $c(x_i)$ for $i \in \mathrm{I}$.

For any subbigebra $\mathfrak{H}$ of $\mathfrak{G}$, $\mathfrak{H}_{(\mathrm{K})}$ is a subbigebra of $\mathfrak{G}_{(\mathrm{K})}$ and $\mathfrak{G} \cap \mathfrak{H}_{(\mathrm{K})} = \mathfrak{H}$; if $\mathfrak{H}_1$, $\mathfrak{H}_2$ are two subbigebras of $\mathfrak{G}$,

$$(\mathfrak{H}_1 \cap \mathfrak{H}_2)_{(\mathrm{K})} = \mathfrak{H}_{1(\mathrm{K})} \cap \mathfrak{H}_{2(\mathrm{K})}$$

(the formula is already true for vector subspaces of $\mathfrak{G}$); the subalgebra of $\mathfrak{G}_{(\mathrm{K})}$ generated by $\mathfrak{H}_{1(\mathrm{K})} + \mathfrak{H}_{2(\mathrm{K})}$ is $\mathfrak{H}_{(\mathrm{K})}$, where $\mathfrak{H}$ is the subalgebra of $\mathfrak{G}$ generated by $\mathfrak{H}_1 + \mathfrak{H}_2$; as

$$\mathfrak{H} = \sup(\mathfrak{H}_1, \mathfrak{H}_2)$$

in the set of subbigebras of $\mathfrak{G}$, we have

$$(\sup(\mathfrak{H}_1, \mathfrak{H}_2))_{(K)} = \sup(\mathfrak{H}_{1(K)}, \mathfrak{H}_{2(K)}).$$

If $\mathfrak{N}$ is a stable invariant subbigebra of $\mathfrak{G}$ and J the two-sided ideal of $\mathfrak{G}$ generated by $\mathfrak{N}^+$, the construction of the privileged basis $(Z_\alpha)$ in Proposition 5 shows that $\mathfrak{N}_{(K)}$ is a stable invariant subbigebra of $\mathfrak{G}_{(K)}$, $J_{(K)}$ the two-sided ideal generated by $\mathfrak{N}^+_{(K)}$, and $\mathfrak{N}_{(K)}$ the largest subbigebra contained in $K \cdot 1 + J_{(K)}$; furthermore, the quotient bigebra $\mathfrak{G}_{(K)}/J_{(K)}$ is naturally isomorphic to $(\mathfrak{G}/J)_{(K)}$, as may at once be seen from the construction of the basis of $\mathfrak{G}/J$ in No. 4. If $\mathfrak{G}' = \mathfrak{G}/J$ and $u: \mathfrak{G} \to \mathfrak{G}'$ is the natural homomorphism, $u_{(K)}: \mathfrak{G}_{(K)} \to \mathfrak{G}_{(K)}$ is the natural homomorphism; furthermore, if $\mathfrak{N}$ is reduced, then, for any subbigebra $\mathfrak{H}'$ of $\mathfrak{G}'$, if $\mathfrak{H}$ is the largest subbigebra contained in $u^{-1}(\mathfrak{H}')$, then $\mathfrak{H}_{(K)}$ is the largest subbigebra of $\mathfrak{G}_{(K)}$ contained in $u^{-1}_{(K)}(\mathfrak{H}'_{(K)})$: indeed, $\mathfrak{N}_{(K)}$ is the largest subbigebra contained in $K \cdot 1 + J_{(K)}$, and as $\mathfrak{H} \supset \mathfrak{N}$, we have $\mathfrak{H}_{(K)} \supset \mathfrak{N}_{(K)}$; on the other hand, from the relation $u(\mathfrak{H}) = \mathfrak{H}'$, we deduce

$$u_{(K)}(\mathfrak{H}_{(K)}) = \mathfrak{H}'_{(K)};$$

the conclusion then follows from the remark after Proposition 9.

From the relation $\mathfrak{H}_{(K)} \cap \mathfrak{G} = \mathfrak{H}$ for any subbigebra $\mathfrak{H}$ of $\mathfrak{G}$, it follows that two subbigebras $\mathfrak{H}_1$, $\mathfrak{H}_2$ of $\mathfrak{G}$ centralize each other if and only if $\mathfrak{H}_{1(K)}$ and $\mathfrak{H}_{2(K)}$ centralize each other in $\mathfrak{G}_{(K)}$. For any subbigebra $\mathfrak{H}$ of $\mathfrak{G}$,

$$(\mathscr{Z}(\mathfrak{H}))_{(K)} = \mathscr{Z}(\mathfrak{H}_{(K)}),$$

for the left-hand side is obviously contained in the right-hand side; the converse follows by taking a basis $(\xi_\lambda)$ of K over $k$; if we write that an element $\sum_\lambda \xi_\lambda U_\lambda$ in $\mathfrak{G}_{(K)}$ commutes with every $Y \in \mathfrak{H}$, the fact that the products $U_\lambda Y$ and $YU_\lambda$ are in $\mathfrak{G}$ implies that we must have $U_\lambda Y = YU_\lambda$ for every $\lambda$ and every $Y \in \mathfrak{H}$, and this proves our assertion.

Finally, let $\mathfrak{N}_1$, $\mathfrak{N}_2$ be two invariant subbigebras of $\mathfrak{G}$; we then have

$$[\mathfrak{N}_1, \mathfrak{N}_2]_{(K)} = [\mathfrak{N}_{1(K)}, \mathfrak{N}_{2(K)}];$$

for by definition, if J is the two-sided ideal of $\mathfrak{G}$ generated by $[\mathfrak{N}_1, \mathfrak{N}_2]^+$, the images of $\mathfrak{N}_{1(K)}$ and $\mathfrak{N}_{2(K)}$ into $\mathfrak{G}_{(K)}/J_{(K)}$ centralize each other, hence

$$[\mathfrak{N}_1, \mathfrak{N}_2]_{(K)} \supset [\mathfrak{N}_{1(K)}, \mathfrak{N}_{2(K)}].$$

On the other hand, if J′ is the two-sided ideal of $\mathfrak{G}_{(K)}$ generated by $[\mathfrak{N}_{1(K)}, \mathfrak{N}_{2(K)}]^+$, J′ contains all brackets $[Y_1, Y_2]$ where $Y_i$ belongs to the two-sided ideal of $\mathfrak{G}$ generated by $\mathfrak{N}_i^+$ $(i = 1, 2)$; therefore J′ contains J, and *a fortiori* contains $J_{(K)}$, which proves the relation $J' = J_{(K)}$.

From now on, suppose $\mathfrak{G}$ has finite dimension, so that I may be taken to be the interval $[1, n]$ and the algebra $\mathfrak{G}^*$ is identified to $k[[t_1, \ldots, t_n]]$ where the $t_i$ are indeterminates; $A = \mathfrak{G}^* \hat{\otimes} \mathfrak{G}^*$ is therefore identified to

$$k[[z_1, \ldots, z_n, x_1, \ldots, x_n]]$$

where the $z_i$ and $x_i$ are indeterminates, and the canonical generic points $z: \mathfrak{G}^* \to A$ and $x: \mathfrak{G}^* \to A$ of $\mathbf{G}_A$ may be identified to the homomorphisms which, respectively, send $t_i$ to $z_i$ and to $x_i$ for $1 \leqslant i \leqslant n$. Consider the mapping $v: s \mapsto zsz^{-1}$, which is an inner automorphism of the group $\mathbf{G}_A$; in particular $v(x)$ is a continuous algebra homomorphism $\mathfrak{G}^* \to A$, which to each $t_i$ associates a formal power series $\varphi_i(z,x) \in A$; each $s \in \mathfrak{G}^*$ may be written $s(t_1, \ldots, t_n)$ (formal power series), and $v(s)$ is the formal power series

$$s(\varphi_1(z, x), \ldots, \varphi_n(z, x))$$

obtained by substitution; if $s$ is the counit of $\mathfrak{G}^*$ (neutral element of $\mathbf{G}_A$), so must be $v(s)$, and conversely, which shows that the $\varphi_i(z, x)$ have no term in the $z_i$ alone. Now, if

$$B = k[[z_1, \ldots, z_n]],$$

A may be identified with the ring $B[[x_1, \ldots, x_n]]$ of formal power series with coefficients in B; as B is an integral domain, we may consider a *perfect field* K containing B, and then A is identified with a subring of $K[[x_1, \ldots, x_n]]$, and the $\varphi_i$ with elements of that ring, which, as we have seen, are power series in the $x_i$ with coefficients in K and without constant term. There is therefore a unique continuous K-homomorphism ${}^t u$ of $K[[t_1, \ldots, t_n]]$ into itself such that

$${}^t u(t_i) = \varphi_i(z, t) \qquad \text{for} \quad 1 \leqslant i \leqslant n.$$

But

$$K[[t_1, \ldots, t_n]] = (\mathfrak{G}_{(K)})^*,$$

and we are going to show that ${}^t u$ is the transposed mapping of a *bigebra* homomorphism of $\mathfrak{G}_{(K)}$ into itself. Indeed, if $s'$, $s''$ are the canonical generic points of $\mathbf{G}_{(K)}$ in the K-algebra $(\mathfrak{G}_{(K)})^* \hat{\otimes}_K (\mathfrak{G}_{(K)})^*$, we have only to show that

$$(s's'') \circ {}^t u = (s' \circ {}^t u)(s'' \circ {}^t u) \qquad \text{(I, §3, No. 2);}$$

but it follows immediately from the definition of the coproduct of $(\mathfrak{G}_{(K)})^*$ that this relation is a consequence of the relation

$$v(y'y'') = v(y')v(y'')$$

where $y'$ and $y''$ are any points of $\mathbf{G}_A$. We will denote the bigebra homomorphism $u$ by Int($z$) and the corresponding formal homomorphism by **Int**($z$). As the preceding argument may be repeated with $v^{-1}: s \mapsto z^{-1}sz$, it yields a bigebra homomorphism Int($z^{-1}$) which is *inverse* to Int($z$), so that the latter is in fact an *automorphism* of $\mathfrak{G}_{(K)}$.

Now consider a reduced subbigebra $\mathfrak{H}$ of $\mathfrak{G}$, and let $j: \mathfrak{H} \to \mathfrak{G}$ be the natural injection. Taking a privileged basis of $\mathfrak{G}$ adapted to $\mathfrak{H}$ (§2, No. 2), we may assume that ${}^t j: \mathfrak{G}^* \to \mathfrak{H}^*$ is such that

$${}^t j(t_i) = 0 \qquad \text{for} \quad m + 1 \leqslant i \leqslant n,$$

and

$${}^t j(t_i) = s_i \qquad \text{for} \quad 1 \leqslant i \leqslant m,$$

if

$$\mathfrak{H}^* = k[[s_1, \ldots, s_m]].$$

Let $A_0$ be the algebra $\mathfrak{G}^* \hat{\otimes} \mathfrak{H}^*$, identified to

$$k[[z_1, \ldots, z_n, y_1, \ldots, y_m]],$$

the canonical generic points $z: \mathfrak{G}^* \to A_0$ and $y: \mathfrak{H}^* \to A_0$ being identified to the homomorphisms which, respectively, send $t_i$ to $z_i$ ($1 \leqslant i \leqslant n$) and $s_i$ to $y_i$ ($1 \leqslant i \leqslant m$). The condition for $\mathfrak{H}$ to be an *invariant* subbigebra is that $z(y \circ {}^t j)z^{-1}$ belongs to $\mathbf{H}_{A_0}$ (I, §3, No. 6). But from the fact that $\mathbf{H}$ is a subgroup of $\mathbf{G}$ it follows that this condition is equivalent, with the preceding notations, to the $n - m$ relations

$$\varphi_i(z_1, \ldots, z_n, y_1, \ldots, y_m, 0, \ldots, 0) = 0 \qquad \text{for} \quad m + 1 \leqslant i \leqslant n.$$

But these relations are equivalent to the condition

$$\text{Int}(z)(\mathfrak{H}_{(\mathrm{K})}) \subset \mathfrak{H}_{(\mathrm{K})} \tag{24}$$

for a perfect field K containing $k[[z_1, \ldots, z_n]]$.

The same argument shows that the condition for $\mathfrak{H}$ to be invariant is also equivalent to

$$\text{Int}(z)^{-1}(\mathfrak{H}_{(\mathrm{K})}) \subset \mathfrak{H}_{(\mathrm{K})} \qquad \text{or} \qquad \text{Int}(z)(\mathfrak{H}_{(\mathrm{K})}) = \mathfrak{H}_{(\mathrm{K})}. \tag{25}$$

**9.** *Solvable and nilpotent reduced groups.* Let $\mathfrak{G}$ be a reduced bigebra of finite dimension, and $\mathfrak{N}_1$, $\mathfrak{N}_2$ two reduced invariant subbigebras of $\mathfrak{G}$; then $[\mathfrak{N}_1, \mathfrak{N}_2]$ is also a reduced subbigebra as is seen from I, §3, No. 17, formula (4) applied to the shift V. Furthermore, we may complete the results of I, §3, No. 17 on commutator subgroups for *reduced* subgroups of reduced groups.

In the first place, let $x_1$, $x_2$ be the canonical generic points of $\mathbf{N}_1$ and $\mathbf{N}_2$ in the algebra $\mathrm{A} = \mathfrak{N}_1^* \hat{\otimes} \mathfrak{N}_2^*$, and let $\mathbf{j}_i$ ($j = 1, 2$) be the natural formal injection $\mathbf{N}_i \to \mathbf{G}$. We have seen that the reduced formal subgroup $[\mathbf{N}_1, \mathbf{N}_2]$ is the smallest *invariant* formal subgroup $\mathbf{N}$ of $\mathbf{G}$ such that $\mathbf{N}_\mathrm{A}$ contains the point

$$y = \mathbf{j}_1(x_1)\mathbf{j}_2(x_2)\mathbf{j}_1(x_1)^{-1}\mathbf{j}_2(x_2)^{-1}.$$

We may now prove that it is also the *smallest reduced formal subgroup* $\mathbf{H}$ such that $\mathbf{H}_\mathrm{A}$ contains $y$. One clearly has $\mathbf{H} \subset \mathbf{N}$, and all we have to show is that $\mathbf{H}$ is *invariant*. We use criterion (24), taking a perfect field K containing $k[[z_1, \ldots, z_n]]$, and denoting by $\mathbf{G}'$, $\mathbf{H}'$, $\mathbf{N}_i'$, $\mathbf{j}_i'$ the formal groups and formal injections deduced from $\mathbf{G}$, $\mathbf{H}$, $\mathbf{N}_i$, $\mathbf{j}_i$ by extending the field of scalars to K; if $\mathrm{A}' = \mathfrak{N}_1'^* \hat{\otimes}_\mathrm{K} \mathfrak{N}_2'^*$, the canonical generic points of $\mathbf{N}_1'$ and $\mathbf{N}_2'$ in $\mathrm{A}'$ are the continuous extensions $x_1'$, $x_2'$ of $x_1$, $x_2$, and similarly the point $\mathbf{j}_i'(x_i)$ in $\mathbf{G}_{\mathrm{A}'}'$ is the continuous extension of $\mathbf{j}_i(x_i)$ ($j = 1, 2$), hence

$$y' = \mathbf{j}_1'(x_1')\mathbf{j}_2'(x_2')\mathbf{j}_1'(x_1')^{-1}\mathbf{j}_2'(x_2')^{-1}$$

is the continuous extension of $y$. The assumption that $\mathbf{N}_i$ is invariant in

$\mathbf{G}$ implies by (24) that $\mathbf{Int}(z)(\mathbf{j}_i'(x_i'))$ is a point of $(\mathbf{N}_i')_{A'}$, hence of the form $\mathbf{j}_i'(w_i')$, where $w_i'$ is a specialization of $x_i'$; but then $\mathbf{Int}(z)\,(y')$ is a specialization of $y'$ since the two points $x_1'$, $x_2'$ are independent. Therefore $\mathbf{Int}(z)(y')$ is a point of $\mathbf{H}'_{A'}$, and as $\mathbf{H}'$ is the smallest reduced subgroup of $\mathbf{G}'$ such that $\mathbf{H}'_{A'}$ contains $y'$, we have

$$\mathrm{Int}(z)^{-1}(\mathfrak{H}') \supset \mathfrak{H}',$$

which proves that $\mathbf{H}$ is invariant by (24).

Next we show that for any reduced invariant subgroup $\mathbf{N}$ of $\mathbf{G}$,

$$\mathscr{D}(\mathbf{N}) = [\mathbf{N}, \mathbf{N}]. \tag{26}$$

Again we have to prove that $\mathscr{D}(\mathbf{N})$ is *invariant* in $\mathbf{G}$. With the same notations, we have

$$\mathbf{Int}(z)\,(\mathbf{N}') = \mathbf{N}',$$

and by definition $\mathbf{Int}(z)^{-1}(\mathscr{D}(\mathbf{N}'))$ is therefore an invariant subgroup of $\mathbf{N}'$ such that $\mathbf{N}'/\mathbf{Int}(z)^{-1}(\mathscr{D}(\mathbf{N}'))$ is commutative. But then by definition this implies that

$$\mathbf{Int}(z)^{-1}(\mathscr{D}(\mathbf{N}')) \supset \mathscr{D}(\mathbf{N}'),$$

and as $\mathscr{D}(\mathbf{N}') = \mathscr{D}(\mathbf{N})'$, this proves (26).

For reduced formal groups of finite dimension, we may now, in virtue of (26), define as usual the *derived series* $(\mathscr{D}^n(\mathbf{G}))$ inductively by

$$\mathscr{D}^0(\mathbf{G}) = \mathbf{G}, \qquad \mathscr{D}^n(\mathbf{G}) = \mathscr{D}(\mathscr{D}^{n-1}(\mathbf{G}))$$

and the *descending central series* $(\mathscr{C}^n(\mathbf{G}))$ by

$$\mathscr{C}^0(\mathbf{G}) = \mathbf{G}, \qquad \mathscr{C}^n(\mathbf{G}) = [\mathbf{G}, \mathscr{C}^{n-1}(\mathbf{G})].$$

These groups are invariant in $\mathbf{G}$ by (26); as they are reduced and $\mathbf{G}$ is finite dimensional, the two sequences $(\mathscr{D}^n(\mathbf{G}))$ and $(\mathscr{C}^n(\mathbf{G}))$ are stationary, and we write $\mathscr{D}^\infty(\mathbf{G})$ and $\mathscr{C}^\infty(\mathbf{G})$ their smallest elements. From I, §3, No. 17, formulas (4) and (5), we deduce by induction that, for any *surjective* homomorphism $\mathbf{u}\colon \mathbf{G} \to \mathbf{G}'$ of formal groups,

$$\mathbf{u}(\mathscr{D}^n(\mathbf{G})) = \mathscr{D}^n(\mathbf{G}'), \qquad \mathbf{u}(\mathscr{C}^n(\mathbf{G})) = \mathscr{C}^n(\mathbf{G}') \tag{27}$$

and for any reduced subgroup **H** of **G**

(28) $$\mathscr{D}^n(\mathbf{H}) \subset \mathscr{D}^n(\mathbf{G}), \qquad \mathscr{C}^n(\mathbf{H}) \subset \mathscr{C}^n(\mathbf{G})$$

for all $n \geqslant 0$.

A reduced formal group **G** is called *solvable* (resp. *nilpotent*) if

$$\mathscr{D}^\infty(\mathbf{G}) = \mathbf{e} \qquad (\text{resp. } \mathscr{C}^\infty(\mathbf{G}) = \mathbf{e}).$$

From (27) and (28) it follows that reduced subgroups and quotient groups of solvable (resp. nilpotent) reduced groups are solvable (resp. nilpotent). The standard arguments of group theory show that solvable groups are the reduced groups having a composition series consisting of reduced invariant subgroups with *commutative* quotients. Furthermore, if **G** is a finite-dimensional reduced group having an invariant reduced subgroup **N** such that **N** and **G**/**N** are solvable, then **G** is solvable.

Finally, for a reduced formal group G of finite dimension, one defines the *ascending central series* $({}^r\mathscr{Z}^n(\mathbf{G}))$ by ${}^r\mathscr{Z}^0(\mathbf{G}) = \mathbf{e}$, and defining ${}^r\mathscr{Z}^n(\mathbf{G})$ as the reduced inverse image in **G** of the reduced center of $\mathbf{G}/{}^r\mathscr{Z}^{n-1}(\mathbf{G})$. As $\mathscr{C}^{n-1}(\mathbf{G})/\mathscr{C}^n(\mathbf{G})$ is reduced and contained in the center of $\mathbf{G}/\mathscr{C}^n(\mathbf{G})$, it is also contained in its reduced center; a standard argument of group theory then proves that the nilpotent reduced groups are exactly those for which the ascending central series terminates at **G**.

**10.** *The covariant bigebra as set of operators.* Let us retain the assumptions and notations of No. 8, so that $\mathfrak{G}^*$ is identified with $k[[t_1, \ldots, t_n]]$ and $\mathfrak{G}^* \hat{\otimes} \mathfrak{G}^*$ with

$$k[[z_1, \ldots, z_n, x_1, \ldots, x_n]] = \mathrm{B}[[x_1, \ldots, x_n]].$$

If K is a perfect field containing $\mathrm{B} = k[[z_1, \ldots, z_n]]$, we may identify the bigebra $\mathfrak{G}$ with a $k$-subbigebra of

$$\mathfrak{G}_{(\mathrm{K})} = \mathfrak{G} \otimes_k \mathrm{K}.$$

Now, for any element $u \in \mathfrak{G}^*$, $c(u)$ belongs to $\mathfrak{G}^* \hat{\otimes} \mathfrak{G}^*$, which we identify with a subset of $(\mathfrak{G}_{(\mathrm{K})})^*$, and therefore for any $Z \in \mathfrak{G} \subset \mathfrak{G}_{(\mathrm{K})}$, $\langle c(u), Z\rangle$ is well defined and an element of B. If we identify B with $\mathfrak{G}^*$ by identifying each $t_i$ with $z_i$, we have thus defined $Z$ as a *linear operator* on $\mathfrak{G}^*$.

More precisely, let $(Z_\alpha)$ be the privileged basis of $\mathfrak{G}$, dual to the pseudobasis $(t^\alpha)$ of $\mathfrak{G}^*$. If we write

(29) $$c(t^\gamma) = \sum_{\alpha,\beta} a_{\alpha\beta\gamma} z^\alpha x^\beta$$

we have by duality $\langle t^\gamma, Z_\alpha Z_\beta \rangle = \langle c(t^\gamma), Z_\alpha \otimes Z_\beta \rangle$, hence

(30) $$Z_\alpha Z_\beta = \sum_\gamma a_{\alpha\beta\gamma} Z_\gamma.$$

From (29) and the preceding definition, it follows that $Z_\beta$, considered as an operator on $\mathfrak{G}^*$, is such that

(31) $$Z_\beta \cdot t^\gamma = \sum_\alpha a_{\alpha\beta\gamma} t^\alpha$$

so that $c(t^\gamma) = \sum_\beta (Z_\beta \cdot t^\gamma) x^\beta$, and in general, for any element $u = \sum_\gamma b_\gamma t^\gamma$,

(32) $$c(u) = \sum_\alpha (Z_\alpha \cdot u) x^\alpha$$

which for Lie groups reduces to the *Taylor formula* of the group. The associativity of the multiplication in $\mathfrak{G}$ (or the coassociativity of $c$) implies that this multiplication is identified to the *composition* of the operators.

In particular, for the elements $X_i$ of the Lie algebra $\mathfrak{g}_0$, $X_i \cdot u$ is the coefficient of $x_i$ in $c(u)$; as $c(uv) = c(u)c(v)$, the operator $X_i$ is a *derivation* of the ring of formal power series $k[[t_1, \ldots, t_n]]$. More precisely, if we note $\mathrm{D}_i$ the derivation $\partial/\partial t_i$ in that ring, and if we write, for $1 \leqslant j \leqslant n$,

(33) $$c(t_j) = t_j + \sum_{i=1}^{n} v_{ji}(t) x_i + \cdots$$

where the unwritten terms have degree $\geqslant 2$ in the $x_i$'s, we get for $X_i$ (as an operator) the expression

(34) $$X_i = \sum_{j=1}^{n} v_{ji}(t) \mathrm{D}_j.$$

CHAPTER III

# Infinitesimal Commutative Groups

## §1. Generalities

**1.** *Commutative covariant bigebras.* The commutative formal groups over a field $k$ obviously form a *full subcategory* of the category of all formal groups over $k$, which is equivalent to the subcategory of *commutative covariant bigebras* (and also of course to the dual of the subcategory of *cocommutative contravariant bigebras*). But there is here a fourth category which naturally intervenes: a commutative covariant bigebra over $k$ has an antipodism as we have seen (I, §2, No. 13), and therefore belongs to the subcategory of the category $\boldsymbol{CAlg}_k$ of $\boldsymbol{Alg}_k$-*cogroups* (I, §1, No. 6) consisting of the bigebras of that category which are *cocommutative.* But the dual of that subcategory is obviously the category of *commutative affine group schemes* over $k$, which is thus in natural duality with the category of *commutative formal groups* over $k$; we will not study that duality, for which we refer to P. Cartier [9].

From now on, as in Chapter II, we will assume that the field $k$ is *perfect* and of characteristic $p > 0$, and we will consider only *infinitesimal* formal commutative groups over $k$ (observe that in the Cartier-Gabriel decomposition theorem (II, §1, No. 4), the semidirect product becomes a direct product for commutative formal groups, and the study of the general commutative formal groups thus splits entirely into the study of etale groups and of infinitesimal groups).

**2.** *The ring of endomorphisms of a commutative bigebra.* In the remainder of the chapter, a *bigebra* $\mathfrak{G}$ will always mean the covariant bigebra of an infinitesimal commutative formal group $\mathbf{G}$ over $k$. For any linearly compact algebra $\mathrm{A} \in \boldsymbol{Alc}_k$, the group $\mathbf{G}_{\mathrm{A}} = \mathrm{Hom}_{\boldsymbol{Alc}_k}(\mathfrak{G}^*, \mathrm{A})$ is commutative, and we will write it *additively*; it can of course be identified with the group $\mathrm{Hom}_{k\text{-cog.}}(\mathrm{C}, \mathfrak{G})$ where $\mathrm{C} = \mathrm{A}^*$ is the $k$-cogebra dual to A. We recall that the sum $u + v$ of two elements in that group is the composite mapping

$$\mathrm{C} \xrightarrow{\ c\ } \mathrm{C} \otimes \mathrm{C} \xrightarrow{\ u\otimes v\ } \mathfrak{G} \otimes \mathfrak{G} \xrightarrow{\ \mu\ } \mathfrak{G} \tag{1}$$

where $c$ is the comultiplication in the cogebra C (I, §1, No. 6).

Now consider two (commutative) bigebras $\mathfrak{G}, \mathfrak{G}'$; the set $\mathrm{Hom}_{k\text{-big.}}(\mathfrak{G}, \mathfrak{G}')$ of all *bigebra* homomorphisms of $\mathfrak{G}$ into $\mathfrak{G}'$ is a subset of the group $\mathrm{Hom}_{k\text{-cog.}}(\mathfrak{G}, \mathfrak{G}')$: it is in fact a *subgroup* of that group. Indeed, with the notations of I, §2, No. 13, the commutativity of $\mathfrak{G}$ implies that $\mu\colon \mathfrak{G} \otimes \mathfrak{G} \to \mathfrak{G}$ is an *algebra* homomorphism; as it is already a cogebra homomorphism, it is a bigebra homomorphism. The same argument applied to $\mathfrak{G}^*$ shows that $\Delta\colon \mathfrak{G} \to \mathfrak{G} \otimes \mathfrak{G}$ is also a *bigebra* homomorphism; as, for any two bigebra homomorphisms $u$, $v$ of $\mathfrak{G}$ into $\mathfrak{G}'$, $u \otimes v$ is a bigebra homomorphism, it follows from formula (1) that $u + v$ is again a bigebra homomorphism. On the other hand, the neutral element of $\mathrm{Hom}_{k\text{-cog.}}(\mathfrak{G}, \mathfrak{G}')$ is the composite $\mathfrak{G} \to k \to \mathfrak{G}'$ of the unit of $\mathfrak{G}'$ and the counit of $\mathfrak{G}$, and both are bigebra homomorphisms. Finally, as the antipodism of $\mathfrak{G}^*$ is an algebra homomorphism, the antipodism $a$ of $\mathfrak{G}$ is both a cogebra and an algebra homomorphism, i.e., a bigebra homomorphism; as the opposite $-u$ in $\mathrm{Hom}_{k\text{-cog.}}(\mathfrak{G}, \mathfrak{G}')$ is $a' \circ u$, where $a'$ is the antipodism of $\mathfrak{G}'$, it is a bigebra homomorphism if $u$ is, and this proves our assertion.

Furthermore, if $\mathfrak{G}_1, \mathfrak{G}'_1$ are two other commutative bigebras, $f\colon \mathfrak{G}_1 \to \mathfrak{G}$, $g\colon \mathfrak{G}' \to \mathfrak{G}'_1$ bigebra homomorphisms, then we have, for $u$, $v$ in $\mathrm{Hom}_{k\text{-big.}}(\mathfrak{G}, \mathfrak{G}')$,

$$(u + v) \circ f = (u \circ f) + (v \circ f) \tag{2}$$

$$g \circ (u + v) = (g \circ u) + (g \circ v). \tag{3}$$

This follows at once from (1) and from the relations $\Delta \circ f = (f \otimes f) \circ \Delta_1$ and $\mu'_1 \circ (g \otimes g) = g \circ \mu'$ (I, §1, No. 5).

In particular, formulas (2) and (3) show that the set of *bigebra endo-*

*morphisms* of a commutative bigebra $\mathfrak{G}$, which we will henceforth simply denote by End($\mathfrak{G}$), has a structure of a (noncommutative) *ring*, for the addition $u + v$ defined above and composition $u \circ v$ as multiplication; the unit element is just the identity mapping $1_{\mathfrak{G}}$.

**3.** In the ring End($\mathfrak{G}$), for any (positive or negative) integer $r$, we may consider the endomorphism $r \cdot 1_{\mathfrak{G}}$, also written $[r]$ or $[r]_{\mathfrak{G}}$. For $r > 0$, it is simply the sum (for the additive group law of End($\mathfrak{G}$)) of $r$ terms equal to $1_{\mathfrak{G}}$. It is easy to obtain its explicit expression for *stable* bigebras $\mathfrak{G}$. For each integer $r > 0$, let $\mathfrak{G}^{\otimes r}$ be the tensor product of $r$ bigebras equal to $\mathfrak{G}$, and define by induction on $r$ the mappings

$$\mu^r : \mathfrak{G}^{\otimes r} \to \mathfrak{G}, \qquad \Delta^r : \mathfrak{G} \to \mathfrak{G}^{\otimes r}$$

as equal, respectively, to $\mu \circ (\mu^{r-1} \otimes 1_{\mathfrak{G}})$ and $(\Delta^{r-1} \otimes 1_{\mathfrak{G}}) \circ \Delta$ for $r \geqslant 3$, with $\mu^2 = \mu$ and $\Delta^2 = \Delta$ by definition. Starting from formula (1), it is then easy to show, by induction on $r$, that for any $r$ elements $u_1, u_2, \ldots, u_r$ in End($\mathfrak{G}$), the sum $u_1 + u_2 + \cdots + u_r$ is equal to

$$\mu^r \circ (u_1 \otimes u_2 \otimes \cdots \otimes u_r) \circ \Delta^r.$$

Now let $(Z_\alpha)$ be a privileged basis for $\mathfrak{G}$. An immediate induction on $r$ shows that

$$\Delta^r(Z_\alpha) = \sum_{(\beta_i)} Z_{\beta_1} \otimes Z_{\beta_2} \otimes \cdots \otimes Z_{\beta_r} \tag{4}$$

where the sum is extended to *all* sequences $(\beta_1, \beta_2, \ldots, \beta_r)$ of multiindices such that $\beta_1 + \beta_2 + \cdots + \beta_r = \alpha$. From the previous expression of $u_1 + u_2 + \cdots + u_r$, applied to the case where all $u_j$ are equal to $1_{\mathfrak{G}}$, we deduce at once that

$$[r](Z_\alpha) = \sum_{(\beta_i)} Z_{\beta_1} Z_{\beta_2} \cdots Z_{\beta_r} \tag{5}$$

with the same convention for the sum. Owing to the commutativity of $\mathfrak{G}$, this can be written in a different form. The symmetric group $\mathfrak{S}_r$ operates on the set $\mathrm{D}_r$ of all sequences $(\beta_1, \ldots, \beta_r)$ of multiindices satisfying $\beta_1 + \cdots + \beta_r = \alpha$, each permutation $\pi \in \mathfrak{S}_r$ transforming the sequence $(\beta_j)$ into $(\beta_{\pi^{-1}(j)})$. Let $\mathrm{E}_r$ be the set of orbits of $\mathrm{D}_r$ under that action, and let $\mathrm{N}(\omega)$ be the number of elements in each orbit $\omega \in \mathrm{E}_r$. Then formula

(5) may be written

$$[r](Z_\alpha) = \sum_{\omega \in E_r} N(\omega) Z_{\beta_1} Z_{\beta_2} \cdots Z_{\beta_r} \tag{6}$$

where for each $\omega \in E_r$, $(\beta_j)$ is an element of the orbit $\omega$.

**4.** The commutativity of the bigebra $\mathfrak{G}$ implies, by the same argument as in II, §1, No. 3, that the mapping $F: Z \mapsto Z^p$ is a *bigebra* homomorphism of $\mathfrak{G}$ into $\mathfrak{G}^{(1)}$, which we will again call the *Frobenius homomorphism*. We now have in $\mathfrak{G}$ two natural mappings, the shift $V: \mathfrak{G} \to \mathfrak{G}^{(-1)}$ and the Frobenius homomorphism $F: \mathfrak{G} \to \mathfrak{G}^{(1)}$, and their composites

$$\mathfrak{G} \xrightarrow{F} \mathfrak{G}^{(1)} \xrightarrow{V} \mathfrak{G} \qquad \text{and} \qquad \mathfrak{G} \xrightarrow{V} \mathfrak{G}^{(-1)} \xrightarrow{F} \mathfrak{G}$$

are endomorphisms of the bigebra $\mathfrak{G}$. They are determined by the following proposition:

**Proposition 1.** *For any commutative bigebra $\mathfrak{G}$, we have the relations*

$$F \circ V = V \circ F = [p]. \tag{7}$$

As everyone of the subbigebras $\mathfrak{s}_r$ of $\mathfrak{G}$ is stable under F, V and any endomorphism of $\mathfrak{G}$, we may suppose that $\mathfrak{G}$ has finite height, hence is stable. We have

$$F(V(Z_\alpha)) = V(F(Z_\alpha)) = 0 \qquad \text{if } \alpha \text{ is not divisible by } p$$

and

$$F(V(Z_\alpha)) = V(F(Z_\alpha)) = (Z_{\alpha/p})^p \qquad \text{if } \alpha \text{ is divisible by } p$$

(II, §2, No. 7, formula (30)). On the other hand, if we take $r = p$ in (6), and consider the *distinct* elements $\gamma_1, \ldots, \gamma_h$ in a sequence $(\beta_j)$ such that $\beta_1 + \beta_2 + \cdots + \beta_p = \alpha$, the number $N(\omega)$ is $p!/(\nu_1! \cdots \nu_h!)$ if $\nu_j$ is the number of elements $\beta_i$ equal to $\gamma_j$ for $1 \leqslant j \leqslant h$. The number $N(\omega)$ is therefore *divisible by p* except when $h = 1$, and that can only occur if all $\beta_i$ are *equal*, hence $\alpha$ is divisible by $p$ and then $N(\omega) = 1$; all the $\beta_i$ are then $\alpha/p$, and formula (6) boils down to

$$[p](Z_\alpha) = 0 \qquad \text{if } \alpha \text{ is not divisible by } p,$$

$$[p](Z_\alpha) = (Z_{\alpha/p})^p \qquad \text{if } \alpha \text{ is divisible by } p,$$

which proves the proposition.

**5.** It is convenient to put on the ring End($\mathfrak{G}$) of a commutative bigebra $\mathfrak{G}$ a *topology* defined as follows. For each $r \geqslant 0$, let $\mathfrak{N}_r$ be the set of all endomorphisms $u$ of $\mathfrak{G}$ such that $u(\mathfrak{s}_r(\mathfrak{G}) \cap \mathfrak{G}^+) = 0$; as the $\mathfrak{s}_r$ are subbigebras of $\mathfrak{G}$ which are stable under any endomorphism of $\mathfrak{G}$, it follows from formula (1) that $\mathfrak{N}_r$ is a *two-sided ideal* of End($\mathfrak{G}$); they therefore constitute a fundamental system of neighborhoods of 0 for a topology on End($\mathfrak{G}$), compatible with the ring structure. Furthermore, for this topology, End($\mathfrak{G}$) is *metrizable* and *complete*. Indeed, it is clear that the intersection of the $\mathfrak{N}_r$ is the element 0 in End($\mathfrak{G}$), i.e., the mapping $\eta \circ e$, for it is the only endomorphism of $\mathfrak{G}$ which vanishes in $\mathfrak{G}^+$. On the other hand, if $(u_r)_{r \geqslant 0}$ is a sequence of endomorphisms of $\mathfrak{G}$ such that

$$v_r = u_{r+1} - u_r \in \mathfrak{N}_r \qquad \text{for all} \quad r \geqslant 0,$$

it follows from formula (1) that $u_{r+1}$ and $u_r$ *coincide in* $\mathfrak{s}_r$, for if $(Z_\alpha)$ is a privileged basis for $\mathfrak{s}_r$, we have $v_r(Z_\alpha) = 0$ except for $\alpha = 0$, hence

$$\mu((u_r \otimes v_r)(\Delta(Z_\alpha))) = \mu(u_r(Z_\alpha) \otimes 1) = u_r(Z_\alpha).$$

There is therefore a unique endomorphism $u$ of $\mathfrak{G}$ which coincides with $u_r$ in each $\mathfrak{s}_r$, and it is clear that $u$ is the limit of the sequence $(u_r)$ for the topology of End($\mathfrak{G}$), which proves that End($\mathfrak{G}$) is complete.

## §2. Free commutative bigebras

**1.** The study of commutative bigebras over a perfect field $k$ of characteristic $p > 0$ is based on the properties of the *free commutative bigebras*, i.e., those which as *k-algebras* are free, or equivalently *algebras of poly-*

*nomials* (in finitely many or infinitely many indeterminates). Observe that such a bigebra $\mathfrak{G}$ must have *infinite dimension*, for $Z \mapsto Z^p$ is a semilinear mapping of the Lie algebra $\mathfrak{g}_0$ of $\mathfrak{G}$ into itself, and if $\mathfrak{g}_0$ was finite dimensional, we would have nontrivial linear relations between the successive powers $X, X^p, X^{p^2}, \ldots$ of an element $X \neq 0$ of $\mathfrak{g}_0$, contrary to assumption.

We are indeed going to show that there are free commutative bigebras having as underlying algebras the algebras of polynomials in a finite number or a denumerable set of indeterminates, and that they are essentially unique.

**Theorem 1.** *Let $r$ denote either an integer $\geqslant 0$, or $+\infty$, and let $\mathfrak{G}_r$ be the algebra of polynomials $k[(\mathrm{T}_h)]_{0 \leqslant h < r+1}$ ($r + 1$ being replaced by $+\infty$ if $r = +\infty$). There is then on $\mathfrak{G}_r$ a structure of commutative bigebra and a privileged basis $(Z_\alpha)$ having the following properties*:

a) *The set $\mathrm{L}_r$ of multiindices is $[0, p^{r+1}[^{(\mathbf{N})}$ ($p^{r+1}$ being replaced by $+\infty$ if $r = +\infty$).*

b) *$X_{hi} = \mathrm{T}_h^{p^i}$ for $0 \leqslant h < r + 1$ and $i \in \mathbf{N}$.*

c) *The privileged basis $(Z_\alpha)$ is canonical* (II, §3, No. 1).

*The privileged basis $(Z_\alpha)$ having these properties is unique. Furthermore, for each $\alpha \in \mathrm{L}_r$, let $v(\alpha)$ be the multiindex such that $v(\alpha)(i) = \alpha(i - 1)$ if $i \geqslant 1$, and $v(\alpha)(0) = 0$ ("lateral shift of $\alpha$"). Then, for every $\alpha \in \mathrm{L}_r$,*

d) $Z_\alpha^p = Z_{v(\alpha)}$.

*Finally, there is a unique graduation $\pi$ on $\mathfrak{G}_r$ such that*

$$\pi(\mathrm{T}_h) = p^h \qquad \textit{for} \quad 0 \leqslant h < r + 1.$$

*Each $Z_\alpha$ is a polynomial in the $\mathrm{T}_h$ without constant term, with coefficients in the prime field $\mathbf{F}_p$, and homogeneous of degree*

$$\pi(Z_\alpha) = \sum_{i \in \mathbf{N}} p^i \alpha(i) \qquad \textit{(which we write } \pi(\alpha)\textit{)}.$$

*The Frobenius subbigebra $\mathfrak{s}_t(\mathfrak{G}_r)$, for $t < r$, is identical to $\mathfrak{G}_t$, and the $Z_\alpha$ of height $t$ form its privileged basis verifying conditions a), b), and c).*

We first note that condition b) entirely determines the $X_\alpha$ (II, §2, No. 6, formula (25); the order chosen of course has no importance here): if

$$\alpha = \alpha_0 + p\alpha_1 + p^2\alpha_2 + \cdots \qquad \text{where the } \alpha_h \text{ have height 0,}$$

then

$$(1) \qquad X_\alpha = \mathrm{T}_0^{\nu_0}\mathrm{T}_1^{\nu_1} \cdots \mathrm{T}_h^{\nu_h} \cdots, \qquad \text{where} \quad \nu_h = \sum_{i \in \mathbf{N}} p^i \alpha_h(i).$$

From II, §3, No. 1, it then follows that condition c) determines the $Z_\alpha$ *uniquely*.

To prove existence of the basis $(Z_\alpha)$, we proceed by induction on $r$; once the theorem has been proved for all integers $r \geqslant 0$, it will be also proved for $r = +\infty$, and in that case we will have shown that $\mathfrak{G}_\infty$ is *reduced*. For $r = 0$, we take for $\mathfrak{g}_0(\mathfrak{G}_0)$ the vector subspace of $k[\mathrm{T}_0]$ having as a basis the

$$X_{0i} = \mathrm{T}_0^{p^i} \qquad (i \in \mathbf{N});$$

the set $\mathrm{L}_0$ is then the subset of $\mathbf{N}^{(\mathbf{N})}$ formed by the multiindices of height 0, and for $\alpha \in \mathrm{L}_0$, we take $Z_\alpha = X_\alpha/\alpha!$, comultiplication being defined by

$$\Delta(\mathrm{T}_0) = 1 \otimes \mathrm{T}_0 + \mathrm{T}_0 \otimes 1.$$

That this gives a bigebra satisfying the required conditions is an immediate consequence of II, §2, Proposition 1.

We now assume the theorem has been proved for $\mathfrak{G}_{r-1}$ and we denote by $\Delta'$ the comultiplication in $\mathfrak{G}_{r-1} = k[\mathrm{T}_0, \ldots, \mathrm{T}_{r-1}]$, by $(Z'_\alpha)$, for $\alpha \in \mathrm{L}_{r-1}$, the privileged basis of the theorem for $\mathfrak{G}_{r-1}$. The first step is to define the comultiplication $\Delta$ in $\mathfrak{G}_r = k[\mathrm{T}_0, \ldots, \mathrm{T}_{r-1}, \mathrm{T}_r]$; as $\mathfrak{G}_r$ is a free $k$-algebra, $\Delta$, as an algebra homomorphism of $\mathfrak{G}_r$ into $\mathfrak{G}_r \otimes_k \mathfrak{G}_r$, is entirely determined by its value for the $\mathrm{T}_h$ with $h \leqslant r$; we take

$$(2) \qquad \Delta(\mathrm{T}_h) = \Delta'(\mathrm{T}_h) \qquad \text{for} \quad 0 \leqslant h \leqslant r - 1;$$

$$(3) \qquad \Delta(\mathrm{T}_r) = 1 \otimes \mathrm{T}_r + \mathrm{T}_r \otimes 1 + \sum_{\beta + \gamma = p^r\varepsilon_0} Z'_\beta \otimes Z'_\gamma.$$

We first have to check that this defines on $\mathfrak{G}_r$ a structure of commutative bigebra, i.e., we have to verify the coassociativity and cocommutativity of the cogebra structure defined by $\Delta$; due to the fact that $\Delta$ is an algebra homomorphism, and that $\mathfrak{G}_r$ is free, it is enough to check these conditions:

$$(4) \qquad (\Delta \otimes 1)(\Delta(\mathrm{T}_h)) = (1 \otimes \Delta)(\Delta(\mathrm{T}_h)),$$

(5) $$\sigma(\Delta(\mathrm{T}_h)) = \Delta(\mathrm{T}_h),$$

on the generators $\mathrm{T}_h$ for $0 \leqslant h \leqslant r$. But (5) obviously follows from (2), (3), and the induction assumption; and the same is true for (4), due to the fact that $(Z'_\alpha)$ is a privileged basis for $\mathfrak{G}_{r-1}$.

The bigebra $\mathfrak{G}_r$ being thus defined, the first result to be proved is

(6) $$\mathfrak{g}_0(\mathfrak{G}_r) = \mathfrak{g}_0(\mathfrak{G}_{r-1});$$

in other words, we have to prove that if an element $Y \in \mathfrak{G}_r$ is primitive then it has to belong to $\mathfrak{G}_{r-1}$; by definition, we may write

(7) $$Y = \mathrm{T}_r^m Y_m + \mathrm{T}_r^{m-1} Y_{m-1} + \cdots + Y_0,$$

where $Y_0, \ldots, Y_m$ are elements of $\mathfrak{G}_{r-1}$ and $Y_m \neq 0$, and we have to show that this is impossible unless $m = 0$. We may assume that $Y$ is *homogeneous* for the degree $\pi$ (since $\Delta$ is a homomorphism of *graded* algebras, as follows from (3)), and then the $Y_j$ may also be taken homogeneous. We first prove that no power $\mathrm{T}_r^i$ in (7) may have a coefficient $Y_i \neq 0$ unless $i$ is a power of $p$. Indeed, suppose there is at least one such term and take the one for which $i$ is the largest possible; then, from (3) it follows that $\Delta(\mathrm{T}_r^i) = (\Delta(\mathrm{T}_r))^i$ contains a term of the form $\mathrm{T}_r^j \otimes \mathrm{T}_r^{i-j}$ with a coefficient $\neq 0$ in $k$, for some $j$ such that $0 < j < i$; such a term cannot be equal to a term coming from $\Delta(\mathrm{T}_r^h)$ with $h > i$ since by assumption all these exponents $h$ are powers of $p$, hence

$$\Delta(\mathrm{T}_r^h) = 1 \otimes \mathrm{T}_r^h + \mathrm{T}_r^h \otimes 1 + V, \qquad \text{where} \quad V \in \mathfrak{G}_{r-1} \otimes \mathfrak{G}_{r-1};$$

for reasons of degree, a term $\mathrm{T}_r^j \otimes \mathrm{T}_r^{i-j}$ with $0 < j < i$ cannot be equal to a term coming from $\Delta(\mathrm{T}_r^h)$ with $h < i$; finally, there is no such term in $Y \otimes 1 + 1 \otimes Y$, and this proves our assertion.

Next we show that if in (7) there is a term $\mathrm{T}_i^r Y_i$ with $i > 0$, then $Y_i$ must be a *scalar* in $k$; suppose again there is at least one term for which $i > 0$ and $Y_i \notin k$, and take the largest exponent $i$ of that kind. Then, in $\Delta(Y)$, this term gives a term $\mathrm{T}_r^i \otimes Y_i$; remembering that all the exponents $i$ are powers of $p$, we see at once that this term cannot be equal to any term coming from $\Delta(\mathrm{T}_r^h Y_h)$ with $h \neq i$; furthermore, since $Y_i \notin k$, there is no term $\mathrm{T}_r^i \otimes Y_i$ in $1 \otimes Y + Y \otimes 1$, and this shows that all the $Y_i$ with the exception of $Y_0$ must be scalars; however, for reasons of homogeneity, there can be no more than *one* such term in the right-hand side of (7).

Multiplying $Y$ by a scalar, we see that the only possible form for $Y$ if $Y \notin \mathfrak{G}_{r-1}$ would be $\mathrm{T}_r^{p^h} + Y_0$. Noting that

$$\Delta(Y) = 1 \otimes Y + Y \otimes 1,$$

we then obtain from (3) and from the relations $Z_\alpha'^p = Z'_{v(\alpha)}$, which are valid in $\mathfrak{G}_{r-1}$ by the inductive assumption

(8)
$$\Delta(Y_0) = 1 \otimes Y_0 + Y_0 \otimes 1 - \sum_{\beta+\gamma=p^r\varepsilon_h} Z'_\beta \otimes Z'_\gamma.$$

But as $\mathfrak{G}_{r-1}$ has height $r-1$, no element of that bigebra may verify (8), from the definition of a privileged basis; this ends the proof of (6).

The dual $\mathfrak{G}_r^*/\mathfrak{G}_{r-1}^0$ of $\mathfrak{G}_{r-1}$ has a pseudobasis $(\bar{x}^\alpha)$ dual to $(Z'_\alpha)$ (for $\alpha \in \mathrm{L}_{r-1}$); it follows from (6) that $\mathfrak{G}_{r-1}^0 \subset \mathfrak{R}^2$, hence a closed vector subspace of $\mathfrak{G}_r^*$ supplementary to $\mathfrak{R}^2$ in $\mathfrak{R}$ is naturally identified to a supplementary subspace to $\mathfrak{R}^2/\mathfrak{G}_{r-1}^0$ in $\mathfrak{R}/\mathfrak{G}_{r-1}^0$; if we denote by $x_i$ the elements corresponding to the $\bar{x}_i$ in this identification, we may construct a privileged basis $(Z_\alpha)$ of $\mathfrak{G}_r$ corresponding to these elements. We then have

$$x^\alpha \in \mathfrak{G}_{r-1}^0 \qquad \text{if} \quad \mathrm{ht}(\alpha) = r,$$

hence

$$\langle x^\alpha, Z'_\beta \rangle = 0 \qquad \text{if} \quad \beta \in \mathrm{L}_{r-1};$$

by definition of $(Z_\alpha)$, we therefore have $Z_\alpha = Z'_\alpha$ for $\alpha \in \mathrm{L}_{r-1}$.

Next, when we compare $\Delta(\mathrm{T}_r^{p^i})$, computed from (3) and the relations $Z_\alpha'^p = Z'_{v(\alpha)}$, and $\Delta(X_{ri})$, we see that $X_{ri} - \mathrm{T}_r^{p^i}$ is a primitive element of $\mathfrak{G}_r$, in other words,

(9)
$$X_{ri} = \mathrm{T}_r^{p^i} + \sum_{j \geqslant 0} b_{ij} X_{0j};$$

it is then possible (II, §1, No. 1) to replace the $x_i$ by

$$x'_i = x_i + \sum_{j \geqslant 0} b_{ji} x_j^{p^r}$$

and for the new privileged basis of $\mathfrak{G}_r$ corresponding to that choice, the $Z_\alpha$ for $\alpha \in \mathrm{L}_{r-1}$ are not changed, and now $X_{ri} = \mathrm{T}_r^{p^i}$ for every $i \in \mathbf{N}$. The method of II, §3, Proposition 2 can next be applied to modify the $x_i$ without changing the $Z_\alpha$ for $|\alpha| \leqslant p^r$, and making the basis $(Z_\alpha)$ *canonical.*

With this basis $(Z_\alpha)$ thus chosen, we prove that the $Z_\alpha$ are *homogeneous* of degree $\pi(\alpha)$. We use induction on $|\alpha|$; the inductive assumption on $\mathfrak{G}_{r-1}$, together with the relations $X_{ri} = \mathrm{T}_r^{p^i}$, imply already that the $Z_\beta$ are homogeneous for $|\beta| \leqslant p^r$. We may write (II, §2, No. 6, Eq. (26))

$$Z_\alpha = \frac{1}{\alpha!} X_\alpha + \sum_{|\beta| < |\alpha|} c_{\alpha\beta} Z_\beta \tag{10}$$

and we have to prove that $c_{\alpha\beta} = 0$ except for the indices such that $\pi(\beta) = \pi(\alpha)$. As

$$\Delta(Z_\alpha) - 1 \otimes Z_\alpha - Z_\alpha \otimes 1 = \sum_{0<\lambda<\alpha} Z_\lambda \otimes Z_{\alpha-\lambda}$$

has degree $\pi(\alpha)$ by the inductive hypothesis, and $\Delta$ is a homomorphism of graded algebras, the sum

$$\sum_{|\beta| < |\alpha|, \pi(\beta) \neq \pi(\alpha)} c_{\alpha\beta}(\Delta(Z_\beta) - 1 \otimes Z_\beta - Z_\beta \otimes 1)$$

is 0; but the elements

$$\Delta(Z_\beta) - 1 \otimes Z_\beta - Z_\beta \otimes 1 = \sum_{0<\lambda<\beta} Z_\lambda \otimes Z_{\beta-\lambda},$$

*for* $|\beta| > 1$, are *linearly independent* in $\mathfrak{G}_r \otimes \mathfrak{G}_r$. Therefore we already have $c_{\alpha\beta} = 0$ for $|\beta| > 1$ and $\pi(\beta) \neq \pi(\alpha)$; on the other hand, if in the right-hand side of (10) we replace each $Z_\beta$ by its expression in terms of the $X_\lambda$ (II, §2, No. 6), we get no term in $X_{0i}$ for $|\beta| > 1$, since the basis $(Z_\alpha)$ is canonical (II, §3, No. 1); for the same reason, we must therefore have $c_{\alpha\varepsilon_i} = 0$, and this shows that $Z_\alpha$ is a homogeneous polynomial in the $\mathrm{T}_h$ for the degree $\pi$. That the coefficients of that polynomial are in the prime field $\mathbf{F}_p$ follows from the fact that the construction of Theorem 1 may be done over any perfect field, in particular, over $\mathbf{F}_p$; however, if $(Z_\alpha)$ is then the basis obtained over $\mathbf{F}_p$, the $Z_\alpha \otimes 1$ form a basis over $k$ which has properties a), b), c), hence is the *unique* basis over $k$ having these properties, and this proves our assertion.

We still have to prove that $Z_\alpha^p = Z_{v(\alpha)}$. We may again use induction on $|\alpha| \geqslant p^r$; as $\mathfrak{G}_r$ is *commutative*, we may write

$$\Delta(Z_\alpha^p - Z_{v(\alpha)}) = \sum_{0 \leqslant \beta \leqslant \alpha} Z_\beta^p \otimes Z_{\alpha-\beta}^p - \sum_{0 \leqslant \beta \leqslant \alpha} Z_{v(\beta)} \otimes Z_{v(\alpha-\beta)},$$

and taking into account the inductive assumption, we see that the right-hand side reduces to

$$1 \otimes (Z_\alpha^p - Z_{v(\alpha)}) + (Z_\alpha^p - Z_{v(\alpha)}) \otimes 1,$$

in other words, $Z_\alpha^p - Z_{v(\alpha)}$ belongs to $\mathfrak{g}_0$. But if we write

$$Z_\alpha = \frac{1}{\alpha!} X_\alpha + \sum_{|\beta|<|\alpha|} e_{\alpha\beta} X_\beta$$

(II, §3, No. 1, formula (3)), the commutativity of $\mathfrak{G}_r$ gives

$$Z_\alpha^p = \frac{1}{\alpha!} X_\alpha^p + \sum_{|\beta|<|\alpha|} e_{\alpha\beta}^p X_\beta^p,$$

and on the other hand

$$Z_{v(\alpha)} = \frac{1}{\alpha!} X_{v(\alpha)} + \sum_{|\beta|<|\alpha|} e_{v(\alpha),v(\beta)} X_{v(\beta)}.$$

But since the basis $(Z_\alpha)$ is canonical and by definition $X_\alpha^p = X_{v(\alpha)}$ for every $\alpha$ and $|v(\alpha)| = |\alpha|$, we see that in the expression of $Z_\alpha^p - Z_{v(\alpha)}$ there are no terms in $X_{0i}$, hence from the relation $Z_\alpha^p - Z_{v(\alpha)} \in \mathfrak{g}_0$ it follows that in fact $Z_\alpha^p = Z_{v(\alpha)}$. Q.E.D.

We will say that the free commutative bigebra $\mathfrak{G}_\infty$ (also written $\mathfrak{G}_\infty(k)$) thus defined is the *Witt bigebra* over $k$; we have proved that $\mathfrak{G}_\infty(k) = \mathfrak{G}_\infty(\mathbf{F}_p) \otimes_{\mathbf{F}_p} k$ up to isomorphism.

**2.** *The universal property of the Witt bigebras.* As usual, we may expect that free commutative bigebras are "universal" in the category of reduced commutative bigebras. We are indeed going to prove the following theorem:

**Theorem 2.** *Let $\mathfrak{G}'$ be a reduced commutative bigebra over $k$, $(X'_{0i})_{i\in \mathrm{I}}$ a basis of the Lie algebra $\mathfrak{g}_0(\mathfrak{G}')$. For each $i \in \mathrm{I}$, let $\mathfrak{G}_i$ be a bigebra isomorphic to $\mathfrak{G}_\infty(k)$, where the indeterminates are written $\mathrm{T}_{hi}$ instead of $\mathrm{T}_h$ for $h \in \mathbf{N}$, and let $\mathfrak{G}$ be the tensor product of the $\mathfrak{G}_i$ for $i \in \mathrm{I}$. Then there exists a surjective bigebra homomorphism $u: \mathfrak{G} \to \mathfrak{G}'$ such that $u(\mathrm{T}_{0i}) = X'_{0i}$ for each $i \in \mathrm{I}$.*

The tensor product of the $\mathfrak{G}_i$, when I is infinite, is as usual defined as the direct limit of the tensor products $\mathfrak{G}_\Phi$, where $\Phi$ runs through the finite subsets of I and $\mathfrak{G}_\Phi$ is the tensor product of the $\mathfrak{G}_i$ for $i \in \Phi$; a canonical basis of $\mathfrak{G}$ is therefore obtained in the following way. Let L be the disjoint union (or set theoretic "sum") of a family $(\mathrm{N}(i))_{i\in \mathrm{I}}$ of sets, each of which $\mathrm{N}(i)$ is a "copy" of $\mathbf{N}$, in other words there is a bijection $n \mapsto n(i)$ of $\mathbf{N}$ onto each $\mathrm{N}(i)$. A multiindex $\lambda \in \mathbf{N}^{(\mathrm{L})}$ considered as a function defined on L, has a restriction $\lambda_i$ to each $\mathrm{N}(i)$ (which is 0 except for a finite number of indices $i \in \mathrm{I}$); we take as canonical basis of $\mathfrak{G}$ the set of the elements

$$Z_\lambda = \bigotimes_{i\in\mathrm{I}} Z_{\lambda_i},$$

tensor product of the bases $(Z_\alpha)$ constructed in Theorem 1, with $\alpha \in \mathbf{N}^{(\mathrm{N}(\mathrm{I}))}$. In addition, we *imbed* the set $\mathbf{N}^{(\mathrm{I})}$ into $\mathbf{N}^{(\mathrm{L})}$ in the following fashion: for each $\alpha \in \mathbf{N}^{(\mathrm{I})}$ we associate to $\alpha$ the element $\lambda(\alpha) \in \mathbf{N}^{(\mathrm{L})}$ whose restriction $\lambda_i$ to $\mathrm{N}(i)$ is given by

$$\lambda_i(0(i)) = \alpha(i) \qquad \text{and} \qquad \lambda_i(n(i)) = 0 \qquad \text{for } n > 1;$$

it is clear that we have thus defined an injection $\alpha \mapsto \lambda(\alpha)$ of $\mathbf{N}^{(\mathrm{I})}$ into $\mathbf{N}^{(\mathrm{L})}$.

To prove Theorem 2, we use the notations of II, §2, No. 9 for the privileged basis of $\mathfrak{G}'$, except that we accent the elements $Z'_\alpha$ of that basis and write $X'_{0i}$ instead of $X_i$ for $i \in \mathrm{I}$ and $\mathfrak{N}'$ instead of $\mathfrak{N}$ for the augmentation ideal of $\mathfrak{G}'^*$. As a matter of fact, we are going to define not one, but a sequence $(Z'^{(n)}_\alpha)$ of privileged bases of $\mathfrak{G}'$ corresponding to the basis $(X'_{0i})_{i\in\mathrm{I}}$ of $\mathfrak{g}_0(\mathfrak{G}')$. Each of these bases will correspond to a choice $(x'^{(n)}_i)$ of representatives in the classes $\bar{x}'_i$ modulo $\mathfrak{N}'^2$ which constitute the pseudo-basis of $\mathfrak{N}'/\mathfrak{N}'^2$ dual to $(X'_{0i})$. At the same time, we will define for each integer $n$ an *algebra* homomorphism $u_n$ of $\mathfrak{G}$ into $\mathfrak{G}'$, such that the two following conditions will be satisfied:

1° $x'^{(n+1)}_i \equiv x'^{(n)}_i$ modulo $\mathfrak{N}'^{n+1}$, hence

$$Z'^{(n+1)}_\alpha = Z'^{(n)}_\alpha \qquad \text{for} \quad |\alpha| \leqslant n;$$

2° $u_n(Z_{\lambda(\alpha)}) = Z'^{(n)}_\alpha$ for $|\alpha| \leqslant n$.

The definition of $u_n$ is done by assigning values to $u_n(\mathrm{T}_{hi})$ for arbitrary values of $h \in \mathbf{N}$ and $i \in \mathrm{I}$; we take

$$u_n(\mathrm{T}_{hi}) = X'^{(n)}_{hi}$$

the construction of the $x'^{(n)}_i$ and the proof of conditions 1° and 2° will be done by induction on $n$.

For $n = 1$, we may take the representatives $x'^{(1)}_i$ arbitrarily, and condition 2° is trivially satisfied. We assume next that the induction has proceeded until $n$ and we consider two cases:

a) $n + 1$ is not a power of $p$; then by definition $u_{n+1}$ and $u_n$ coincide in the Frobenius subbigebra $\mathfrak{s}_r(\mathfrak{G})$ for the smallest integer $r$ such that $p^{r+1} > n + 1$, and their restriction to $\mathfrak{s}_r(\mathfrak{G})$ is a *bigebra* homomorphism of $\mathfrak{s}_r(\mathfrak{G})$ into $\mathfrak{G}'$: as $u_n$ is an algebra homomorphism, we have only to check that, for the comultiplications $\Delta$ and $\Delta'$ in $\mathfrak{G}$ and $\mathfrak{G}'$, we have

$$\Delta'(u_n(\mathrm{T}_{hi})) = (u_n \otimes u_n)(\Delta(\mathrm{T}_{hi})) \qquad \text{for all} \quad h \leqslant r \quad \text{and} \quad i \in \mathrm{I}.$$

But by definition, $\mathrm{T}_{hi}$ is the element $X_{h,0(i)}$ of the basis $(Z_\lambda)$ of $\mathfrak{G}$, corresponding to the multiindex $\rho$ such that $\rho(n(j)) = 0$ if $j \neq i$ or if $j = i$ and $n \geqslant 1$, and $\rho(0(i)) = p^h$; this index can be written $\lambda(p^h\varepsilon_i)$, and the multiindices $\mu$, $\nu$ in $\mathbf{N}^{(\mathrm{L})}$ such that $\mu + \nu = \lambda(p^h\varepsilon_i)$ are exactly the indices $\lambda(\beta)$, $\lambda(\gamma)$ such that $\beta + \gamma = p^h\varepsilon_i$; we therefore can write

$$\Delta(\mathrm{T}_{hi}) = 1 \otimes \mathrm{T}_{hi} + \mathrm{T}_{hi} \otimes 1 + \sum_{\beta+\gamma=p^h\varepsilon_i} Z_{\lambda(\beta)} \otimes Z_{\lambda(\gamma)},$$

and by the inductive assumption,

$$u_n(Z_{\lambda(\beta)}) = Z'^{(n)}_\beta \qquad \text{for} \quad 0 < \beta < p^h\varepsilon_i.$$

The relation to be verified then results from the definition $u_n(\mathrm{T}_{hi}) = X'^{(n)}_{hi}$ and the relation

$$\Delta'(X'^{(n)}_{hi}) = 1 \otimes X'^{(n)}_{hi} + X'^{(n)}_{hi} \otimes 1 + \sum_{\beta+\gamma=p^h\varepsilon_i} Z'^{(n)}_\beta \otimes Z'^{(n)}_\gamma.$$

The assumption on $n$ implies that for all multiindices $\alpha \in \mathbf{N}^{(\mathrm{I})}$ such that $|\alpha| = n + 1$, we have $Z_{\lambda(\alpha)} \in \mathfrak{s}_r(\mathfrak{G})$, hence

$$\Delta'(u_n(Z_{\lambda(\alpha)})) = (u_n \otimes u_n)(\Delta(Z_{\lambda(\alpha)}));$$

using the inductive assumption, this implies that $u_n(Z_{\lambda(\alpha)}) - Z'^{(n)}_\alpha$ is a primitive element in $\mathfrak{G}'$, in other words

$$(11) \qquad u_n(Z_{\lambda(\alpha)}) = Z_\alpha'^{(n)} + \sum_{i \in \mathrm{I}} b_{\alpha i} X'_{0i}$$

with the $b_{\alpha i}$ equal to 0 except for a finite number of indices $i \in \mathrm{I}$.

We then choose the new representatives $x_i'^{(n+1)}$ by the relations

$$x_i'^{(n+1)} = x_i'^{(n)} - \sum_{|\alpha| = n+1} b_{\alpha i} (x'^{(n)})^\alpha$$

which is possible by II, §3, No. 1, and gives for $Z_\alpha'^{(n+1)}$ the value of the right-hand side of (11) for $|\alpha| = n + 1$.

b) $n + 1 = p^{r+1}$. The arguments of a) still apply for those multi-indices $\alpha$ such that $|\alpha| = p^{r+1}$, but which are *of height r*. We may therefore carry on the same change of representatives, but of course we only obtain the relations

$$u_n(Z_{\lambda(\alpha)}) = Z_\alpha'^{(n+1)}$$

for the $\alpha$ *of height r* with $|\alpha| = n + 1$. We then define $u_{n+1}$ by the conditions

$$u_{n+1}(\mathrm{T}_{hi}) = X_{hi}'^{(n+1)} \qquad \text{for all pairs } (h, i),$$

and as $X_{hi}'^{(n+1)} = X_{hi}'^{(n)}$ for $h \leqslant r$, $u_{n+1}$ and $u_n$ coincide in $\mathfrak{s}_r(\mathfrak{G})$, and therefore we have

$$u_{n+1}(Z_{\lambda(\alpha)}) = Z_\alpha'^{(n+1)} \qquad \text{for } all\ \alpha \text{ with } |\alpha| = n + 1.$$

The induction may thus proceed; we then define $u(Z_{\lambda(\alpha)})$ as equal to $u_n(Z_{\lambda(\alpha)})$ for all $n \geqslant |\alpha|$, and $u$ satisfies the conditions of Theorem 2.

**Corollary.** *Under the same assumptions as in Theorem 2, let $X_0$ be an arbitrary element of the Lie algebra $\mathfrak{g}_0(\mathfrak{G}')$. Then there exists a bigebra homomorphism $v: \mathfrak{G}_\infty \to \mathfrak{G}'$ such that $v(\mathrm{T}_0) = X'_0$.*

Indeed, we may suppose that there is a basis $(X_{0i})_{i \in \mathrm{I}}$ of $\mathfrak{g}_0(\mathfrak{G}')$ such that $0 \in \mathrm{I}$ and $X'_{00} = X'_0$, and that $\mathfrak{G}_\infty$ is the subbigebra of $\mathfrak{G}$ generated by

$$\mathrm{T}_{00} = \mathrm{T}_0, \mathrm{T}_{10}, \ldots, \mathrm{T}_{h0}, \ldots.$$

The result is then obtained by taking for $v$ the restriction of the homomorphism $u$ to $\mathfrak{G}_\infty$.

**3.** *The Witt ring.* This section and the next are devoted to the elucidation of the structure of the topological ring $\mathrm{End}(\mathfrak{G}_\infty(k))$ for the free commutative bigebra defined in Theorem 1. We first consider the *subring* $\mathrm{W}(k)$ of $\mathrm{End}(\mathfrak{G}_\infty(k))$ consisting of those endomorphisms which respect the graduation $\pi$ of $\mathfrak{G}_\infty(k)$ (in other words, they must be endomorphisms of the *graded* algebra $\mathfrak{G}_\infty(k)$). We shall say that $\mathrm{W}(k)$ is the *Witt ring* of the perfect field $k$.

We have seen in Theorem 1 that we may write

$$(12) \qquad Z_\alpha = \mathrm{E}_\alpha(\mathrm{T}_0, \mathrm{T}_1, \ldots.)$$

where $\mathrm{E}_\alpha$ is a polynomial with coefficients in the prime field $\mathbf{F}_p$, homogeneous (for $\pi$) and of degree

$$\pi(\alpha) = \sum_{i \in \mathrm{I}} p^i \alpha(i).$$

For any scalar $\xi \in k$ we therefore have

$$(13) \qquad \mathrm{E}_\alpha(\xi \mathrm{T}_0, \xi^p \mathrm{T}_1, \xi^{p^2} \mathrm{T}_2, \ldots) = \xi^{\pi(\alpha)} Z_\alpha$$

and as $\pi(\beta + \gamma) = \pi(\beta) + \pi(\gamma)$, it follows from that relation that, for $\xi \neq 0$, $(\xi^{\pi(\alpha)} Z_\alpha)$ is still a privileged basis for $\mathfrak{G}_\infty$; the algebra homomorphism of $\mathfrak{G}_\infty$ into itself which transforms each $\mathrm{T}_h$ into $\xi^{p^h}\mathrm{T}_h$ for every $h \geqslant 0$ is thus *graded*, and a *bigebra* endomorphism of $\mathfrak{G}_\infty$, which is bijective if $\xi \neq 0$, and the zero element of $\mathrm{End}(\mathfrak{G}_\infty)$ if $\xi = 0$. If we write $e(\xi)$ that element of $\mathrm{W}(k)$, it follows from its definition that

$$(14) \qquad e(\xi\eta) = e(\xi)e(\eta) = e(\eta)e(\xi),$$

in other words, $\xi \mapsto e(\xi)$ is a homomorphism for the *multiplicative* laws of $k$ and $\mathrm{W}(k)$, sending the unit element of $k$ onto $1_{\mathfrak{G}_\infty}$. But $\xi \mapsto e(\xi)$ is *not* a homomorphism for the *additive* laws of $k$ and $\mathrm{W}(k)$, since $[p] = p \cdot 1_{\mathfrak{G}_\infty} \neq 0$ by Proposition 1 of §1. In the following, we shall write $\omega$ for the endomorphism $[p] = \mathrm{F} \circ \mathrm{V}$ of $\mathfrak{G}_\infty$; as

$$[p](Z_\alpha) = (Z_{\alpha/p})^p,$$

that endomorphism belongs to W($k$) and *commutes* with all the $e(\xi)$.

**Proposition 1.** *The Witt ring* W($k$) *is a commutative domain of integrity of characteristic* 0, *equipped with a discrete valuation* $w$ *such that* $w(\omega) = 1$; *the topology defined by* $w$ *is identical to the one induced by the topology of* End($\mathfrak{G}_\infty$) (§1, No. 5), *and* W($k$) *is complete for that topology. The residual field of* W($k$) *is naturally identified to* $k$, *and every element of* W($k$) *may be written uniquely as a series*

$$u = e(\xi_0) + e(\xi_1^{p^{-1}})\omega + \cdots + e(\xi_n^{p^{-n}})\omega^n + \cdots. \tag{15}$$

If $u \in \mathrm{End}(\mathfrak{G}_\infty)$ is in the closure of W($k$), for each $r \geqslant 0$, there is an element $u_r \in \mathrm{W}(k)$ such that $u$ coincides with $u_r$ in $\mathfrak{s}_r(\mathfrak{G}_\infty)$; this implies that $u$ is a *graded* algebra endomorphism, hence belongs to W($k$), which is therefore closed in End($\mathfrak{G}_\infty$), hence complete for the induced topology. Furthermore, from the definition of $\omega$ and from the properties of the shift V, it follows that the two-sided principal ideal of W($k$) generated by $\omega^r$ is contained in the ideal $\mathfrak{N}_{r-1}$ of End($\mathfrak{G}_\infty$) (§1, No. 5). Let $u \in \mathrm{W}(k)$, and suppose we have proved the existence of elements $\xi_i \in k$ $(0 \leqslant i \leqslant r)$ such that

$$v = u - (e(\xi_0) + e(\xi_1^{p^{-1}})\omega + \cdots + e(\xi_r^{p^{-r}})\omega^r)$$

belongs to the ideal $\mathfrak{N}_r$, so that $v(\mathrm{T}_h) = 0$ for $0 \leqslant h \leqslant r$, hence $v(Z_\alpha) = 0$ for each multiindex $\alpha$ of height $\leqslant r$. As $v$ is a bigebra endomorphism, this implies

$$\Delta(v(\mathrm{T}_{r+1})) = 1 \otimes v(\mathrm{T}_{r+1}) + v(\mathrm{T}_{r+1}) \otimes 1,$$

in other words $v(\mathrm{T}_{r+1}) \in \mathfrak{g}_0(\mathfrak{G}_\infty)$; but as $v$ is a *graded* bigebra endomorphism, this is only possible if

$$v(\mathrm{T}_{r+1}) = \xi_{r+1}\mathrm{T}_0^{p^{r+1}} \qquad \text{for some scalar} \quad \xi_{r+1} \in k.$$

From this result we deduce that $v - e(\xi_{r+1}^{p^{-r-1}})\omega^{r+1}$ belongs to the ideal $\mathfrak{N}_{r+1}$, and the induction may proceed. As the topology of W($k$) is Hausdorff and complete, this proves the existence of the series (15) for any $u \in \mathrm{W}(k)$

and also proves that W($k$) is commutative. To prove uniqueness, we have only to show that if

(16)

$$e(\xi_r^{p^{-r}})\omega^r + e(\xi_{r+1}^{p^{-r-1}})\omega^{r+1} + \cdots = e(\eta_r^{p^{-r}})\omega^r + e(\eta_{r+1}^{p^{-r-1}})\omega^{r+1} + \cdots$$

then $\xi_r = \eta_r$; but this is clear if we observe that both sides of (16) coincide, respectively, with $e(\xi_r^{p^{-r}})\omega^r$ and $e(\eta_r^{p^{-r}})\omega^r$ in $\mathfrak{s}_r(\mathfrak{G}_\infty)$, and when we take the value at $T_r$ of these last two endomorphisms, we obtain $\xi_r = \eta_r$. Furthermore, the value of the left-hand side of (16) at $T_r$ is 0 if and only if $\xi_r = 0$, hence the intersection $\mathfrak{N}_r \cap W(k)$ is exactly equal to the principal ideal $(\omega^{r+1})$.

If we define $w(u)$ as the smallest integer $r$ for which $\xi_r \neq 0$ in (15) if $u \neq 0$, and $+\infty$ if $u = 0$, it follows at once, from (14) and from the fact that $\xi \mapsto e(\xi)$ is injective, that $w$ is a *discrete valuation* on W($k$) such that $w(\omega) = 1$, and that $k$ is the residual field for that valuation. As W($k$) is a valuation ring, it is an integral domain, and as $k$ has characteristic $p$, W($k$) can only have characteristic $p$ or 0; but as $p \cdot 1_{\mathfrak{G}_\infty} = \omega$ is not 0, W($k$) has characteristic 0. Q.E.D.

We may observe here that as $\omega = p \cdot 1_{\mathfrak{G}_\infty}$, the valuation ring W($k$) is *unramified*, which directly proves the existence of such a ring having a given residual field $k$, perfect and of characteristic $p > 0$.

**Corollary.** (i) *There is a unique system of polynomials in* $2h + 2$ *indeterminates* $F_h(X_0, \ldots, X_h; T_0, \ldots, T_h)$ *with coefficients in* $\mathbf{F}_p$, *such that, for every* $u \in W(k)$ *given by* (15) *and for each* $h \geqslant 0$,

$$u(T_h) = F_h(\xi_0, \ldots, \xi_h; T_0, \ldots, T_h). \tag{17}$$

*Furthermore, for the degree* $\pi$ *on* $k[X_0, \ldots, X_h; T_0, \ldots, T_h]$ *defined by* $\pi(X_i) = \pi(T_i) = p^i$, $F_h$ *is homogeneous and of degree* $p^h$ *separately in the* $\xi_i$ *and in the* $T_i$.

(ii) *There exist two systems of polynomials*

$$S_h(T_0, \ldots, T_h; U_0, \ldots, U_h), \qquad P_h(T_0, \ldots, T_h; U_0, \ldots, U_h)$$

*in* $2h + 2$ *indeterminates, with coefficients in* $\mathbf{F}_p$, *such that, for any pair of elements*

$$u = \sum_n e(\xi_n^{p^{-n}})\omega^n, \qquad v = \sum_n e(\eta_n^{p^{-n}})\omega^n \qquad \textit{of} \quad \mathrm{W}(k),$$

*if*

$$u + v = \sum_n e(\zeta_n^{p^{-n}})\omega^n, \qquad uv = \sum_n e(\theta_n^{p^{-n}})\omega^n,$$

*we have*

(18)

$$\zeta_h = \mathrm{S}_h(\xi_0, \ldots, \xi_h; \eta_0, \ldots, \eta_h), \qquad \theta_h = \mathrm{P}_h(\xi_0, \ldots, \xi_h; \eta_0, \ldots, \eta_h)$$

*for every $h \geqslant 0$. The polynomials $\mathrm{S}_h$ and $\mathrm{P}_h$ are uniquely determined by these properties; furthermore, for the degree $\pi$ defined by $\pi(\mathrm{T}_i) = \pi(\mathrm{U}_i) = p^i$, $\mathrm{S}_h$ is homogeneous of degree $p^h$ in the $2h + 2$ indeterminates $\mathrm{T}_i$, $\mathrm{U}_i$, and $\mathrm{P}_h$ is homogeneous and of degree $p^h$ separately in the $\mathrm{T}_i$ and in the $\mathrm{U}_i$.*

(i) If

$$u_h = e(\xi_0) + e(\xi_1^{p^{-1}})\omega + \cdots + e(\xi_h^{p^{-h}})\omega^h,$$

we have $u(\mathrm{T}_h) = u_h(\mathrm{T}_h)$; we may therefore assume that all $\xi_i$ of index $i \geqslant h + 1$ are 0. We then have (§1, No. 3),

$$u_h = \mu^{h+1} \circ (e(\xi_0) \otimes e(\xi_h^{p^{-h}})\omega \otimes \cdots \otimes e(\xi_h^{p^{-h}})\omega^h) \circ \Delta^{h+1},$$

and therefore $u_h(\mathrm{T}_h)$ is the sum of the terms

(19) $$\xi_0^{\pi(\beta_0)}\xi_1^{\pi(\beta_1)} \cdots \xi_h^{\pi(\beta_h)} Z_{\beta_0} Z_{\beta_1}^p \cdots Z_{\beta_h}^{p^h}$$

where $(\beta_i)_{0 \leqslant i \leqslant h}$ runs through all sequences such that

$$\beta_0 + p\beta_1 + \cdots + p^h\beta_h = p^h\varepsilon_0.$$

It follows from (12) that each of these terms is a polynomial in $\xi_0, \ldots, \xi_h$, $\mathrm{T}_0, \ldots, \mathrm{T}_h$ with coefficients in $\mathbf{F}_p$, separately homogeneous (for $\pi$) of degree $p^h$ in the $\xi_i$ and the $\mathrm{T}_i$. The uniqueness of the polynomials $\mathrm{F}_h$ comes from the fact that one may take for $k$ an infinite extension of $\mathbf{F}_p$, since a polynomial in $h + 1$ indeterminates with coefficients in such a field is entirely determined by its values in $k^{h+1}$.

Observe in addition that the only term (19) which may contain $\xi_h$ corresponds to the case in which

$$\beta_0 = \beta_1 = \cdots = \beta_{h-1} = 0 \qquad \text{and} \qquad \beta_h = \varepsilon_0,$$

hence is equal to $\xi_h \mathrm{T}_0^{p^h}$, so that we may write

(20) $\mathrm{F}_h(\mathrm{X}_0, \ldots, \mathrm{X}_h; \mathrm{T}_0, \ldots, \mathrm{T}_h) = \mathrm{X}_h \mathrm{T}_0^{p^h} + \mathrm{F}'_h(\mathrm{X}_0, \ldots, \mathrm{X}_{h-1}; \mathrm{T}_0, \ldots, \mathrm{T}_h)$

where the homogeneous (for $\pi$) polynomial $\mathrm{F}'_h$ does not contain $\mathrm{X}_h$.

(ii) From (i) and relation (12) we deduce that, for any multiindex $\alpha \in \mathbf{N}^{(\mathbf{N})}$,

$$u(Z_\alpha) = \mathrm{G}_\alpha(\xi_0, \ldots, \xi_h, \ldots; \mathrm{T}_0, \ldots, \mathrm{T}_h, \ldots)$$

where $\mathrm{G}_\alpha$ is homogeneous of degree $\pi(\alpha)$, separately in the $\xi_i$ and in the $\mathrm{T}_i$, and has coefficients in $\mathbf{F}_p$. But by definition (§1, No. 2)

(21)

$$\begin{aligned}(u + v)(\mathrm{T}_h) &= \sum_{\beta+\gamma = p^h\varepsilon_0} u(Z_\beta)v(Z_\gamma) \\ &= \sum_{\beta+\gamma = p^h\varepsilon_0} \mathrm{G}_\beta(\xi_0, \ldots, \xi_h; \mathrm{T}_0, \ldots, \mathrm{T}_h)\mathrm{G}_\gamma(\eta_0, \ldots, \eta_h; \mathrm{T}_0, \ldots, \mathrm{T}_h)\end{aligned}$$

where, for reasons of degrees, $\mathrm{G}_\beta$ and $\mathrm{G}_\gamma$ do not contain terms in $\xi_i$ or $\mathrm{T}_i$ for $i > h$. Using (20), we see that, in the left-hand side of (21), the coefficient of $\mathrm{T}_0^{p^h}$ has the form

$$\zeta_h + \Phi_h(\zeta_0, \ldots, \zeta_{h-1}),$$

where $\Phi_h$ is a polynomial with coefficients in $\mathbf{F}_p$, homogeneous (for $\pi$) of degree $p^h$; if we write that this coefficient is equal to the coefficient of $\mathrm{T}_0^{p^h}$ in the right-hand side of (21), we see, by induction on $h$, that we have the first relation (18), with $\mathrm{S}_h$ homogeneous of degree $p^h$.

Similarly,

$$v(\mathrm{T}_h) = \mathrm{F}_h(\eta_0, \ldots, \eta_h; \mathrm{T}_0, \ldots, \mathrm{T}_h) = \sum_\alpha c_\alpha(\eta_0, \ldots, \eta_h)X_\alpha,$$

where $\alpha$ runs through the multiindices of degree $\pi(\alpha) = p^h$, and $c_\alpha$ is a polynomial with coefficients in $\mathbf{F}_p$, homogeneous (for $\pi$) of degree $p^h$; as we may write, for each $\alpha$,

$$u(X_\alpha) = H_\alpha(\xi_0, \ldots, \xi_h, \ldots; T_0, \ldots, T_h, \ldots),$$

where $H_\alpha$ has properties similar to the $G_\alpha$, we see that the coefficient of $T_0^{p^h}$ in $u(v(T_h))$ is a polynomial in the $\xi_i$ and $\eta_i$, separately homogeneous of degree $p^h$ in the $\xi_i$ and $\eta_i$; on the other hand that coefficient has by (20) the form

$$\theta_h + \Phi_h(\theta_1, \ldots, \theta_{h-1}),$$

and we conclude as above that we have the second relation (18), with $P_h$ homogeneous of degree $p^h$ separately in the $\xi_i$ and $\eta_i$. The uniqueness of the polynomials $S_h$ and $P_h$ is proved as in (i). This ends the proof of the corollary.

From the commutativity of $W(k)$, it follows that the polynomials $S_h$ and $P_h$ do not change by exchanging $T_i$ and $U_i$ for $0 \leqslant i \leqslant h$; furthermore, the associativity of addition and multiplication in $W(k)$ and the distributivity of multiplication on addition imply "associativity" identities:

$$\begin{aligned} &S_h(T_0, \ldots, T_h; S_0(U_0; V_0), \ldots, S_h(U_0, \ldots, U_h; V_0, \ldots, V_h)) \\ &\qquad = S_h(S_0(T_0; U_0), \ldots, S_h(T_0, \ldots, T_h; U_0, \ldots, U_h); V_0, \ldots, V_h) \\ &P_h(T_0, \ldots, T_h; P_0(U_0; V_0), \ldots, P_h(U_0, \ldots, U_h; V_0, \ldots, V_h)) \\ &\qquad = P_h(P_0(T_0; U_0), \ldots, P_h(T_0, \ldots, T_h; U_0, \ldots, U_h); V_0, \ldots, V_h) \end{aligned}$$

and "distributivity" identities:

$$\begin{aligned} &P_h(T_0, \ldots, T_h; S_0(U_0; V_0), \ldots, S_h(U_0, \ldots, U_h; V_0, \ldots, V_h)) \\ &\qquad = S_h(P_0(T_0; U_0), \ldots, P_h(T_0, \ldots, T_h; U_0, \ldots, U_h); P_0(T_0; V_0), \ldots, \\ &\qquad\quad P_h(T_0, \ldots, T_h; V_0, \ldots, V_h)) \end{aligned}$$

which are obtained by taking as above for $k$ an infinite extension of $\mathbf{F}_p$. Similarly, the existence of the inverse for addition in $W(k)$ implies the existence of a system of homogeneous polynomials $I_h(T_0, \ldots, T_h)$ of degree $p^h$ and with coefficients in $\mathbf{F}_p$, such that

$$S_h(T_0, \ldots, T_h; I_0(T_0), \ldots, I_h(T_0, \ldots, T_h)) = T_h \qquad \text{for every} \quad h \geqslant 0.$$

For any commutative algebra A over $\mathbf{F}_p$, the definition of W($k$) may be generalized, the *Witt ring* W(A) being the set of infinite sequences $(\xi_0, \ldots, \xi_h, \ldots)$ of elements of A (the "Witt vectors" over A), with the ring structure defined by

$$(\xi_i) + (\eta_i) = (\zeta_i), \qquad (\xi_i)(\eta_i) = (\theta_i),$$

where the $\zeta_i$ and $\theta_i$ are defined by (18). This definition is slightly different from the usual one, the polynomials $\mathrm{S}_h$ and $\mathrm{P}_h$ defined here being obtained from the original Witt polynomials by a "change of variables," which is described explicitly in [19] and [20].

**4.** *Structure of the ring* $\mathrm{End}(\mathfrak{G}_\infty(k))$. Let us first consider the Witt bigebra $\mathfrak{G}_\infty(\mathbf{F}_p)$ over the prime field $\mathbf{F}_p$; as the automorphism $\xi \mapsto \xi^p$ of $\mathbf{F}_p$ is the identity, the Frobenius homomorphism F and the shift V in $\mathfrak{G}_\infty(\mathbf{F}_p)$ are *linear* mappings. This enables one to define, for an arbitrary perfect field $k$ of characteristic $p$, *endomorphisms* of the Witt bigebra

$$\mathfrak{G}_\infty(k) = \mathfrak{G}_\infty(\mathbf{F}_p) \otimes_{\mathbf{F}_p} k,$$

by the formulas

$$\pi = \mathrm{F} \otimes 1 \tag{22}$$

$$\mathbf{t} = \mathrm{V} \otimes 1 \tag{23}$$

which are therefore such that $\pi(Z_\alpha) = Z_\alpha^p$ and $\mathbf{t}(Z_\alpha) = Z_{\alpha/p}$ for the basis of Theorem 1, and therefore coincide on that basis with the Frobenius homomorphism and the shift, respectively, but of course are distinct from these mappings when $k \neq \mathbf{F}_p$. From §1, Proposition 1 it follows that

$$\pi\mathbf{t} = \mathbf{t}\pi = \omega \tag{24}$$

in $\mathrm{End}(\mathfrak{G}_\infty(k))$. On the other hand, for $\xi \in k$, we have

$$e(\xi)(\pi(\mathrm{T}_h)) = \xi^{p^{h+1}}\mathrm{T}_h^p \qquad \text{and} \qquad \pi(e(\xi)(\mathrm{T}_h)) = \xi^{p^h}\mathrm{T}_h^p;$$

if $\sigma$ is the canonical automorphism of the Witt ring W($k$) which, to each Witt vector $(\xi_i)$ associates the Witt vector $(\xi_i^p)$, we see that, for every $x \in \mathrm{W}(k)$, we have in $\mathrm{End}(\mathfrak{G}_\infty)$

$$x\pi = \pi x^{\sigma}, \tag{25}$$

and one proves similarly that

$$\mathbf{t}x = x^{\sigma}\mathbf{t} \tag{26}$$

for $x \in \mathrm{W}(k)$.

**Theorem 3.** *Every element in the topological ring* $\mathrm{End}(\mathfrak{G}_{\infty}(k))$ *may be written uniquely as*

$$u = \sum_{n=0}^{\infty} x_n \mathbf{t}^n + \sum_{n=1}^{\infty} \pi^n y_n \tag{27}$$

*where the* $x_n$ *and* $y_n$ *are elements of* $\mathrm{W}(k)$ *and the sequence* $(y_n)$ *tends to* 0 *in* $\mathrm{W}(k)$. *Conversely, for any pair of sequences* $(x_n)$, $(y_n)$ *of elements of* $\mathrm{W}(k)$ *such that* $\lim_{n\to\infty} y_n = 0$, *the series on the right-hand side of* (27) *are convergent in* $\mathrm{End}(\mathfrak{G}_{\infty})$. *The ring* $\mathrm{End}(\mathfrak{G}_{\infty})$ *has no divisors of zero.*

We shall prove that any element $u \in \mathrm{End}(\mathfrak{G}_{\infty})$ may be written uniquely as

$$u = \sum_{n=0}^{\infty} z_n \mathbf{t}^n \tag{28}$$

with

$$z_n = \sum_{j=0}^{m_n} \pi^j e(\xi_{nj}) \tag{29}$$

where the $\xi_{nj} \in k$. From (24), (25), and (26), these formulas imply (27) with

$$x_n = e(\xi_{n0}) + e(\xi_{n+1,1}^{p^{-1}})\omega + \cdots + e(\xi_{n+h,h}^{p^{-h}})\omega^h + \cdots, \tag{30}$$

$$y_n = e(\xi_{0n}) + e(\xi_{1,n}^{p})\omega + \cdots + e(\xi_{h,n}^{p^h})\omega^h + \cdots. \tag{31}$$

As for each $h \geqslant 0$, there are only a finite number of indices $n$ such that $\xi_{hn} \neq 0$, these formulas will show that $\lim_{n\to\infty} y_n = 0$; conversely, it follows from Proposition 1 that the $x_n$ (resp. $y_n$) can be uniquely written in the form (30) (resp. (31)), and the assumption $\lim_{n\to\infty} y_n = 0$ then implies that, for each $h \geqslant 0$, the number of indices $n$ such that $\xi_{hn} \neq 0$ is finite; one can thus go back from (27) to (28) and (29), and the uniqueness of the $\xi_{nj}$ implies the uniqueness of the $x_n$ and $y_n$.

To prove existence and uniqueness of the $\xi_{nj}$, we use induction on $n$ and assume the $\xi_{nj}$ of index $n \leqslant r$ have been defined and are such that if $u_r$ is the sum of the terms of index $n \leqslant r$ in (28), then (with the notations of §1, No. 5), $u - u_r \in \mathfrak{N}_r$. The same argument as in the proof of Proposition 1 then shows that if $v = u - u_r$, we have $v(\mathrm{T}_h) = 0$ for $0 \leqslant h \leqslant r$, and $v(\mathrm{T}_{r+1}) \in \mathfrak{g}_0(\mathfrak{G}_\infty)$, which means that

$$v(\mathrm{T}_{r+1}) = \sum_{j=0}^{m} \xi_{r+1,j}\mathrm{T}_0^{p^j};$$

but this implies that $v - \sum_{j=0}^{m} \pi^j e(\xi_{r+1,j})\mathbf{t}^{r+1}$ belongs to the ideal $\mathfrak{N}_{r+1}$ and that the $\xi_{r+1,j}$ having that property are uniquely determined. This proves, by induction on $r$, the existence of the $\xi_{nj}$ satisfying (28) and (29); on the other hand, if we have a development (28), it is clear that $u - \sum_{n=0}^{r} z_n\mathbf{t}^n$ belongs to $\mathfrak{N}_r$, hence the preceding argument also proves the uniqueness of the $\xi_{nj}$ by induction on $r$.

When the $x_n$ are arbitrary and the $y_n$ tend to 0, the convergence of the two series on the right-hand side of (27) is obvious, since $\mathbf{t}^n$ tends to 0 when $n$ tends to $+\infty$.

Finally, suppose

$$u = z_r\mathbf{t}^r + z_{r+1}\mathbf{t}^{r+1} + \cdots, \qquad u' = z'_s\mathbf{t}^s + z'_{s+1}\mathbf{t}^{s+1} + \cdots$$
$$\text{with} \quad z_r \neq 0, \quad z'_s \neq 0;$$

if we prove that $z_r z_s'^{\sigma^r} \neq 0$, it will follow that in the development of $uu'$ in powers of $\mathbf{t}$, the term of lowest degree (in $\mathbf{t}^{r+s}$) will have a coefficient $\neq 0$ according to (26), and therefore $uu' \neq 0$. We thus have to check that if

$$z = \pi^i e(\xi_i) + \pi^{i+1}e(\xi_{i+1}) + \cdots, \qquad z' = \pi^j e(\xi'_j) + \pi^{j+1}e(\xi'_{j+1}) + \cdots$$
$$\text{with} \quad \xi_i \neq 0, \quad \xi'_j \neq 0,$$

then $zz' \neq 0$; but from (25) it follows that the term of lowest degree (in $\pi^{i+j}$) in $zz'$ has $e(\xi_i^{p^i}\xi_j) \neq 0$ as coefficient, which ends the proof.

## §3. Modules of hyperexponential vectors

**1.** *Hyperexponential vectors in a reduced commutative bigebra.* Among the polynomials $E_\alpha$ defined in §2, No. 3, formula (12), we will be especially interested in those for which $\alpha = m\varepsilon_0$, and we will simply write $E_m$ instead of $E_{m\varepsilon_0}$; we will say that the $E_m$ (for $m$ an integer $\geqslant 0$) are the *hyperexponential polynomials* with respect to the system of indeterminates $\mathbf{T} = (T_0, T_1, \ldots, T_h, \ldots)$. We shall first see that these polynomials are *characterized* by the following properties:

1° $E_m$ *has its coefficients in* $\mathbf{F}_p$ *and is homogeneous (for the degree* $\pi$) *of degree* $m$.

2° $E_{p^h}(\mathbf{T}) = T_h$.

3° *If* $\mathbf{U} = (U_0, U_1, \ldots, U_h, \ldots)$ *is a second system of indeterminates, and if, for each* $h \geqslant 0$, *we write*

$$W_h = \sum_{m=0}^{p^h} E_m(\mathbf{T})E_{p^h-m}(\mathbf{U}). \tag{1}$$

*then, for each* $m \geqslant 0$, *if we write* $\mathbf{W} = (W_0, W_1, \ldots, W_h, \ldots)$, *we have*

$$E_m(\mathbf{W}) = \sum_{i=0}^{m} E_i(\mathbf{T})E_{m-i}(\mathbf{U}). \tag{2}$$

Indeed, conditions 1° and 2° are trivially satisfied by the $E_m$; furthermore, it follows from (12) that in $\mathfrak{G}_\infty$

$$\Delta(T_h) = \sum_{m=0}^{p^h} E_m(\mathbf{T}) \otimes E_{p^h-m}(\mathbf{T})$$

which can also be written

$$\Delta(T_h) = \sum_{m=0}^{p^h} (E_m(\mathbf{T}) \otimes 1)(1 \otimes E_{p^h-m}(\mathbf{T}))$$

or equivalently

(3) $$\Delta(\mathrm{T}_h) = \sum_{m=0}^{p^h} \mathrm{E}_m(\mathbf{T} \otimes 1)\mathrm{E}_{p^h-m}(1 \otimes \mathbf{T})$$

where we note $\mathbf{T} \otimes 1$ (resp. $1 \otimes \mathbf{T}$) the system $(\mathrm{T}_0 \otimes 1, \mathrm{T}_1 \otimes 1, \ldots)$ (resp. $(1 \otimes \mathrm{T}_0, 1 \otimes \mathrm{T}_1, \ldots)$); similarly, noting that the $Z_\alpha$ form a privileged basis, we obtain, for each $m \geqslant 0$,

(4) $$\mathrm{E}_m(\Delta(\mathrm{T}_0), \Delta(\mathrm{T}_1), \ldots) = \sum_{i=0}^{m} \mathrm{E}_i(\mathbf{T} \otimes 1)\mathrm{E}_{m-i}(1 \otimes \mathbf{T}).$$

If we observe that, in the tensor product $\mathbf{F}_p[\mathbf{T}] \otimes \mathbf{F}_p[\mathbf{T}]$, the family of elements $\mathrm{T}_i \otimes 1$ and $1 \otimes \mathrm{T}_i$ (for all $i \geqslant 0$) is algebraically free, we obtain relation (2) when the $\mathrm{W}_h$ are defined by (1).

Conversely, suppose a system $(\mathrm{E}'_m)$ of polynomials satisfies the three conditions 1°, 2°, and 3°, and let us prove by induction on $m$ that $\mathrm{E}'_m = \mathrm{E}_m$. It follows from (4) applied both to $\mathrm{E}_m$ and to $\mathrm{E}'_m$ and from the inductive hypothesis that

$$\mathrm{E}_m(\mathbf{T}) - \mathrm{E}'_m(\mathbf{T}) \in \mathfrak{g}_0(\mathfrak{G}_\infty),$$

hence, by homogeneity, $\mathrm{E}_m(\mathbf{T}) - \mathrm{E}'_m(\mathbf{T})$ must be a monomial of type $c \cdot \mathrm{T}_0^{p^i}$ for some $i$; as $\mathrm{E}_m$ and $\mathrm{E}'_m$ are homogeneous of degree $m$, this would be possible only if $m = p^i$; but then it follows from 2° that $c = 0$, which ends our proof.

Let now $\mathfrak{G}$ be a *reduced* commutative bigebra over $k$; we will say that a sequence $\mathbf{Y} = (Y_0, Y_1, \ldots, Y_h, \ldots)$ of elements of $\mathfrak{G}$ is a *hyperexponential vector* of $\mathfrak{G}$ if there exists a bigebra homomorphism $u: \mathfrak{G}_\infty \to \mathfrak{G}$ such that $u(\mathrm{T}_h) = Y_h$ for all $h \geqslant 0$ (a condition which we will simply write as $u(\mathbf{T}) = \mathbf{Y}$); such a homomorphism is of course unique, and there is therefore a natural bijection of $\mathrm{Hom}_{k\text{-big.}}(\mathfrak{G}_\infty, \mathfrak{G})$ onto the set $\mathscr{E}(\mathfrak{G})$ of the hyperexponential vectors of $\mathfrak{G}$. As $\mathrm{Hom}_{k\text{-big.}}(\mathfrak{G}_\infty, \mathfrak{G})$ is naturally equipped with a structure of *commutative group* (§1, No. 2), this structure may be transferred to $\mathscr{E}(\mathfrak{G})$ by the preceding bijection. From the definition of addition in $\mathrm{Hom}_{k\text{-big.}}(\mathfrak{G}_\infty, \mathfrak{G})$ (§1, No. 2) it follows that if $\mathbf{Y} = (Y_i)$, $\mathbf{Y}' = (Y'_i)$ are two hyperexponential vectors in $\mathfrak{G}$, then $\mathbf{Y}'' = \mathbf{Y} + \mathbf{Y}' = (Y''_i)$ is given by

(5) $$Y''_h = \sum_{m=0}^{p^h} \mathrm{E}_m(\mathbf{Y})\mathrm{E}_{p^h-m}(\mathbf{Y}') \qquad \text{for every} \quad h \geqslant 0.$$

**2.** *Modules of hyperexponential vectors.* From now on, we will denote by $\mathscr{A}$, or $\mathscr{A}(k)$, the *subring* of $\mathrm{End}(\mathfrak{G}_\infty(k))$ consisting of the elements

$$u = x_0 + x_1\mathbf{t} + x_2\mathbf{t}^2 + \cdots + x_m\mathbf{t}^m + \cdots \tag{6}$$

with coefficients $x_m \in \mathrm{W}(k)$, i.e., the elements for which, in formula (27) of §2, Theorem 3, the $y_n$ are all 0. It follows from §2, Theorem 3, that $\mathscr{A}$ is a *noncommutative* topological ring, complete for the topology induced by that of $\mathrm{End}(\mathfrak{G}_\infty)$, and without zero divisors.

For any reduced commutative bigebra $\mathfrak{G}$, the set $\mathscr{E}(\mathfrak{G})$ of hyperexponential vectors of $\mathfrak{G}$ is naturally equipped with a structure of *right $\mathscr{A}$-module*: if $\mathbf{Y} = u(\mathbf{T})$ is an element of $\mathscr{E}(\mathfrak{G})$, we define, for any $v \in \mathscr{A}$,

$$\mathbf{Y} \cdot v = u(v(\mathbf{T})) \tag{7}$$

which by definition belongs to $\mathscr{E}(\mathfrak{G})$; the verification of the axioms of $\mathscr{A}$-modules is trivial, due to formulas (2) and (3) of §1, No. 2. More generally, we may replace in that definition the subring $\mathscr{A}$ by the full ring $\mathrm{End}(\mathfrak{G}_\infty)$. This remark enables one to prove immediately the first part of the following proposition:

**Proposition 1.** *For the $\mathscr{A}$-module structure of $\mathscr{E}(\mathfrak{G})$,*

$$\mathscr{E}(\mathfrak{G}) \cdot \omega \subset \mathscr{E}(\mathfrak{G}) \cdot \mathbf{t} \tag{8}$$

*and the relation $\mathbf{Y} \cdot \mathbf{t} = 0$ for an element $\mathbf{Y} \in \mathscr{E}(\mathfrak{G})$ is equivalent to $\mathbf{Y} = 0$.*

Indeed, the first statement follows at once from the fact that in $\mathrm{End}(\mathfrak{G}_\infty)$ we have $\omega = \boldsymbol{\pi}\mathbf{t}$ (§2, formula (24)). To prove the second statement, we may assume $\mathbf{Y} = u(\mathbf{T})$; the relation $u(\mathbf{t}(\mathbf{T})) = 0$ implies by definition that

$$u(\mathbf{t}(\mathrm{T}_h)) = u(\mathrm{T}_{h-1}) = 0 \qquad \text{for every} \quad h \geqslant 1,$$

hence $Y_h = 0$ for every $h \geqslant 0$.

**Proposition 2.** *Let* M *be a set of hyperexponential vectors of a reduced commutative bigebra $\mathfrak{G}$. Then the subalgebra* B *of $\mathfrak{G}$ generated by the components of all hyperexponential vectors in* M *is a reduced subbigebra of $\mathfrak{G}$.*

Indeed, from the definition of hyperexponential vectors and from formula (5), we deduce that for any $\mathbf{Y} = (Y_0, Y_1, \ldots, Y_h, \ldots)$ in $\mathscr{E}(\mathfrak{G})$, we have

$$\Delta Y_h = \sum_{0 \leqslant m \leqslant p^h} \mathrm{E}_m(\mathbf{Y} \otimes 1)\mathrm{E}_{p^h - m}(1 \otimes \mathbf{Y}), \tag{9}$$

hence $\Delta Y_h$ belongs to $\mathrm{B} \otimes \mathrm{B}$, which proves that B is a cogebra, hence a bigebra (I, §3, No. 3). Furthermore, as a bigebra homomorphism commutes with the shifts (II, §2, No. 7) and $\mathrm{V}(\mathrm{T}_{h+1}) = \mathrm{T}_h$ in $\mathfrak{G}_\infty$, we also have $\mathrm{V}(Y_{h+1}) = Y_h$ in $\mathfrak{G}$, which proves that $\mathrm{V}(\mathrm{B}) = \mathrm{B}$ and B is reduced.

We may define on $\mathrm{Hom}_{k\text{-big.}}(\mathfrak{G}_\infty, \mathfrak{G})$ (or on $\mathscr{E}(\mathfrak{G})$, which amounts to the same thing) a *topology*, by the same process as in §1, No. 5: for each integer $r \geqslant 0$, $\mathrm{P}_r$ is the set of all bigebra homomorphisms $u: \mathfrak{G}_\infty \to \mathfrak{G}$ such that $u$ is 0 in $\mathfrak{s}_r(\mathfrak{G}_\infty) \cap \mathfrak{G}_\infty^+$; it is clear that the $\mathrm{P}_r$ are *sub-$\mathscr{A}$-modules* of $\mathrm{Hom}_{k\text{-big.}}(\mathfrak{G}_\infty, \mathfrak{G})$, and if one takes them as neighborhoods of 0, one obtain a structure of *topological $\mathscr{A}$-module* on $\mathrm{Hom}_{k\text{-big.}}(\mathfrak{G}_\infty, \mathfrak{G})$; the verification of the axioms is trivial.

**3.** *Functorial properties of modules of hyperexponential vectors.* Consider now two reduced commutative bigebras $\mathfrak{G}$, $\mathfrak{G}'$. For every $v \in \mathrm{Hom}_{k\text{-big.}}(\mathfrak{G}, \mathfrak{G}')$, $\tilde{v}: u \mapsto v \circ u$ is a homomorphism of right $\mathscr{A}$-modules of $\mathrm{Hom}_{k\text{-big.}}(\mathfrak{G}_\infty, \mathfrak{G})$ into $\mathrm{Hom}_{k\text{-big.}}(\mathfrak{G}_\infty, \mathfrak{G}')$, which we may also consider as a right $\mathscr{A}$-module homomorphism $\tilde{v}: \mathscr{E}(\mathfrak{G}) \to \mathscr{E}(\mathfrak{G}')$. Furthermore, if $\mathfrak{G}''$ is a third reduced commutative bigebra and $v' \in \mathrm{Hom}_{k\text{-big.}}(\mathfrak{G}', \mathfrak{G}'')$, $v'' = v' \circ v$, we have $\tilde{v}'' = \tilde{v}' \circ \tilde{v}$. When $\mathscr{E}(\mathfrak{G})$ and $\mathscr{E}(\mathfrak{G}')$ are equipped with the topologies defined above, the definition of $\tilde{v}$ shows that it is a *continuous* homomorphism of $\mathscr{A}$-modules.

**Proposition 3.** *If $\mathfrak{G}$ and $\mathfrak{G}'$ are two reduced commutative bigebras, the $\mathscr{A}$-module $\mathscr{E}(\mathfrak{G} \otimes \mathfrak{G}')$ is naturally isomorphic to the direct sum $\mathscr{E}(\mathfrak{G}) \oplus \mathscr{E}(\mathfrak{G}')$.*

Let

$$i: \mathfrak{G} \to \mathfrak{G} \otimes \mathfrak{G}', \qquad j: \mathfrak{G}' \to \mathfrak{G} \otimes \mathfrak{G}'$$

be the natural injections,

$$\rho: \mathfrak{G} \otimes \mathfrak{G}' \to \mathfrak{G}, \qquad \sigma: \mathfrak{G} \otimes \mathfrak{G}' \to \mathfrak{G}'$$

the natural surjections (I, §3, No. 12). One readily verifies that, for any bigebra homomorphism $u\colon \mathfrak{G}_\infty \to \mathfrak{G} \otimes \mathfrak{G}'$, one has

$$u = i \circ \rho \circ u + j \circ \sigma \circ u$$

in the group $\mathrm{Hom}_{k\text{-big.}}(\mathfrak{G}_\infty, \mathfrak{G} \otimes \mathfrak{G}')$. This implies that

$$(v, w) \mapsto i \circ v + j \circ w$$

is a bijection of the group

$$\mathrm{Hom}_{k\text{-big.}}(\mathfrak{G}_\infty, \mathfrak{G}) \oplus \mathrm{Hom}_{k\text{-big.}}(\mathfrak{G}_\infty, \mathfrak{G}')$$

onto the group $\mathrm{Hom}_{k\text{-big.}}(\mathfrak{G}_\infty, \mathfrak{G} \otimes \mathfrak{G}')$, compatible with products on the right by elements of $\mathrm{End}(\mathfrak{G}_\infty)$; hence the result.

**4.** *Equivalence of the category of finite dimensional reduced commutative bigebras and of the category of their* $\mathscr{A}$*-modules of hyperexponential vectors.* Let $\mathfrak{G}$ be a reduced commutative bigebra, and let $(X_{0i})_{i\in \mathrm{I}}$ be a basis of its Lie algebra $\mathfrak{g}_0(\mathfrak{G})$. The result of §2, Theorem 2 may be expressed in the following way: for each index $i \in \mathrm{I}$, there exists a bigebra homomorphism $u_i\colon \mathfrak{G}_\infty \to \mathfrak{G}$ such that $u_i(\mathrm{T}_0) = X'_{0i}$; furthermore, if for every $m > 0$, we write

$$Z'_{m\varepsilon_i} = u_i(Z_{m\varepsilon_0}) = u_i(\mathrm{E}_m(\mathbf{T}))$$

and, for $\alpha \in \mathbf{N}^{(\mathrm{I})}$,

$$Z'_\alpha = \prod_{i\in \mathrm{I}} Z'_{\alpha(i)\varepsilon_i},$$

then $(Z'_\alpha)$ is a privileged basis of $\mathfrak{G}$.

**Proposition 4.** *Suppose* I *is finite. Then the hyperexponential vectors*

$$\mathbf{X}'_i = (X'_{0i}, X'_{1i}, \ldots, X'_{hi}, \ldots)$$

*form a system of generators of the* $\mathscr{A}$*-module* $\mathscr{E}(\mathfrak{G})$.

Let $u$ be an element of $\mathrm{Hom}_{k\text{-big.}}(\mathfrak{G}_\infty, \mathfrak{G})$; suppose we have proved that, for an integer $r \geqslant 0$, there exist elements

$$v_{ir} = \sum_{h=0}^{r} e(\xi_{hi})\mathbf{t}^h$$

of $\mathscr{A}$ such that the homomorphism

$$w = u - \sum_{i \in \mathrm{I}} u_i \circ v_{ir}$$

in $\mathrm{Hom}_{k\text{-big.}}(\mathfrak{G}_\infty, \mathfrak{G})$ satisfies the condition $w(\mathrm{T}_h) = 0$ for $0 \leqslant h \leqslant r$. This implies $w(Z_{m\varepsilon_0}) = 0$ for $1 \leqslant m < p^{r+1}$, hence

$$\Delta(w(\mathrm{T}_{r+1})) = (w \otimes w)(\Delta(\mathrm{T}_{r+1})) = 1 \otimes w(\mathrm{T}_{r+1}) + w(\mathrm{T}_{r+1}) \otimes 1,$$

in other words $w(\mathrm{T}_{r+1}) \in \mathfrak{g}_0(\mathfrak{G})$; therefore there exist elements $\xi_{r+1,i} \in k$ such that

$$w(\mathrm{T}_{r+1}) = \sum_{i \in \mathrm{I}} \xi_{r+1,i} X'_{0i}.$$

This proves that the homomorphisms $w$ and $\sum_{i \in \mathrm{I}} u_i \circ e(\xi_{r+1,i})\mathbf{t}^{r+1}$ coincide in $\mathfrak{s}_{r+1}(\mathfrak{G}_\infty)$, and the elements

$$v_{i,r+1} = \sum_{h=0}^{r+1} e(\xi_{hi})\mathbf{t}^h$$

are such that the homomorphism

$$w' = u - \sum_{i \in \mathrm{I}} u_i \circ v_{i,r+1}$$

satisfies $w'(\mathrm{T}_h) = 0$ for $0 \leqslant h \leqslant r + 1$. It is then clear that each of the sequences $(v_{ir})_{r \geqslant 0}$ converges to an element $v_i \in \mathscr{A}$, and that $u = \sum_{i \in \mathrm{I}} u_i \circ v_i$, which ends the proof.

*Remark.* If I is infinite, the same argument shows that one may still write $u = \sum_{i \in \mathrm{I}} u_i \circ v_i$ with $\lim v_i = 0$, the limit being taken for the filter of complements of finite sets of I; the sum is taken for the topology on $\mathrm{Hom}_{k\text{-big.}}(\mathfrak{G}_\infty, \mathfrak{G})$ defined in No. 2.

It follows from the definitions (No. 2) that the $\mathscr{A}$-module $\mathscr{E}(\mathfrak{G}_\infty)$ is isomorphic to $\mathrm{End}(\mathfrak{G}_\infty)$ considered as a right $\mathscr{A}$-module; we will denote by $\mathscr{E}'(\mathfrak{G}_\infty)$ the *sub-$\mathscr{A}$-module* of $\mathscr{E}(\mathfrak{G}_\infty)$ generated by

$$\mathbf{T} = (\mathrm{T}_0, \mathrm{T}_1, \ldots, \mathrm{T}_h, \ldots),$$

which is of course isomorphic to $\mathscr{A}$ considered as a right $\mathscr{A}$-module.

For an integer $n > 0$, the sub-$\mathscr{A}$-module of $\mathscr{E}(\mathfrak{G}_\infty^{\otimes n})$ generated by the $\mathbf{T}_i = (\mathrm{T}_{hi})_{h \geqslant 0}$ (with the notations of Theorem 2 of §2) has the $\mathbf{T}_i$ as a *basis* and is therefore isomorphic to $(\mathscr{E}'(\mathfrak{G}_\infty))^n$ with which we identify it. With respect to the basis $(\mathbf{T}_i)_{1 \leqslant i \leqslant n}$, every endomorphism of the $\mathscr{A}$-module $(\mathscr{E}'(\mathfrak{G}_\infty))^n$ is therefore defined by a square matrix $V = (v_{ji})$ of order $n$ with elements in the ring $\mathscr{A}$.

**Theorem 1.** (i) *If $\mathfrak{G}$ is a reduced commutative bigebra of finite dimension n, the right $\mathscr{A}$-module $\mathscr{E}(\mathfrak{G})$ is isomorphic to the quotient of the $\mathscr{A}$-module $(\mathscr{E}'(\mathfrak{G}_\infty))^n$ by a submodule* M, *which is the image of $(\mathscr{E}'(\mathfrak{G}_\infty))^n$ by an endomorphism of $(\mathscr{E}'(\mathfrak{G}_\infty))^n$ whose matrix with respect to $(\mathbf{T}_i)_{1 \leqslant i \leqslant n}$ has the form $I\omega - V\mathbf{t}$, where $I$ is the unit matrix and $V$ a square matrix of order n with elements in $\mathscr{A}$.*

(ii) *Conversely, for any square matrix V of order n with elements in $\mathscr{A}$, if* M *is the image of $(\mathscr{E}'(\mathfrak{G}_\infty))^n$ by the endomorphism of that $\mathscr{A}$-module whose matrix with respect to $(\mathbf{T}_i)$ is $I\omega - V\mathbf{t}$, then there exists a reduced commutative bigebra $\mathfrak{G}$ of dimension n, such that the right $\mathscr{A}$-module $\mathscr{E}(\mathfrak{G})$ is isomorphic to $(\mathscr{E}'(\mathfrak{G}_\infty))^n/\mathrm{M}$.*

(i) The homomorphisms $u_i$ defined at the beginning of this section can be considered as composed of a surjective homomorphism $u\colon \mathfrak{G}_\infty^{\otimes n} \to \mathfrak{G}$ such that

$$u(Z_\alpha) = Z'_\alpha \qquad \text{for} \quad \alpha \in \mathbf{N}^{\mathrm{I}},$$

and of the natural injections $\mathfrak{G}_\infty \to \mathfrak{G}_\infty^{\otimes n}$ sending $\mathbf{T}$ onto $\mathbf{T}_i$ for $i \in \mathrm{I}$. By No. 3, we deduce from $u$ an $\mathscr{A}$-module homomorphism $(\mathscr{E}'(\mathfrak{G}_\infty))^n \to \mathscr{E}(\mathfrak{G})$ which we will again write $\tilde{u}$, and which is *surjective* by Proposition 4; we are going to determine its *kernel* M.

We note that, for each $i \in \mathrm{I}$, $u_i \circ \pi$ is a bigebra homomorphism of $\mathfrak{G}_\infty$ into $\mathfrak{G}$, hence $(X'^p_{0i}, X'^p_{1i}, \ldots, X'^p_{hi}, \ldots)$ is a hyperexponential vector of $\mathscr{E}(\mathfrak{G})$; Proposition 4 shows that there exist elements $v_{ji} \in \mathscr{A}$ such that, for $i \in \mathrm{I}$,

$$(10) \qquad (X'^p_{0i}, \ldots, X'^p_{hi}, \ldots) = \sum_j \mathbf{X}'_j v_{ji}.$$

But from Proposition 1 it follows that we have equivalent relations by multiplying both sides on the right by $\mathbf{t}$, which gives, by §2, formula (24),

$$(11) \qquad \tilde{u}(\mathbf{T}_i\omega) = \sum_j \tilde{u}(\mathbf{T}_j v_{ji}\mathbf{t})$$

and therefore the elements $\mathbf{T}_i\omega - \sum_j \mathbf{T}_j v_{ji}\mathbf{t}$, or, equivalently, the vectors $(I\omega - V\mathbf{t}) \cdot \mathbf{T}_i$ with $V = (v_{ji})$, belong to the kernel of $\tilde{u}$ (as usual vectors of $(\mathscr{E}'(\mathfrak{G}_\infty))^n$ are identified with one column matrices.)

Conversely, suppose that $\tilde{u}(\sum_i \mathbf{T}_i w_i) = 0$ for a family $(w_i)_{i \in \mathrm{I}}$ of elements of $\mathscr{A}$; we will prove that, for each $i \in \mathrm{I}$, there exists a sequence $(z'_{mi})_{m \geqslant 0}$ of elements of $\mathrm{W}(k)$ such that if, for each integer $m$, we write

$$(12) \qquad w'_{mi} = \sum_{h<m} z'_{hi}\mathbf{t}^h$$

then, for each pair $(i, m)$, there exists an element $w''_{mi} \in \mathscr{A}$ for which

$$(13) \qquad \sum_i \mathbf{T}_i w_i = (I\omega - V\mathbf{t}) \cdot (\sum_i \mathbf{T}_i w'_{mi}) + \sum_i \mathbf{T}_i w''_{mi}\mathbf{t}^m.$$

As it is obvious from (12) that the sequence $(w'_{mi})_{m \geqslant 0}$ converges to an element $w'_i \in \mathscr{A}$, this will prove that

$$\sum_i \mathbf{T}_i w_i = (I\omega - V\mathbf{t}) \cdot (\sum_i \mathbf{T}_i w'_i),$$

and therefore will establish that $\mathrm{M}$ is the image of $(\mathscr{E}'(\mathfrak{G}_\infty))^n$ by the endomorphism of matrix $I\omega - V\mathbf{t}$.

We use induction on $m$, as the assertion is trivial for $m = 0$. Starting from relation (13), write

$$w''_{mi} = z_{0i} + z_{1i}\mathbf{t} + \cdots + z_{hi}\mathbf{t}^h + \cdots \qquad \text{with} \quad z_{hi} \in \mathrm{W}(k);$$

by definition, the hyperexponential vector

$$\mathbf{Y} = (Y_0, Y_1, \ldots, Y_h, \ldots) = \sum_i \mathbf{T}_i w''_{mi}\mathbf{t}^m$$

of $\mathfrak{G}_\infty^{\otimes n}$ is such that $Y_h = 0$ for $h < m$, and if we write

$$z_{0i} = \sum_{h=0}^{\infty} \zeta_{h0i}\omega^h,$$

we have $Y_m = \sum_i \zeta_{00i}\mathrm{T}_{0i}$. The assumption $\tilde{u}(\mathbf{Y}) = 0$ yields the relation $\sum_i \zeta_{00i}X'_{0i} = 0$, hence $\zeta_{00i} = 0$ for $i \in \mathrm{I}$; we may therefore write $z_{0i} = z'_{mi}\omega$ for some $z'_{mi} \in \mathrm{W}(k)$, hence

$$\sum_i \mathbf{T}_i w''_{mi}\mathbf{t}^m = (I\omega - V\mathbf{t}) \cdot (\sum_i \mathbf{T}_i z'_{mi}\mathbf{t}^m) + \sum_i \mathbf{T}_i w''_{m+1,i}\mathbf{t}^{m+1}$$

for suitable elements $w''_{m+1,i} \in \mathscr{A}$; the induction may thus proceed and this ends the proof of (i).

(ii) Let $\mathfrak{N}$ be the subbigebra of $\mathfrak{G}_\infty^{\otimes n}$ generated by the components of all hyperexponential vectors $\mathbf{Y} \in \mathrm{M}$; Proposition 2 shows that $\mathfrak{N}$ is a reduced subbigebra of $\mathfrak{G}_\infty^{\otimes n}$, which is invariant (as all subbigebras of a commutative bigebra). Let P be the ideal of $\mathfrak{G}_\infty^{\otimes n}$ generated by $\mathfrak{N}^+$, $\mathfrak{G}$ the quotient bigebra $\mathfrak{G}_\infty^{\otimes n}/\mathrm{P}$, and $u\colon \mathfrak{G}_\infty^{\otimes n} \to \mathfrak{G}$ the natural homomorphism; let us write $Z'_\alpha = u(Z_\alpha)$ for every $\alpha \in \mathbf{N}^{(\mathrm{I})}$, and in particular

$$X'_{hi} = u(\mathrm{T}_{hi}) \qquad \text{for} \quad h \geqslant 0, \quad i \in \mathrm{I}.$$

We are going to prove that the $Z'_\alpha$ form a *privileged basis* for $\mathfrak{G}$. Write as usual $\alpha = \alpha_0 + p\alpha_1 + \cdots + p^r\alpha_r$, where the $\alpha_j$ have height 0, and

$$X'_\alpha = \prod_{h,i} X'^{\alpha_h(i)}_{hi};$$

as $u$ is a bigebra homomorphism and the subbigebras $\mathfrak{s}_r(\mathfrak{G})$ are finite-dimensional vector spaces over $k$, it will be enough (by II, §2, No. 7, Proposition 2) to show that, for every integer $r \geqslant 0$, the $X'_\alpha$ such that $\mathrm{ht}(\alpha) \leqslant r$ form a *basis* for the vector space $\mathfrak{s}_r(\mathfrak{G})$.

Let us write $\mathbf{Y}_i = (I\omega - V\mathbf{t}) \cdot \mathbf{T}_i$, and let $Y_{hi}$ be the component of index $h$ of the hyperexponential vector $\mathbf{Y}_i$; it follows from the definitions that the component of index $h$ of $V \cdot \mathbf{T}_i$ is a polynomial with respect to the $\mathrm{T}_{ij}$, of degree $\leqslant p^h$ (for the graduation $\pi$), which therefore can contain the $\mathrm{T}_{hj}$ for $1 \leqslant j \leqslant n$ only in the *first degree*, and does not contain any $\mathrm{T}_{lj}$ with $l > h$. We may thus write

$$(14)\qquad Y_{hi} = \mathrm{T}_{h-1,i} - \sum_{j=1}^{n} \beta_{hij}\mathrm{T}_{h-1,j}^{p} - \mathrm{G}_{hi}(\mathrm{T}_{01}, \ldots, \mathrm{T}_{0n}, \ldots, \mathrm{T}_{h-2,1}, \ldots, \mathrm{T}_{h-2,n})$$

where the $\beta_{hij}$ are scalars in $k$ and $\mathrm{G}_{hi}$ is a polynomial with coefficients in $k$.

The ideal P is therefore generated by the polynomials on the right-hand side of (14), and for every $r \geqslant 0$, the ideal $\mathrm{P}_r = \mathrm{P} \cap \mathfrak{s}_r(\mathfrak{G}_\infty^{\otimes n})$ of the algebra $\mathfrak{s}_r(\mathfrak{G}_\infty^{\otimes n})$ is generated by the right-hand sides of the relations (14) for $h - 1 \leqslant r$, as follows from the fact that the image of $\mathfrak{s}_r(\mathfrak{G}_\infty^{\otimes n})$ by $u$ is $\mathfrak{s}_r(\mathfrak{G})$ (II, §2, No. 4) since $\mathfrak{N}$ is reduced. We will use the following elementary lemma:

**Lemma 1.** *Let* C *be a commutative ring (with unit), and let* $\mathfrak{b}$ *be the ideal of the polynomial ring* $\mathrm{C}[\mathrm{T}_1, \ldots, \mathrm{T}_n]$ *generated by the n polynomials*

$$\mathrm{Q}_i = \mathrm{T}_i^m - \sum_{j=1}^{n} b_{ij}\mathrm{T}_j - c_i \qquad (1 \leqslant i \leqslant n)$$

*where m is an integer* $> 1$, *the* $b_{ij}$ *and* $c_i$ *elements of* C. *Then, if* $\xi_i$ *is the class of* $\mathrm{T}_i$ mod $\mathfrak{b}$, *the monomials* $\xi_1^{\mu_1} \cdots \xi_n^{\mu_n}$ *for* $0 \leqslant \mu_i < m$, *constitute a basis of the* C-*module* $\mathrm{C}[\mathrm{T}_1, \ldots, \mathrm{T}_n]/\mathfrak{b}$.

This lemma will successively be applied for $m = p$ and $\mathrm{C} = k$, $\mathrm{C} = \mathfrak{s}_0(\mathfrak{G}), \ldots, \mathrm{C} = \mathfrak{s}_{r-1}(\mathfrak{G})$, and will then prove by induction that the $X'_\alpha$ such that $\mathrm{ht}(\alpha) \leqslant r$ constitute a basis for $\mathfrak{s}_r(\mathfrak{G})$.

To prove the lemma, consider the polynomials

$$\mathrm{R}_{\mu\nu} = \mathrm{T}_1^{\mu_1} \cdots \mathrm{T}_n^{\mu_n}\mathrm{Q}_1^{\nu_1} \cdots \mathrm{Q}_n^{\nu_n}$$

for multiindices $\mu$, $\nu$ such that $0 \leqslant \mu_i < m$ for every $i$, $\nu_j \geqslant 0$ for all $j$. The term of highest total degree in $\mathrm{R}_{\mu\nu}$ is obviously $\mathrm{T}_1^{m\nu_1+\mu_1} \cdots \mathrm{T}_n^{m\nu_n+\mu_n}$, and it follows at once that if $(\mu, \nu) \neq (\mu', \nu')$, these terms in $\mathrm{R}_{\mu\nu}$ and $\mathrm{R}_{\mu'\nu'}$ cannot be the same; this immediately proves that the $\mathrm{R}_{\mu\nu}$ are linearly independent over C. On the other hand, as one may write

$$\mathrm{T}_i^m = \mathrm{Q}_i + \sum_{j=1}^{n} b_{ij}\mathrm{T}_j + c_i,$$

one proves at once by induction on $r$ that any monomial in the $\mathrm{T}_i$, of total degree $\leqslant r$, is a linear combination of the $R_{\mu\nu}$ such that $m|\nu| + |\mu| \leqslant r$. Therefore the $R_{\mu\nu}$ constitute a *basis* of the C-module $C[T_1, \ldots, T_n]$; the lemma is an immediate consequence of that fact.

Let us now end the proof of part (ii) of Theorem 1. From the definition of $u$ it follows that we have the relations (11), hence also the relations (10) which are equivalent to (11). The argument made in (i) then proves that the kernel of the $\mathscr{A}$-module homomorphism

$$\tilde{u} : (\mathscr{E}'(\mathfrak{G}_\infty))^n \to \mathscr{E}(\mathfrak{G})$$

deduced from $u$ is the image of $(\mathscr{E}'(\mathfrak{G}_\infty))^n$ by the matrix $I\omega - V\mathbf{t}$, i.e., the given submodule M. As $u$ is surjective by Proposition 4, this ends the proof.

**Theorem 2.** *Let $\mathfrak{G}$, $\mathfrak{G}'$ be two reduced commutative bigebras, $\mathfrak{G}$ being finite dimensional. The natural mapping $v \mapsto \tilde{v}$ defined in No. 3 is an isomorphism of the group* $\mathrm{Hom}_{k\text{-big.}}(\mathfrak{G}, \mathfrak{G}')$ *onto the group*

$$\mathrm{Hom}_{\mathscr{A}\text{-mod.}}(\mathscr{E}(\mathfrak{G}), \mathscr{E}(\mathfrak{G}')).$$

We note first that if $\tilde{v} = 0$, this implies in particular that $v \circ u_i = 0$ for the homomorphisms $u_i \colon \mathfrak{G}_\infty \to \mathfrak{G}$ defined at the beginning of this section; hence, with the same notations, $v(Z'_\alpha) = 0$ for all $\alpha \neq 0$, which means by definition that $v = 0$. The mapping $v \mapsto \tilde{v}$ being thus injective, we have to prove that it is surjective. Suppose given an $\mathscr{A}$-module homomorphism $f\colon \mathscr{E}(\mathfrak{G}) \to \mathscr{E}(\mathfrak{G}')$; if $u\colon \mathfrak{G}_\infty^{\otimes n} \to \mathfrak{G}$ is defined as in Theorem 1, we obtain by composition an $\mathscr{A}$-module homomorphism

$$g = f \circ \tilde{u} : (\mathscr{E}'(\mathfrak{G}_\infty))^n \to \mathscr{E}(\mathfrak{G}').$$

If we write

$$\mathbf{Y}_i = g(\mathbf{T}_i) = (Y_{hi})_{h \geqslant 0} \qquad \text{for} \quad 1 \leqslant i \leqslant n = \dim(\mathfrak{G}),$$

the $\mathbf{Y}_i$ are hyperexponential vectors in $\mathscr{E}(\mathfrak{G}')$, and therefore it follows from the definition of hyperexponential vectors that the $k$-algebra homomorphism $w\colon \mathfrak{G}_\infty^{\otimes n} \to \mathfrak{G}'$ defined by $w(T_{hi}) = Y_{hi}$ for $h \geqslant 0$ and $1 \leqslant i \leqslant n$ is a *bigebra* homomorphism such that $\tilde{w} = g$. Furthermore, it follows from the definition of $g$ that this homomorphism is 0 in the kernel M of $\tilde{u}$;

with the notations of the proof of Theorem 1, $w$ is therefore 0 for all elements of $\mathfrak{N}^+$, hence also for all elements of the ideal P generated by $\mathfrak{N}^+$; this implies the factorization $w = v \circ u$ for a bigebra homomorphism $v: \mathfrak{G} \to \mathfrak{G}'$ such that, by definition, $\tilde{v}$ and $f$ coincide on the hyperexponential vectors $\mathbf{X}_i' = (X_{0i}', \ldots, X_{hi}', \ldots)$ of $\mathscr{E}(\mathfrak{G})$; as these vectors generate $\mathscr{E}(\mathfrak{G})$ by Proposition 4, we have $\tilde{v} = f$. Q.E.D.

We may express Theorems 1 and 2 by saying that the category of reduced finite-dimensional commutative bigebras over $k$ is *equivalent* to the category of the *$\mathscr{A}$-modules of hyperexponential vectors* of these bigebras. In §§4 and 5, we will characterize intrinsically the category of these $\mathscr{A}$-modules.

With the notations of Theorem 2, suppose both $\mathfrak{G}$ and $\mathfrak{G}'$ have finite dimensions $n$ and $m$. We may write, by Theorem 1,

$$\mathscr{E}(\mathfrak{G}) = (\mathscr{E}'(\mathfrak{G}_\infty))^n/\mathrm{M}, \qquad \mathscr{E}(\mathfrak{G}') = (\mathscr{E}'(\mathfrak{G}_\infty))^m/\mathrm{M}',$$

where M (resp. M′) is the image of $(\mathscr{E}'(\mathfrak{G}_\infty))^n$ (resp. $(\mathscr{E}'(\mathfrak{G}_\infty))^m$) by an endomorphism having as matrix $I_n\omega - V\mathbf{t}$ (resp. $I_m\omega - V'\mathbf{t}$), where $V$ (resp. $V'$) is a square matrix of order $n$ (resp. $m$) with elements in $\mathscr{A}$. Any $\mathscr{A}$-module homomorphism $f: \mathscr{E}(\mathfrak{G}) \to \mathscr{E}(\mathfrak{G}')$ may be considered as obtained from a (nonuniquely determined) homomorphism

$$\Phi: (\mathscr{E}'(\mathfrak{G}_\infty))^n \to (\mathscr{E}'(\mathfrak{G}_\infty))^m$$

such that $\Phi(\mathrm{M}) \subset \mathrm{M}'$, by passage to the quotients. To each such homomorphism $\Phi$ corresponds uniquely an $m \times n$ matrix $W$ with elements in $\mathscr{A}$, and the relation $\Phi(\mathrm{M}) \subset \mathrm{M}'$ is equivalent to the existence of another $m \times n$ matrix $W_1$ with elements in $\mathscr{A}$, satisfying the relation

$$W \cdot (I_n\omega - V\mathbf{t}) = (I_m\omega - V'\mathbf{t}) \cdot W_1. \tag{15}$$

With these notations, the bigebra $\mathfrak{G}^{(r)}$ ($r$ positive or negative integer) corresponds to the matrix $V^{\sigma^r}$, where $\sigma$ is the automorphism of the ring $\mathscr{A}$ such that, for any $u = x_0 + x_1\mathbf{t} + \cdots + x_m\mathbf{t}^m + \cdots$ with $x_j \in \mathrm{W}(k)$, then $u^\sigma = x_0^\sigma + x_1^\sigma\mathbf{t} + \cdots + x_m^\sigma\mathbf{t}^m + \cdots$. It is then easy to verify that the shift $\mathrm{V}: \mathfrak{G} \to \mathfrak{G}^{(-1)}$ corresponds to the matrix $W = I_n\mathbf{t}$ in the preceding description, and the Frobenius homomorphism $\mathrm{F}: \mathfrak{G} \to \mathfrak{G}^{(1)}$ to the matrix $W = V$; the endomorphism $[p] = p \cdot 1_{\mathfrak{G}}: \mathfrak{G} \to \mathfrak{G}$ corresponds to the matrix $W = I_n\omega$, or to the matrix $W = V\mathbf{t} = \mathbf{t}V^{\sigma^{-1}}$.

We may also say that the $\mathscr{A}$-module $\mathscr{E}(\mathfrak{G}^{(r)})$ is naturally identified with $(\mathscr{E}(\mathfrak{G}))^{\sigma^r}$, i.e., to the additive group $\mathscr{E}(\mathfrak{G})$ in which the product of an element $\mathbf{Y}$ and an operator $u \in \mathscr{A}$ is the product $\mathbf{Y} \cdot u^{\sigma^r}$ for the structure of $\mathscr{A}$-module of $\mathscr{E}(\mathfrak{G})$ previously defined. To the shift $\mathrm{V} : \mathfrak{G} \to \mathfrak{G}^{(-1)}$ corresponds then the homothetic mapping $\tilde{\mathrm{V}} : \mathbf{Y} \to \mathbf{Yt}$, which is an $\mathscr{A}$-homomorphism of $\mathscr{E}(\mathfrak{G})$ into $(\mathscr{E}(\mathfrak{G}))^{\sigma^{-1}}$.

**5.** *Isogenies and $\mathscr{R}$-modules associated to reduced commutative bigebras.* We will write $\mathscr{R}(k) = \mathscr{R} \supset \mathscr{A}$ the ring of noncommutative power series

$$x_h\mathbf{t}^h + x_{h+1}\mathbf{t}^{h+1} + \cdots$$

with coefficients in $\mathrm{W}(k)$, where now the leading exponent $h$ may be *negative*, the commutation rule for negative exponents of $\mathbf{t}$ being

$$x\mathbf{t}^{-1} = \mathbf{t}^{-1}x^{\sigma} \tag{16}$$

for each $x \in \mathrm{W}(k)$. Every element of $\mathscr{R}$ may be written $u\mathbf{t}^{-m}$ with $u \in \mathscr{A}$ and $m \geqslant 0$; this at once implies that $\mathscr{R}$ has *no zero divisors*; furthermore, the neighborhoods of 0 *in* $\mathscr{A}$ are neighborhoods of 0 *in* $\mathscr{R}$ for a topology compatible with the ring structure of $\mathscr{R}$, for which $\mathscr{R}$ is Hausdorff and *complete*.

**Proposition 5.** *Let* M *be any right $\mathscr{A}$-module. The kernel of the natural mapping $z \mapsto z \otimes 1$ of* M *into* $\mathrm{M} \otimes_{\mathscr{A}} \mathscr{R}$ *is the set of $z \in \mathrm{M}$ for which there exists an integer $m \geqslant 0$ such that $z\mathbf{t}^m = 0$.*

If $z\mathbf{t}^m = 0$, then $z \otimes 1 = (z\mathbf{t}^m) \otimes \mathbf{t}^{-m} = 0$. Conversely, suppose that $z \otimes 1 = 0$. Then there is a finitely generated left sub-$\mathscr{A}$-module $\mathscr{R}_1$ of $\mathscr{R}$ such that already $z \otimes 1 = 0$ in $\mathrm{M} \otimes_{\mathscr{A}} \mathscr{R}_1$ [5, p. II–117]; as $\mathscr{R}_1$ is contained in a submodule $\mathscr{A}\mathbf{t}^{-m}$ of $\mathscr{R}$, we may assume that $\mathscr{R}_1 = \mathscr{A}\mathbf{t}^{-m}$. But $u \mapsto u\mathbf{t}^m$ is an isomorphism of the left $\mathscr{A}$-module $\mathscr{A}\mathbf{t}^{-m}$ onto $\mathscr{A}$ considered as left $\mathscr{A}$-module; hence

$$z \otimes \mathbf{t}^m = (z\mathbf{t}^m) \otimes 1 = 0 \quad \text{in} \quad \mathrm{M} \otimes_{\mathscr{A}} \mathscr{A},$$

i.e., $z\mathbf{t}^m = 0$ since $y \mapsto y \otimes 1$ is an isomorphism of M onto $\mathrm{M} \otimes_{\mathscr{A}} \mathscr{A}$.

**Corollary.** *The left $\mathscr{A}$-module $\mathscr{R}$ is flat.*

Let $f\colon \mathrm{M} \to \mathrm{M}'$ an injective $\mathscr{A}$-homomorphism of right $\mathscr{A}$-modules; it is enough to show that

$$f \otimes 1_{\mathscr{R}} : \mathrm{M} \otimes_{\mathscr{A}} \mathscr{R} \to \mathrm{M}' \otimes_{\mathscr{A}} \mathscr{R}$$

is also injective. Any element of $\mathrm{M} \otimes_{\mathscr{A}} \mathscr{R}$ may be written $z \otimes \mathbf{t}^{-m}$ for a $z \in \mathrm{M}$ and an integer $m \geqslant 0$; if its image by $f \otimes 1_{\mathscr{R}}$ is 0, then

$$f(z) \otimes \mathbf{t}^{-m} = (f(z) \otimes 1)\mathbf{t}^{-m} = 0,$$

hence $f(z) \otimes 1 = 0$, and by Proposition 5 there is a $q \geqslant 0$ such that $f(z)\mathbf{t}^q = 0$, or equivalently $f(z\mathbf{t}^q) = 0$. But as $f$ is injective, this implies $z\mathbf{t}^q = 0$, hence $z \otimes 1 = 0$, and *a fortiori*

$$z \otimes \mathbf{t}^{-m} = (z \otimes 1)\mathbf{t}^{-m} = 0.$$

For any reduced commutative finite-dimensional bigebra $\mathfrak{G}$ over $k$, we will write $\mathscr{E}_{(\mathscr{R})}(\mathfrak{G})$ the right $\mathscr{R}$-module $\mathscr{E}(\mathfrak{G}) \otimes_{\mathscr{A}} \mathscr{R}$; from Propositions 5 and 1 it follows that $\mathscr{E}(\mathfrak{G})$ is naturally identified with a *sub-$\mathscr{A}$-module* of $\mathscr{E}_{(\mathscr{R})}(\mathfrak{G})$. Theorem 1 shows that $\mathscr{E}(\mathfrak{G})$ is isomorphic to $\mathscr{A}^n/\mathrm{M}$, where M is the image of $\mathscr{A}^n$ by an endomorphism having a matrix of type $I\omega - V\mathbf{t}$; the corollary to Proposition 5 proves that $\mathscr{E}_{(\mathscr{R})}(\mathfrak{G})$ is isomorphic to $\mathscr{R}^n/\mathrm{M}_{(\mathscr{R})}$, where $\mathrm{M}_{(\mathscr{R})}$ is the image of $\mathscr{R}^n$ by the endomorphism of $\mathscr{R}^n$ having the *same* matrix $I\omega - V\mathbf{t}$ (whose elements are of course considered as elements of $\mathscr{R}$).

**Proposition 6.** *Let $\mathfrak{G}$, $\mathfrak{G}'$ be two reduced finite-dimensional commutative bigebras. For any homomorphism of $\mathscr{R}$-modules*

$$f : \mathscr{E}_{(\mathscr{R})}(\mathfrak{G}) \to \mathscr{E}_{(\mathscr{R})}(\mathfrak{G}'),$$

*there exists an integer $m \geqslant 0$ and a bigebra homomorphism $u\colon \mathfrak{G} \to \mathfrak{G}'^{(-m)}$ such that*

$$\tilde{u}_{(\mathscr{R})} = f \circ (\mathrm{V}^m)\tilde{}_{(\mathscr{R})} = (\mathrm{V}^m)\tilde{}_{(\mathscr{R})} \circ f. \tag{17}$$

Indeed, as $\mathscr{E}(\mathfrak{G})$ is a finitely generated $\mathscr{A}$-module (Proposition 4), there is an integer $m \geqslant 0$ such that, by composition of the homothetic mapping $z' \mapsto z'\mathbf{t}^m$ of $\mathscr{E}(\mathfrak{G}')$ and of the restriction $f_0$ of $f$ to $\mathscr{E}(\mathfrak{G})$, we obtain an $\mathscr{A}$-homomorphism of $\mathscr{E}(\mathfrak{G})$ into $\mathscr{E}(\mathfrak{G}'^{(-m)})$ (see the last remark of No. 4); hence it may be written $\tilde{u}$ where $u: \mathfrak{G} \to \mathfrak{G}'^{(-m)}$ is a bigebra homomorphism (Theorem 2); this proves the relations (17).

**Proposition 7.** *Let $\mathfrak{G}$, $\mathfrak{G}'$ be two reduced finite-dimensional commutative bigebras. In order that a bigebra homomorphism $u: \mathfrak{G} \to \mathfrak{G}'$ be surjective* (resp. *that its reduced kernel be* 0, resp. *that it be an isogeny*), *a necessary and sufficient condition is that the $\mathscr{R}$-module homomorphism $\tilde{u}_{(\mathscr{R})}: \mathscr{E}_{(\mathscr{R})}(\mathfrak{G}) \to \mathscr{E}_{(\mathscr{R})}(\mathfrak{G}')$ be surjective* (resp. *injective*, resp. *bijective*).

We begin with the condition that $u$ be an isogeny. If $u$ is an isogeny, there exists an isogeny $v: \mathfrak{G}' \to \mathfrak{G}^{(-m)}$ such that $v \circ u = \mathrm{V}^m$ and $u \circ v = \mathrm{V}^m$ (II, §3, No. 6, Proposition 13); hence $\tilde{v} \circ \tilde{u}$ (resp. $\tilde{u} \circ \tilde{v}$) is the homothetic mapping $z \mapsto z\mathbf{t}^m$ in $\mathscr{E}(\mathfrak{G})$ (resp. $z' \mapsto z'\mathbf{t}^m$ in $\mathscr{E}(\mathfrak{G}')$), and $\tilde{v}_{(\mathscr{R})} \circ \tilde{u}_{(\mathscr{R})}$ (resp. $\tilde{u}_{(\mathscr{R})} \circ \tilde{v}_{(\mathscr{R})}$) the same mapping in $\mathscr{E}_{(\mathscr{R})}(\mathfrak{G})$ (resp. $\mathscr{E}_{(\mathscr{R})}(\mathfrak{G}')$). But as $\mathbf{t}^m$ is invertible in $\mathscr{R}$, $\tilde{u}_{(\mathscr{R})} \circ \tilde{v}_{(\mathscr{R})}$ and $\tilde{v}_{(\mathscr{R})} \circ \tilde{u}_{(\mathscr{R})}$ are bijective, hence $\tilde{u}_{(\mathscr{R})}$ is an isomorphism.

To prove the converse, suppose we have an isomorphism

$$g : \mathscr{E}_{(\mathscr{R})}(\mathfrak{G}') \to \mathscr{E}_{(\mathscr{R})}(\mathfrak{G})$$

inverse to $\tilde{u}_{(\mathscr{R})}$. Then, by Proposition 6, there is an integer $m \geqslant 0$ and a bigebra homomorphism $v: \mathfrak{G}' \to \mathfrak{G}^{r-mr}$ such that

$$\tilde{v}_{(\mathscr{R})} = g \circ (\mathrm{V}^m)\tilde{}_{(\mathscr{R})} = (\mathrm{V}^m)\tilde{}_{(\mathscr{R})} \circ g;$$

hence

$$\tilde{u}_{(\mathscr{R})} \circ \tilde{v}_{(\mathscr{R})} = (\mathrm{V}^m)\tilde{}_{(\mathscr{R})} \qquad \text{and} \qquad \tilde{v}_{(\mathscr{R})} \circ \tilde{u}_{(\mathscr{R})} = (\mathrm{V}^m)\tilde{}_{(\mathscr{R})}.$$

As $u$, $v$, and V are bigebra homomorphisms, this also gives by restriction to $\mathscr{E}(\mathfrak{G}')$ and $\mathscr{E}(\mathfrak{G})$, respectively, $\tilde{u} \circ \tilde{v} = \tilde{\mathrm{V}}^m$ and $\tilde{v} \circ \tilde{u} = \tilde{\mathrm{V}}^m$, hence, by Theorem 2, $u \circ v = \mathrm{V}^m$ and $v \circ u = \mathrm{V}^m$, which proves that $u$ is an isogeny (II, §3, No. 6).

To prove that when $u$ is surjective (resp. has a reduced kernel equal to 0) then $\tilde{u}_{(\mathscr{R})}$ is surjective (resp. injective) we may, taking into account the preceding result and the canonical decomposition (17) of II, §3, No. 6, assume that $u$ is the natural surjection of $\mathfrak{G}$ onto the quotient bigebra of $\mathfrak{G}$ by a bi-ideal generated by the augmentation ideal $\mathfrak{N}^+$ of a reduced subbigebra $\mathfrak{N}$ of $\mathfrak{G}$ (resp. the natural injection into $\mathfrak{G}'$ of a reduced subbigebra $\mathfrak{H}'$ of $\mathfrak{G}'$). But then $\tilde{u}$ is already a surjection (resp. an injection) of $\mathscr{E}(\mathfrak{G})$ into $\mathscr{E}(\mathfrak{G}')$; this is obvious when $u$ is a natural injection, from the definition of hyperexponential vectors, and when $u$ is a natural surjection, the conclusion follows from the construction of a privileged basis for a quotient bigebra (II, §3, No. 4) and from Proposition 4.

Finally, suppose that $u$ is not surjective (resp. that its reduced kernel is not reduced to 0); then there exists a reduced commutative bigebra $\mathfrak{G}_1'$ (resp. $\mathfrak{G}_1$) and a bigebra homomorphism $v' \neq 0$ of $\mathfrak{G}'$ into $\mathfrak{G}_1'$ (resp. a bigebra homomorphism $v \neq 0$ of $\mathfrak{G}_1$ into $\mathfrak{G}$) such that $v' \circ u = 0$ (resp. $u \circ v = 0$). But then $\tilde{v}'_{(\mathscr{R})} \circ \tilde{u}_{(\mathscr{R})} = 0$ (resp. $\tilde{u}_{(\mathscr{R})} \circ \tilde{v}_{(\mathscr{R})} = 0$), and $\tilde{v}' \neq 0$ (resp. $\tilde{v} \neq 0$) by Theorem 2; as $\mathbf{t}$ is not a zero divisor, we also have $\tilde{v}'_{(\mathscr{R})} \neq 0$ (resp. $\tilde{v}_{(\mathscr{R})} \neq 0$); hence $\tilde{u}_{(\mathscr{R})}$ is not surjective (resp. not injective).

We may express Proposition 7 by saying that the *quotient* category of the category of reduced finite-dimensional commutative bigebras, obtained by taking as objects the *classes* of isogenous bigebras and for morphisms the bigebras homomorphisms *up to isogeny*, is *equivalent* to a full subcategory of the category of finitely generated $\mathscr{R}$-modules.

## §4. Distinguished modules over a Hilbert-Witt ring

**1.** *Properties of Hilbert-Witt rings.* As before, $k$ is a perfect field of characteristic $p > 0$. Let K be the field of fractions of the Witt ring W($k$) (§2, No. 3); for each integer $e \geqslant 1$, we will write $K_e$ the completely ramified extension of K generated by a root $\omega_e$ of the polynomial $T^e - \omega$ of K[T], and $W_e$ the integral closure of W($k$) in $K_e$; clearly $W_e$ is a free W($k$)-module having $\{1, \omega_e, \ldots, \omega_e^{e-1}\}$ as a basis; the automorphism $\sigma$ of W($k$) is uniquely extended to $W_e$ by the convention $\omega_e^\sigma = \omega_e$. The ring $W_e$ is a complete valuation ring, whose residual field is $k$ and whose maximal ideal is generated by $\omega_e$; we will write $w_e$ the corresponding discrete normed valuation, so that $w_e(\omega_e) = 1$; every element $u \in W_e$ can be uniquely written $\sum_{n=1}^{\infty} \xi_n \omega_e^n$ with $\xi_n \in k$ ($k$ being identified with its image by $\xi \mapsto e(\xi)$ in W($k$)); if $u \neq 0$, $w_e(u)$ is the smallest integer $n$ such that $\xi_n \neq 0$.

Note that, for any integer $d \geqslant 1$, the completely ramified extension of $K_e$ generated by a root of the polynomial $T^d - \omega_e$ of $K_e$[T] is naturally identified with $K_{ed}$, the integral closure of $W_e$ in that field being identified with $W_{ed}$, so that $W_{ed}$ is a free $W_e$-module having $\{1, \omega_{ed}, \ldots, \omega_{ed}^{d-1}\}$ as a basis.

In *this paragraph*, the number $e$ will be *fixed* once and for all, and we will therefore write W, $\omega$, and $w$ instead of $W_e$, $\omega_e$, and $w_e$. We will write A (resp. R) the W-module of *formal power series*

$$u = x_h t^h + x_{h+1} t^{h+1} + \cdots$$

in an indeterminate $t$, with coefficients $x_n \in W$, where the exponent $h$ is $\geqslant 0$ (resp. arbitrary). The W-module A (resp. R) is made into a ring by the multiplication law

$$(1) \qquad (xt^m)(yt^n) = (xy^{\sigma^m})t^{m+n} \qquad \text{for} \quad x, y \text{ in W}.$$

It is immediate to prove that R (hence also $A \subset R$) is a noncommutative ring *without zero divisors*, and it is known [7, Chapter 3, §2, No. 10, Theorem 2] that A is a left and right *noetherian* ring. Every element of R may be written $at^h$ or $t^h a'$, with $a$, $a'$ in A and $h \leqslant 0$; the automorphism $\sigma$ of W is extended to R (hence also to A) by the convention

$$\left(\sum_{n=h}^{\infty} x_n t^n\right)^{\sigma} = \sum_{n=h}^{\infty} x_n^{\sigma} t^n.$$

We will say that A is the *Hilbert-Witt ring* relative to the field $k$ and the integer $e$, and R the *localized Hilbert-Witt ring* relative to $k$ and $e$. The same arguments as in §3, No. 5 show that for any right A-module M, the kernel of the natural homomorphism $M \to M \otimes_A R$ is the set of $x \in M$ such that $xt^m = 0$ for some integer $m \geqslant 0$, and that the (left or right) A-module R is *flat*.

Let $a = a_h t^h + a_{h+1} t^{h+1} + \cdots$ an element $\neq 0$ in R, with $a_h \neq 0$; we will write

$$(2) \qquad v(a) = w(a_h)$$

and we will say that this integer $\geqslant 0$ is the *stathm* of $a$; its definition may be extended to the whole ring R by writing $v(0) = +\infty$. It is clear that

$$(3) \qquad v(ab) = v(a) + v(b)$$

for any pair of elements $a$, $b$ in R. In order that $a \in R$ be *invertible* in R, a necessary and sufficient condition is that $a_h$ be invertible in W, and this is equivalent to $v(a) = 0$.

With the same notations, observe that the sequence of numbers $w(a_n)$ for $n \geqslant h$ (some of which may be $+\infty$) has a *finite minimum* if $a \neq 0$; we will write $c(a)$ the *smallest integer* $i$ such that $w(a_{h+i})$ is equal to that minimum, and we will say that $c(a)$ is the *costathm* of $a$; its definition is extended to the whole ring R by writing $c(0) = +\infty$. We have

$$c(ab) = c(a) + c(b) \tag{4}$$

for any pair of elements $a$, $b$ in R. This is obvious if one of the elements $a$, $b$ is 0; if not, let

$$i = c(a), \qquad j = c(b),$$

$$a = a_h t^h + a_{h+1} t^{h+1} + \cdots, \quad b = b_l t^l + b_{l+1} t^{l+1} + \cdots$$
$$\text{with} \quad a_h \neq 0, \quad b_l \neq 0;$$

in $ab$, the coefficient of $t^{h+l+m}$ for $m < i + j$ is sum of the products $a_{h+\alpha} b_{l+\beta}^{\sigma^{h+\alpha}}$ with $\alpha + \beta = m$, hence either $\alpha < i$, or $\beta < j$, and therefore

$$w(a_{h+\alpha} b_{l+\beta}^{\sigma^{h+\alpha}}) > w(a_{h+i}) + w(b_{l+j}),$$

and *a fortiori*

$$w(\sum_{\alpha+\beta=m} a_{h+\alpha} b_{l+\beta}^{\sigma^{h+\alpha}}) > w(a_{h+i}) + w(b_{l+j});$$

on the other hand, the coefficient of $t^{h+l+i+j}$ is sum of $a_{h+i} b_{l+j}^{\sigma^{h+i}}$ and of terms $a_{h+\alpha} b_{l+\beta}^{\sigma^{h+\alpha}}$ where either $\alpha < i$ or $\beta < j$; this shows that $c(ab) = i + j$. The relation $c(a) = 0$ means that $a$ has the form $\omega^{v(a)} u$, where $u$ is *invertible in* R (hence has the form $t^{-h} u_0$, where $h \geqslant 0$ and $u_0$ is invertible *in* A).

**Proposition 1.** *The ring* R *is left and right euclidean for the stathm* $v$; *in other words, for any pair of elements* $a$, $b$ *of* R *such that* $b \neq 0$, *there exist elements* $q_1, q_2, r_1, r_2$ *of* R *such that*

$$a = bq_1 + r_1 \quad \textit{and} \quad r_1 = 0 \quad \textit{or} \quad v(r_1) < v(b) \tag{5}$$

$$a = q_2 b + r_2 \quad \textit{and} \quad r_2 = 0 \quad \textit{or} \quad v(r_2) < v(b). \tag{6}$$

We may assume $a \neq 0$, otherwise we would take $q_1 = q_2 = r_1 = r_2 = 0$. Let

$$a = a_h t^h + a_{h+1} t^{h+1} + \cdots, \qquad b = b_l t^l + b_{l+1} t^{l+1} + \cdots;$$

the sum of the terms $a_i t^i$ (resp. $b_j t^j$) such that $w(a_i) \geqslant v(a)$ (resp. $w(b_j) \geqslant v(b)$) has the form $\omega^n a_0 t^h$ (resp. $\omega^m b_0 t^l$), where $a_0$ and $b_0$ are invertible in A and $n = v(a)$, $m = v(b)$; furthermore, from the definition of $v$, we may write

$$a = \omega^n a_0 t^h + a' t^{h+1} \qquad \text{with} \quad a' \in \mathrm{A}$$

and either $a' = 0$ or $v(a') < n$, and similarly

$$b = \omega^m b_0 t^l + b' t^{l+1} \qquad \text{with} \quad b' \in \mathrm{A}$$

and either $b' = 0$ or $v(b') < m$. If $m > n$, we satisfy to (5) and (6) by taking $q_1 = q_2 = 0$, $r_1 = r_2 = a$. Otherwise, consider the element

$$z_0 = a - c_0 t^{h-l} b, \qquad \text{where} \quad c_0 = \omega^{n-m} a_0 (b_0^{-1})^{\sigma^{h-l}};$$

we may also write $z_0 = (a' - c_0 b'^{\sigma^{h-l}}) t^{h+1}$.
We argue by contradiction, supposing that there is no pair $(x, y)$ of elements of R satisfying the relation $a = xb + y$ with $y = 0$ or $v(y) < v(b)$. Then we must have $v(z_0) \geqslant v(b) = m$, and we could repeat on $z_0$ the same operation as on $a$, so that we would define by induction an infinite sequence $(c_j)_{j \geqslant 0}$ of elements of A such that, for the element

$$z_j = a - (c_0 + c_1 t + \cdots + c_j t^j) t^{h-l} b,$$

we have $v(z_j) \geqslant m$ for every $j$. But then the series

$$c_0 + c_1 t + \cdots + c_j t^j + \cdots$$

is convergent in A; if $x$ is its sum, we have $a = x t^{h-l} b$, for $z_j$ is divisible by $t^{h+j}$ for every $j \geqslant 0$, hence $a - x t^{h-l} b$ is divisible by all powers of $t$, and therefore is 0. But this is contrary to our initial assumption, hence the latter was absurd and this proves the existence of $q_2$ and $r_2$ satisfying (6); the proof of (5) is similar.

We should observe here that the elements $q_i$ and $r_i$ defined in Proposition 1 ($j = 1, 2$) *are not uniquely determined* in general.

The usual argument of the theory of euclidean rings, which consists in taking in an ideal an element of smallest stathm, shows that the left and right ideals of R are *principal ideals*. We will actually prove a more precise result:

**Proposition 2.** *If $a$, $b$ are two elements of* R *such that $a \neq 0$ and $b \neq 0$, there exist two elements $x$, $y$ in* R *such that*

$$aR + bR = xR, \qquad (aR) \cap (bR) = yR \tag{7}$$

*and which satisfy the relations*

$$v(a) + v(b) = v(x) + v(y) \tag{8}$$

$$c(a) + c(b) = c(x) + c(y). \tag{9}$$

This is a known result of O. Ore [43] for euclidean rings; we rapidly recall the proof to be complete. Suppose for instance that $v(a) \geqslant v(b)$, and determine by induction elements $q_i$, $r_i$ of R such that

$$\begin{aligned}
a &= bq_0 + r_0, & v(r_0) &< v(b)\\
b &= r_0q_1 + r_1, & v(r_1) &< v(r_0)\\
r_0 &= r_1q_2 + r_2, & v(r_2) &< v(r_1)\\
&\cdots\cdots\cdots\cdots\cdots\cdots\cdots\cdots\cdots\cdots\cdots\cdots\\
r_{n-1} &= r_nq_{n+1} + x, & v(x) &< v(r_n)\\
r_n &= xq_{n+2}.
\end{aligned}$$

Then it is clear that

$$aR + bR = bR + r_0R = r_0R + r_1R = \cdots = r_nR + xR = xR.$$

On the other hand, we may determine elements $m_j$ ($1 \leqslant j \leqslant n + 1$) and $q'_j$ ($0 \leqslant j \leqslant n + 1$) such that

$$
\begin{aligned}
m_1 &= r_{n-1}q_{n+2} = r_n q'_{n+1}, & m_1\mathrm{R} &= (r_{n-1}\mathrm{R}) \cap (r_n\mathrm{R}) \\
m_2 &= r_{n-2}q'_{n+1} = r_{n-1}q'_n, & m_2\mathrm{R} &= (r_{n-2}\mathrm{R}) \cap (r_{n-1}\mathrm{R}) \\
&\ \vdots & &\ \vdots \\
m_{n+1} &= bq'_2 = r_0 q'_1, & m_{n+1}\mathrm{R} &= (b\mathrm{R}) \cap (r_0\mathrm{R}) \\
aq'_1 &= bq'_0 = y, & y\mathrm{R} &= (a\mathrm{R}) \cap (b\mathrm{R}).
\end{aligned} \tag{10}
$$

To see this, we start with the relations

$$r_{n-1}q_{n+2} = r_n q_{n+1}q_{n+2} + xq_{n+2} = r_n(q_{n+1}q_{n+2} + 1),$$

and define

$$q'_{n+1} = q_{n+1}q_{n+2} + 1 \qquad \text{and} \qquad m_1 = r_{n-1}q_{n+2} = r_n q'_{n+1}.$$

We obviously have $m_1\mathrm{R} \subset (r_{n-1}\mathrm{R}) \cap (r_n\mathrm{R})$; conversely, if

$$f \in (r_{n-1}\mathrm{R}) \cap (r_n\mathrm{R})$$

we may write $f = r_{n-1}s = r_n t$, i.e., $(r_n q_{n+1} + x)s = r_n t$, or $r_n(q_{n+1}s - t) = -xs$ and finally

$$xq_{n+2}(q_{n+1}s - t) = -xs;$$

but as R has no zero divisor and $x \neq 0$, this implies $s = q_{n+2}(t - q_{n+1}s)$, hence

$$f = r_{n-1}q_{n+2}(t - q_{n+1}s) \in m_1\mathrm{R},$$

and we have proved the first relation (10). Using induction, suppose next we have proved the first $i$ relations (10), and let

$$m_{i+1} = r_{n-i-1}q'_{n-i+2};$$

we also have

$$m_{i+1} = r_{n-i}q_{n-i+1}q'_{n-i+2} + r_{n-i+1}q'_{n-i+2} = r_{n-i}q'_{n-i+1},$$

where

$$q'_{n-i+1} = q'_{n-i+3} + q_{n-i+1}q'_{n-i+2}$$

by definition. We thus have

$$m_{i+1}\mathrm{R} \subset (r_{n-i-1}\mathrm{R}) \cap (r_{n-i}\mathrm{R});$$

conversely, suppose $f = r_{n-i-1}s = r_{n-i}t$ is in that intersection; we have

$$(r_{n-i}q_{n-i+1} + r_{n-i+1})s = r_{n-i}t,$$

or equivalently

$$r_{n-i}(t - q_{n-i+1}s) = r_{n-i+1}s.$$

But then the inductive assumption implies the existence of a $z \in \mathrm{R}$ such that

$$r_{n-i+1}s = m_i z = r_{n-i+1}q'_{n-i+2}z;$$

as R has no zero divisor, we obtain $s = q'_{n-i+2}z$, hence $f = m_{i+1}z \in m_{i+1}\mathrm{R}$, which ends the proof of (10) by induction.

Furthermore, from the first equation (10), we deduce

$$v(m_1) + v(x) = v(r_{n-1}) + v(q_{n+2}) + v(x) = v(r_{n-1}) + v(r_n).$$

Suppose we have proved the relation

$$v(m_i) + v(x) = v(r_{n-i}) + v(r_{n-i+1});$$

then we have $v(m_{i+1}) = v(r_{n-i-1}) + v(q'_{n-i+2})$, hence

$$v(m_{i+1}) + v(r_{n-i+1}) = v(r_{n-i-1}) + v(m_i)$$

and therefore

$$\begin{aligned} v(m_{i+1}) + v(r_{n-i+1}) + v(x) &= v(r_{n-i-1}) + v(m_i) + v(x) \\ &= v(r_{n-i-1}) + v(r_{n-i}) + v(r_{n-i+1}); \end{aligned}$$

hence finally

$$v(m_{i+1}) + v(x) = v(r_{n-i-1}) + v(r_{n-i})$$

which proves (8) by induction. The proof of (9) is identical, simply replacing $v$ by $c$.

**2.** *Finitely generated modules over a localized Hilbert-Witt ring.* We first recall the standard theory of modules over a ring without zero divisor R in which *all left and right ideals are principal* [31]. Such a ring is left and right noetherian; *irreducible* (on the right) elements of R are the elements $q \in \mathrm{R}$ such that the right ideal $q\mathrm{R}$ is *maximal*. The first result is that any element $a \neq 0$ in R may be written $a = q_1 q_2 \cdots q_k x$, where the $q_i$ are irreducible and $x$ is invertible: this is shown by writing in succession

$$a = q_1 a_1, \quad a_1 = q_2 a_2, \quad \ldots,$$

and observing that the increasing sequence of left ideals $(\mathrm{R}a_i)$ must be stationary, hence one must arrive at a maximal one. Furthermore the right ideals $q_i\mathrm{R}$ corresponding to such a decomposition are *unique up to a permutation*: indeed, the sequence

$$\mathrm{R} \supset q_1\mathrm{R} \supset q_1 q_2 \mathrm{R} \supset \cdots \supset q_1 q_2 \cdots q_k \mathrm{R} = a\mathrm{R}$$

yields a *Jordan-Hölder sequence* for the R-module $\mathrm{R}/a\mathrm{R}$, for we have $q_1 q_2 \cdots q_j \mathrm{R}/q_1 q_2 \cdots q_j q_{j+1}\mathrm{R}$ isomorphic to the *simple* R-module $\mathrm{R}/q_{j+1}\mathrm{R}$ since $q_1 q_2 \cdots q_j$ is not a zero divisor in R. At the same time, this proves that, for $a \neq 0$, the monogenic R-module $\mathrm{R}/a\mathrm{R}$ has *finite length*. The modules $\mathrm{R}/a\mathrm{R}$ and $\mathrm{R}/b\mathrm{R}$ are isomorphic if and only if $a = bu$ where $u$ is invertible.

Under the same assumption on R, the usual proof of the fact that a submodule of a *free* right R-module is also free can proceed just as in the case of commutative principal ideal rings. Furthermore, for a finitely generated free right R-module L, the number of elements in a basis of L is independent of that basis, for it is equal to the length of the R-module $\mathrm{L}/q\mathrm{L}$ for any irreducible element $q$ in R; we will say that this number is the *dimension* of L.

We now make the stronger assumption that R is a left and right *euclidean* ring, i.e., that there is defined on R a *stathm* $v$, which is a mapping of R into the set consisting of the integers $\geqslant 0$ and $+\infty$, satisfying relation (3), for which Proposition 1 is true, and such that the invertible elements of R are exactly those for which $v(x) = 0$. Then it is known [31] that the usual arguments of the theory of *elementary divisors* carry over to that case, and we recall them again briefly. Let L be a free finitely generated right

R-module, M a submodule of L (which is necessarily finitely generated, since R is noetherian). Let $L^*$ be the dual of L, and consider the stathms of $\langle x^*, y\rangle$ for $y \in M$ and $x^* \in L^*$; they have a *minimum* value $d_1$, hence there exists $y_1 \in M$ and $x_1^* \in L^*$ such that $\langle x_1^*, y_1\rangle = \alpha_1$ with $v(\alpha_1) = d_1$. This definition first implies that, for any $y \in M$, $\langle x_1^*, y\rangle$ is a *right multiple* of $\alpha_1$, and for any $x^* \in L^*$, $\langle x^*, y_1\rangle$ is a *left multiple of* $\alpha_1$. Indeed, if for instance $\beta = \langle x_1^*, y\rangle$ was not a right multiple of $\alpha_1$, a right g.c.d. $\delta$ of $\alpha_1$ and $\beta$ would be such that $v(\delta) < d_1$ and $\delta = \alpha_1\lambda + \beta\mu$, hence

$$\langle x_1^*, y_1\lambda + y\mu\rangle = \delta,$$

which contradicts the definition of $d_1$; a similar argument yields the second assertion, by using a left g.c.d. of $\langle x^*, y_1\rangle$ and $\langle x_1^*, y_1\rangle$ (recall that $L^*$ is a left R-module). In particular, all coordinates of $y_1$ for any basis of L are left multiples of $\alpha_1$, and therefore there exists $e_1 \in L$ such that $y_1 = e_1\alpha_1$, hence $\langle x_1^*, e_1\rangle = 1$, and one immediately checks that L is a direct sum of $e_1R$ and of $L_1 = \mathrm{Ker}(x_1^*)$; furthermore, for each $y \in M$, we may write $\langle x_1^*, y\rangle = \alpha_1\gamma$, hence $\langle x_1^*, y - y_1\gamma\rangle = 0$, and therefore M is a direct sum of $y_1R = e_1\alpha_1R$ and of $L_1 \cap M$. The R-module $L_1$ being free, one has then only to use induction on the dimension $n$ of L: one may assume that there is a basis $(e_i)_{2\leqslant i\leqslant n}$ of $L_1$ and a basis $(e_i\alpha_i)_{2\leqslant i\leqslant m}$ of $L_1 \cap M$ such that $m \leqslant n$ and $\alpha_{i+1}$ is a left multiple of $\alpha_i$ for $2 \leqslant i \leqslant m - 1$. One must then finally show that $\alpha_2$ is a left multiple of $\alpha_1$; now, if $(e_i^*)$ is the basis of $L^*$ dual to $(e_i)$, we have

$$\langle e_1^*, e_1\alpha_1\rangle = \alpha_1, \qquad \langle e_2^*, e_2\alpha_2\rangle = \alpha_2;$$

if $\alpha_2$ were not a left multiple of $\alpha_1$, there would exist $\lambda$, $\mu$ in R such that $\delta = \lambda\alpha_1 + \mu\alpha_2$ would satisfy $v(\delta) < v(\alpha_1) = d_1$; but

$$\langle \lambda e_1^* + \mu e_2^*, e_1\alpha_1 + e_2\alpha_2\rangle = \lambda\alpha_1 + \mu\alpha_2 = \delta,$$

and this again contradicts the definition of $\alpha_1$.

Of course, this theory of elementary divisors immediately yields the structure of all *finitely generated* right R-modules, since such a module is isomorphic to L/M with suitable choice of L and M; but (with the preceding notations) it is clear that L/M is a direct sum of $n - m$ modules isomorphic to R, and of $m$ monogenic R-modules $R/\alpha_iR$ with $\alpha_i \neq 0$. We will say that a right R-module N is a *torsion module* if every element $\neq 0$ in N has a nonzero annihilator in R; a finitely generated right torsion

R-module is thus a *direct sum of monogenic* R-*modules* $R/\beta_i R$ with $\beta_i \neq 0$.

Furthermore, for *any* decomposition of such a module N into a direct sum of monogenic R-modules $R/\gamma_j R$ $(1 \leqslant j \leqslant r)$, *the sum of the stathms* $\sum_j v(\gamma_j)$ *is the same*: indeed, if we write

$$\gamma_j = \gamma_{1j}\gamma_{2j} \cdots \gamma_{k_j,j}, \qquad \text{where the } \gamma_{ij} \text{ are irreducible,}$$

the simple R-modules $R/\gamma_{ij}R$ are the quotients of a Jordan-Hölder sequence for N, hence are well determined by N up to isomorphism and up to a permutation; as

$$\sum_i v(\gamma_j) = \sum_{i,j} v(\gamma_{ij}),$$

this proves our assertion. We will say that the integer $\sum_j v(\gamma_j)$ is the *rank* of the torsion module N and we will write it rk(N). If P is a submodule of N, we obtain a Jordan-Hölder sequence of N by juxtaposition of a Jordan-Hölder sequence of P and a Jordan-Hölder sequence of N/P, hence

$$\text{(11)} \qquad \operatorname{rk}(N) = \operatorname{rk}(P) + \operatorname{rk}(N/P),$$

and in particular the relation rk(P) = rk(N) implies P = N.

When R is in addition a *localized Hilbert-Witt ring*, the same argument as above shows that *the sum of the costathms* $\sum_j c(\gamma_j)$ is independent of the decomposition of N into a direct sum of monogenic R-modules $R/\gamma_j R$; we will say that the integer $\sum_j c(\gamma_j)$ is the *corank* of N and we will write it crk(N). For any submodule P of N, we have as above

$$\text{(12)} \qquad \operatorname{crk}(N) = \operatorname{crk}(P) + \operatorname{crk}(N/P).$$

Observe in addition that the relation crk(N) = 0 means that N is a direct sum of modules of type $R/\omega^m R$, or equivalently that *there exists an integer* $r \geqslant 0$ *such that* $\omega^r N = 0$.

We will use the following general lemma:

**Lemma 1.** *Let* B *be any ring, $a$, $b$ two elements of* B *which are not zero divisors, and consider the right* B-*module* $M = B/abB$ *and its submodule*

$N = aB/abB$, *which is isomorphic to* $B/bB$. *In order that* N *be a direct factor in* M, *a necessary and sufficient condition is that there exist elements* $x, y$ *of* B *such that* $xa + by = 1$. *Then* $(1 - ax)B/abB$ *is a supplementary submodule of* N *in* M, *isomorphic to* $B/aB$.

Let $\varepsilon$ be the natural image of 1 in M, so that we have $M = \varepsilon B$, $N = \varepsilon aB$, and $abB$ is the annihilator of $\varepsilon$; the annihilator of $\varepsilon a$ is therefore $bB$, since $a$ is not a zero divisor, and therefore N is isomorphic to $B/bB$. Suppose $M = N \oplus P$, and write $\varepsilon = \xi + \eta$ with $\xi \in N$, $\eta \in P$; as $M = \varepsilon M = \xi B + \eta B$ and $\xi B \subset N$, $\eta B \subset P$, we have $P = \eta B$ and the relation $\eta u = 0$ for $u \in B$ is equivalent to $\varepsilon u \in N$, or also to $u \in aB$, hence P is isomorphic to $B/aB$. Furthermore, we may write $\eta = \varepsilon d$ with $d \in B$, and $d$ is determined modulo $abB$; as $\varepsilon = \xi + \eta$, there exists an $x \in B$ such that

$$1 - ax - d \in abB$$

and by adding to $d$ an element of $abB$, we may assume that $d = 1 - ax$; then we have just seen that $\varepsilon(1 - ax)a = 0$, in other words,

$$(1 - ax)a = aby \qquad \text{for some} \quad y \in B,$$

which implies $1 = xa + by$ since $a$ is not a zero divisor. Conversely, if we have such a relation, and if we write $d = 1 - ax$, we may write $\varepsilon = \varepsilon ax + \varepsilon d$, and this implies $M = N + \varepsilon dB$; furthermore, if

$$\varepsilon az = \varepsilon dt \in N \cap \varepsilon dB \qquad \text{for} \quad z, t \text{ in } B,$$

we have

$$az - dt = abw \qquad \text{for a} \quad w \in B,$$

i.e., $az - t - axt = abw$, hence $t \in aB$ and

$$dt \in daB = a(1 - xa)B = abyB;$$

therefore $\varepsilon dt = 0$, in other words $N \cap \varepsilon dB = 0$. Q.E.D.

Returning to the study of torsion right R-modules for a localized Hilbert-Witt ring R, let us note that $\omega$ is an irreducible element of R (since $v(\omega) = 1$) belonging to its *center*. Every $a \in R$ can be written $a = \omega^m b$ with $m \geqslant 0$, $b$ not divisible by $\omega$, and $\omega^m$ is divisible by no irreducible element other than $\omega$, since $c(\omega^m) = 0$. We have therefore $\omega^m x + by = 1$ for suitable elements $x, y$ of R by Proposition 2; Lemma 1 then shows that $R/aR$ is isomorphic to the direct sum of $R/\omega^m R$ and $R/bR = N$; furthermore, we have $N = N\omega$. We have thus proved:

**Proposition 3.** *If* R *is a localized Hilbert-Witt ring, every finitely generated torsion right* R*-module* M *can be uniquely written* $M = N \oplus P$, *where* $N = N\omega$ *and* P *is a direct sum of submodules of type* $R/\omega^m R$.

Observe in addition that $Q = R/\omega^m R$ is an *indecomposable* R-module, for its only submodules are the $\omega^h R/\omega^m R$, isomorphic to $R/\omega^{m-h}R$ $(0 \leqslant h \leqslant m)$, since $\omega$ is the only irreducible element dividing $\omega^m$, and Lemma 1 proves that such a submodule cannot be a direct factor of Q for $1 \leqslant h \leqslant m - 1$.

**3.** *Distinguished modules over a Hilbert-Witt ring* (cf. I. Barsotti [1]). We keep the notations of No. 1. In the Hilbert-Witt ring A, $\mathfrak{m} = At = tA$ is a two-sided ideal *contained in the radical of* A: indeed, for any $a \in A$, the sum

$$a' = 1 + at + (at)^2 + \cdots + (at)^n + \cdots$$

has a meaning in A, since it is equal to the formal power series

$$1 + at + a^{1+\sigma}t^2 + \cdots + a^{1+\sigma+\sigma^2+\cdots+\sigma^{n-1}}t^n + \cdots;$$

obviously $(1 - at)a' = a'(1 - at) = 1$, hence $1 - at$ is invertible, and one shows in the same way that $1 - ta$ is invertible.

For any right A-module M, the right A-module $M\mathfrak{m}$ is also equal to $Mt$, as follows from formula (1).

**Proposition 4.** *Let* M, N *be two right* A*-modules,* $u\colon M \to N$ *an* A*-homomorphism. If* N *is finitely generated and if the homomorphism* $M/Mt \to N/Nt$ *deduced from* $u$ *is surjective, then* $u$ *is surjective.*

Since $t$ is contained in the radical, this is simply Nakayama's lemma [6, §6, No. 3, Corollary 2 of Proposition 6].

**Corollary.** *If* M *is a finitely generated right* A*-module,*

$$\bigcap_{n \geqslant 0} \mathrm{M}t^n = 0.$$

Let $\mathrm{N} = \bigcap_{n \geqslant 0} \mathrm{M}t^n$; as A is noetherian, N is finitely generated; furthermore $\mathrm{N} = \mathrm{N}t$ by definition. One then applies Proposition 4 to the homomorphism $0 \to \mathrm{N}$.

We may express this corollary by saying that on M the *filtration* of the $\mathrm{M}t^n$ is *separated.*

**Proposition 5.** *Let* M *and* N *two finitely generated right* A*-modules such that the homothetic mappings* $x \mapsto xt$ *in* M *and in* N *are injective. In order that a homomorphism* $u\colon \mathrm{M} \to \mathrm{N}$ *be injective* (resp. *surjective*, resp. *bijective*), *a sufficient condition is that the homomorphism* $u_0\colon \mathrm{M}/\mathrm{M}t \to \mathrm{N}/\mathrm{N}t$ *deduced from u have the same property.*

Taking Proposition 4 into account, all we have to show is that, if $u_0$ is injective, $u$ is injective. The assumptions imply that for every integer $r > 0$,

$$u_r : \mathrm{M}t^r/\mathrm{M}t^{r+1} \to \mathrm{N}t^r/\mathrm{N}t^{r+1}$$

is injective. We then have only to apply ([7], Chapter III, §2, No. 8, Corollary 1 of Theorem 1), since the filtrations $(\mathrm{M}t^n)$ and $(\mathrm{N}t^n)$ are separated.

Observe that Proposition 4 and 5 are still true when we only suppose that $u$ is a *semilinear* mapping of M into N, relative to an automorphism of A which leaves $t$ invariant.

We will say that a right A-module M is *distinguished* if it is finitely generated and satisfies the following two conditions:

$$\mathrm{M}\omega \subset \mathrm{M}t. \tag{13}$$

(14) *The homothetic mapping* $x \mapsto xt$ *in* M *is injective.*

As the kernel of the natural mapping $\mathrm{M} \to \mathrm{M} \otimes_{\mathrm{A}} \mathrm{R} = \mathrm{M}_{(\mathrm{R})}$ is the set of $x \in \mathrm{M}$ such that $xt^r = 0$ for some integer $r > 0$, condition (14) is equivalemt to saying that this natural mapping is *injective*.

In what follows, we write $\pi = t^{-1}\omega = \omega t^{-1}$ (element of R).

**Proposition 6.** *If* M *is a distinguished* A-*module*, $\mathrm{M}_{(\mathrm{R})}$ *is a torsion* R-*module*.

If we identify M to a sub-A-module of $\mathrm{M}_{(\mathrm{R})}$, condition (13) may also be written $\mathrm{M}\pi \subset \mathrm{M}$. Suppose that $\mathrm{M}_{(\mathrm{R})}$ contains an element $x_0 \neq 0$ whose annihilator is reduced to 0. We may suppose that $x_0 \in \mathrm{M}$. Let $\mathfrak{a}$ be the set of all elements $r \in \mathrm{R}$ such that $x_0 r \in \mathrm{M}$; as by assumption $r \mapsto x_0 r$ is an injection, $\mathfrak{a}$ is a right A-module, isomorphic to a submodule of M, hence it is finitely generated; furthermore, as $\mathrm{M}\pi \subset \mathrm{M}$, $\mathfrak{a}$ contains all powers $\pi^n$ of $\pi$ ($n \geqslant 1$), and of course contains A. Consider in $\mathfrak{a}$ the increasing sequence of right A-modules $(\mathrm{A} + \pi\mathrm{A} + \pi^2\mathrm{A} + \cdots + \pi^n\mathrm{A})$: this sequence is stationary since A is noetherian and $\mathfrak{a}$ finitely generated, hence there is an $n > 0$ such that

$$\pi^n = \sum_{i=0}^{n-1} \pi^i a_i \qquad \text{with} \quad a_i \in \mathrm{A}.$$

However this implies that $\omega^n \in t\mathrm{A}$, which is absurd.

The distinguished A-modules may therefore be characterized as the *finitely generated sub-*A*-modules of the torsion* R*-modules, which are also modules over the subring* $\mathrm{A}[\pi]$ *of* R *generated by* A *and* $\pi$.

For each right A-module M, $\mathrm{M}/\mathrm{M}t$ may be considered as a module over $\mathrm{A}/\mathfrak{m} = \mathrm{W}$; if in addition M is distinguished, $\mathrm{M}/\mathrm{M}t$ is also a module over the field $k = \mathrm{W}/\omega\mathrm{W}$, since $\mathrm{M}\omega \subset \mathrm{M}t$. Every basis of the $k$-vector space $\mathrm{M}/\mathrm{M}t$ is also a system of generators of the A-module $\mathrm{M}/\mathrm{M}t$. It then follows from Proposition 4 that if $(x_i)_{1 \leqslant i \leqslant n}$ is a system of elements of M such that the classes $\bar{x}_i$ of the $x_i$ in $\mathrm{M}/\mathrm{M}t$ constitute a basis of $\mathrm{M}/\mathrm{M}t$ over $k$, then the $x_i$ generate M. We will say that the dimension of the $k$-vector space $\mathrm{M}/\mathrm{M}t$ is the *rank* of the distinguished A-module M, and we will write it rk(M); it is also the minimum number of generators of M; we shall see in §5, No. 6 that rk(M) is equal to *the rank* $\mathrm{rk}(\mathrm{M}_{(\mathrm{R})})$ defined in No. 2.

**Proposition 7.** *In order that a distinguished* A-*module* M *have rank* $n$, *a necessary and sufficient condition is that* M *be isomorphic to a quotient*

*$A^n/u(A^n)$, where $u$ is an endomorphism of $A^n$ whose matrix with respect to the canonical basis of $A^n$ has the form $I_n\omega - Vt$, where $V$ is any square matrix of order $n$ with elements in A.*

Let $u$ be an endomorphism of $A^n$ of the type described in Proposition 7, and let $N = A^n/u(A^n)$; we first prove that N is a distinguished A-module of rank $n$. We may write

$$N/Nt = A^n/(A^n t + u(A^n)),$$

and from the definition of $u$ it follows that

$$A^n t + u(A^n) = A^n t + A^n\omega = (At + A\omega)^n;$$

now $At + A\omega$ is a two-sided maximal ideal of A, and $A/(At + A\omega) = k$; therefore

$$N/Nt = (A/(At + A\omega))^n = k^n$$

is an $n$-dimensional vector space over $k$. On the other hand, for any $z \in A^n$, we have

$$z\omega \equiv Vt \cdot z = (V \cdot z^\sigma)t \qquad \text{mod } u(A^n)$$

and therefore $N\omega \subset Nt$. Finally, if $y \in A^n$ is such that $yt \in u(A^n)$, then we may write

$$yt = z\omega - (Vt) \cdot z = z\omega - V \cdot z^\sigma t \qquad \text{for some} \quad z \in A^n,$$

and this implies that $z \in A^n t$; let $z = z't$ with $z' \in A^n$; then $yt = u(z')t$, hence $y = u(z')$; this shows that the homothetic mapping $x \mapsto xt$ in N is injective and proves that N is a distinguished A-module of rank $n$.

Conversely, let M be such an A-module, and take the generators $x_i$ ($1 \leqslant i \leqslant n$) as above. The relation $M\omega \subset Mt$ implies $n$ equations:

$$(15) \qquad x_i\omega = \sum_{j=1}^{n} x_j v_{ji} t \qquad \text{with} \quad v_{ji} \in A.$$

Let $V$ be the matrix $(v_{ji})$, $u$ the endomorphism of $A^n$ having as matrix with respect to the canonical basis $(e_i)_{1 \leqslant i \leqslant n}$ of $A^n$ the matrix $I_n\omega - Vt$, and let $N = A^n/u(A^n)$. Consider the surjective homomorphism $f: A^n \to M$ such that $f(e_i) = x_i$ for $1 \leqslant i \leqslant n$; it follows from (15) that $u(A^n)$ is contained in the kernel of $f$, hence we obtain from $f$ a surjective homomorphism $g: N \to M$. The homomorphism $g_0: N/Nt \to M/Mt$ corresponding to $g$ is then a surjective homomorphism of $k$-vector spaces; but $N/Nt$ and $M/Mt$ are both $k$-vector spaces of dimension $n$, hence $g_0$ is bijective, and so is $g$ by Proposition 5, since both N and M are distinguished.

**Proposition 8.** *Let* M *be a distinguished ring* A*-module of rank* $m$, N *a sub-*A*-module of* M.

(i) *The quotient module* M/N *is distinguished if and only if* $N \cap Mt = Nt$; *then* N *is distinguished and* $\mathrm{rk}(N) + \mathrm{rk}(M/N) = \mathrm{rk}(M)$.

(ii) *For every distinguished submodule* N *of* M, $\mathrm{rk}(N) \leqslant \mathrm{rk}(M)$; *in addition the following conditions are equivalent*:

a) $\mathrm{rk}(N) = \mathrm{rk}(M)$.
b) *There exists an integer* $r > 0$ *such that* $Mt^r \subset N$.
c) M/N *has finite length*.

(iii) *If* N *is a distinguished submodule of* M, *there exists a system of* $m$ *generators of* M,

$$x_1, \ldots, x_{d_1}, x_{d_1+1}, \ldots, x_{d_1+d_2}, \ldots\ldots, \\ x_{d_1+\cdots+d_{h-1}+1}, \ldots, x_{d_1+\cdots+d_h}, \ldots, x_m$$

*and a strictly increasing sequence of integers*

$$0 \leqslant s_1 < s_2 < \cdots < s_h$$

*such that*

$$x_1 t^{s_1}, \ldots, x_{d_1} t^{s_1}, x_{d_1+1} t^{s_2}, \ldots, x_{d_1+d_2} t^{s_2}, \ldots, \\ x_{d_1+\cdots+d_{h-1}} t^{s_h}, \ldots, x_{d_1+\cdots+d_h} t^{s_h}$$

*is a system of generators of* N, *with* $d_1 + d_2 + \cdots + d_h = \mathrm{rk}(N)$; *furthermore the integers* $s_i$ *and* $d_i$ $(1 \leqslant i \leqslant h)$ *are independent of the choice of the* $x_j$, *and if* $\mathrm{rk}(N) = \mathrm{rk}(M)$, *then* $s_1 d_1 + \cdots + s_h d_h = \mathrm{length}(M/N)$.

(i) Suppose M/N is distinguished and $x \in N \cap Mt$; then $x = yt$ with

$y \in \mathrm{M}$; if $\varphi\colon \mathrm{M} \to \mathrm{M/N}$ is the natural homomorphism, $\varphi(y)t = 0$, hence $\varphi(y) = 0$ by assumption, and this means that $y \in \mathrm{N}$, in other words $\mathrm{N} \cap \mathrm{M}t = \mathrm{M}t$; as $\mathrm{N}\omega \subset \mathrm{M}\omega \subset \mathrm{M}t$, we have $\mathrm{N}\omega \subset \mathrm{N} \cap (\mathrm{M}t) = \mathrm{N}t$ and N is distinguished. Conversely, the relation $\mathrm{N}t = \mathrm{N} \cap \mathrm{M}t$ implies that N is distinguished, as we have just seen; furthermore, the relation $\varphi(y)t = 0$ for $y \in \mathrm{M}$ implies $yt \in \mathrm{N} \cap \mathrm{M}t = \mathrm{N}t$, hence $y \in \mathrm{N}$ and $\varphi(y) = 0$; finally, as $\mathrm{M}\omega \subset \mathrm{M}t$, we clearly have $\varphi(\mathrm{M})\omega \subset \varphi(\mathrm{M})t$, and this shows that $\varphi(\mathrm{M}) = \mathrm{M/N}$ is distinguished.

In addition, if $x_1, \ldots, x_n$ is a minimal system of generators of N, their images $\bar{x}_i$ modulo $\mathrm{N}t$ form a basis (over $k$) of $\mathrm{N}/\mathrm{N}t$, and this $k$-vector space is identified with $(\mathrm{N} + \mathrm{M}t)/\mathrm{M}t$ since $\mathrm{N}t = \mathrm{N} \cap \mathrm{M}t$; there exist therefore $n - m$ elements $\bar{x}_i$ $(n + 1 \leqslant i \leqslant m)$ of $\mathrm{M}/\mathrm{M}t$ which, together with the $\bar{x}_i$ for $1 \leqslant i \leqslant n$, constitute a basis for $\mathrm{M}/\mathrm{M}t$ over $k$; hence, if $x_i$ $(n + 1 \leqslant i \leqslant m)$ is an element of $\bar{x}_i$, the $x_i$ for $1 \leqslant i \leqslant m$ constitute a minimal system of generators of M. As the classes modulo N of the $x_i$ for $n + 1 \leqslant i \leqslant m$ generate M/N, we have $\mathrm{rk(M/N)} \leqslant m - n$; on the other hand, if $y_1, \ldots, y_r$ are elements of M whose classes modulo N generate M/N, then $x_1, \ldots, x_n, y_1, \ldots, y_r$ generate M, hence $m \leqslant n + r$, whence $\mathrm{rk(M/N)} \geqslant m - n$, which ends the proof of (i). We note in addition that the relation $\mathrm{rk(N)} = \mathrm{rk(M)}$ is only possible (when M/N is distinguished) if $\mathrm{M} = \mathrm{N}$, since for a distinguished module P, the relation $\mathrm{rk(P)} = 0$ is equivalent to $\mathrm{P} = 0$.

(ii) and (iii). If $\mathrm{N} \subset \mathrm{M}$ is distinguished, so is $\mathrm{N}' = \mathrm{M} \cap \mathrm{NR}$: indeed, we have $\mathrm{N}'t = \mathrm{M}t \cap \mathrm{NR}$ (since $\mathrm{R}t = \mathrm{R}$) hence

$$\mathrm{N}'\omega \subset \mathrm{M}\omega \cap \mathrm{NR} \subset \mathrm{M}t \cap \mathrm{NR} \subset \mathrm{N}'t.$$

In addition $\mathrm{N}' \cap \mathrm{M}t = \mathrm{N}'t$. By (i) we have therefore

$$\mathrm{rk(N')} + \mathrm{rk(M/N')} = \mathrm{rk(M)}. \tag{16}$$

If $(z_i)_{1 \leqslant i \leqslant n'}$ is a system of generators of N′, the relation $\mathrm{N}' \subset \mathrm{NR}$ shows that there exists an integer $r > 0$ such that $z_i \in \mathrm{N}t^{-r}$ for each $i$, hence $\mathrm{N}'t^r \subset \mathrm{N}$. The module N′/N is then a quotient of $\mathrm{N}'/\mathrm{N}'t^r$; we know that $\mathrm{N}'/\mathrm{N}'t$, which is a $k$-vector space of finite dimension, has finite length; as $\mathrm{N}'t^q/\mathrm{N}'t^{q+1}$ is isomorphic to $\mathrm{N}'/\mathrm{N}'t$ for each $q > 0$, we conclude that $\mathrm{N}'/\mathrm{N}'t^r$ has finite length, hence also N′/N. Furthermore, let $s_1$ be the largest integer $s$ such that $\mathrm{N} \subset \mathrm{N}'t^s$; then $(\mathrm{N} + \mathrm{N}'t^{s_1+1})/\mathrm{N}'t^{s_1+1}$ is a subspace of the $k$-vector space $\mathrm{N}'t^{s_1}/\mathrm{N}'t^{s_1+1}$, hence the image by the canonical isomorphism $\mathrm{N}'/\mathrm{N}'t \to \mathrm{N}'t^{s_1}/\mathrm{N}'t^{s_1+1}$ of a subspace of $\mathrm{N}'/\mathrm{N}'t$;

let $(x_i)_{1 \leqslant i \leqslant d_1}$ be a system of elements of N′ such that the classes modulo $N't^{s_1+1}$ of the elements $x_i t^{s_1}$ constitute a basis of $(N + N't^{s_1+1})/N't^{s_1+1}$; we may assume that $x_i t^{s_1} \in N$ for $1 \leqslant i \leqslant d_1$; if $P_1$ is the submodule of N generated by the $x_i t^{s_1}$, it follows from (i) that $P_1$ is a distinguished submodule of $N't^{s_1}$, and that $N'_1 = N't^{s_1}/P_1$ is distinguished. We may then apply inductively the same process to $N'_1$ and to $N_1 = N/P_1 \supset N'_1 t^{r-s_1}$, which is distinguished in $N'_1$, to obtain the system of generators of N′ consisting of the $x_i$ with $1 \leqslant i \leqslant d_1 + d_2 + \cdots + d_h$, with the property described in Proposition 8; one then completes that system with $m - (d_1 + \cdots + d_h)$ elements $x_i$ to obtain a system of generators of M, by the process described in (i), N′ being such that M/N′ is distinguished. The invariance of the numbers $s_i$ and $d_i$ follows from the consideration of the dimensions of the $k$-vector spaces $(N \cap N't^q)/(N \cap N't^{q+1})$: that dimension is $d_1 + \cdots + d_i$ for $s_i \leqslant q < s_{i+1}$. This also shows that $s_1 d_1 + \cdots + s_h d_h = \text{length}(N'/N)$.

The preceding argument obviously proves that rk(N) = rk(N′) $\leqslant$ rk(M), and that the relation rk(N) = rk(M) is equivalent to N′ = M, by the remark made at the end of (i). In particular, it proves that condition a) in (ii) implies b) and c); we have also in passing shown how b) implies c).

It remains to prove that c) implies a); otherwise, we would have $N' \neq M$ and $Q = M/N'$, which is a quotient of M/N, would have finite length. However as $Q \neq 0$ is distinguished, we would have $Q/Qt \neq 0$, hence also

$$Qt^q/Qt^{q+1} \neq 0 \qquad \text{for every} \quad q > 0,$$

and this shows that Q has infinite length.

**Corollary 1.** *If* M *and* M′ *are two distinguished* A*-modules such that* $M_{(R)}$ *and* $M'_{(R)}$ *are isomorphic* R*-modules, then* M *and* M′ *have the same rank.*

Using an R-isomorphism, we may assume that $M_{(R)} = M'_{(R)}$; as M and M′ are finitely generated and generate $M_{(R)}$, and every element of $M_{(R)}$ has the form $xt^{-m}$ for $x \in M$ and $m \geqslant 0$, we may assume, replacing if necessary M′ by $M't^m$ for a suitable integer $m \geqslant 0$, that $M' \subset M$. The same argument, where M and M′ are exchanged, also shows that we may assume the existence of an $r \geqslant 0$ such that $Mt^r \subset M'$; the corollary then follows from part (ii) of Proposition 8.

Let M be a distinguished A-module, and consider the sub-A-module $M\omega$ of M; as $M\omega^2 \subset Mt\omega = M\omega t$, $M\omega$ is distinguished. We will say that M is *equidimensional* if $\mathrm{rk}(M\omega) = \mathrm{rk}(M)$. We will define the *corank* of M and write crk(M) the number $+\infty$ if M is not equidimensional, and the number length$(M/M\omega)$ if M is equidimensional. In order that M be equidimensional, a necessary and sufficient condition is that the homothetic mapping $x \mapsto x\omega$ of M into itself be *injective*: indeed, if this property holds, then $M\omega$ is isomorphic to M and $\mathrm{rk}(M\omega) = \mathrm{rk}(M)$. Conversely, if there exists $x \in M$ such that $x\omega = 0$ and $x \neq 0$, we may assume (multiplying if necessary $x$ by some $t^{-r}$) that $x \notin Mt$, hence that $x$ belongs to a minimal system of generators of M; but the relation $x\omega = 0$ then implies that $M\omega$ is generated by at most $\mathrm{rk}(M) - 1$ elements, hence $\mathrm{rk}(M\omega) < \mathrm{rk}(M)$.

**Corollary 2.** *If* M *and* M′ *are two distinguished* A*-modules such that* $M_{(R)}$ *and* $M'_{(R)}$ *are isomorphic* R*-modules, then* $\mathrm{crk}(M) = \mathrm{crk}(M')$.

As in Corollary 1, we may assume that $M' \subset M$; if $\mathrm{rk}(M\omega) < \mathrm{rk}(M)$, then as $M'\omega \subset M\omega$, we have

$$\mathrm{rk}(M'\omega) < \mathrm{rk}(M) = \mathrm{rk}(M') \qquad \text{(by Corollary 1)}$$

hence

$$\mathrm{crk}(M') = \mathrm{crk}(M) = +\infty.$$

If on the other hand M is equidimensional, then $x \mapsto x\omega$ is injective in M, hence in M′, and M′ is equidimensional; furthermore $M\omega/M'\omega$ is isomorphic to $M/M'$, and therefore

$$\begin{aligned}\mathrm{length}(M/M') + \mathrm{crk}(M') &= \mathrm{length}(M/M'\omega)\\ &= \mathrm{length}(M/M\omega) + \mathrm{length}(M\omega/M'\omega)\\ &= \mathrm{crk}(M) + \mathrm{length}(M/M')\end{aligned}$$

which proves the corollary.

We will see in §5, No. 5 that when M is equidimensional crk(M) is equal to *the corank* $\mathrm{crk}(M_{(R)})$ defined in No. 2.

## §5. The case of an algebraically closed field.

The first goal of this paragraph is to determine explicitly the structure of the *finitely generated torsion* R-*modules*, when R is a localized Hilbert-Witt ring over an *algebraically closed field k* of characteristic $p > 0$. Until further notice, $k$ will therefore always be assumed to be algebraically closed, and the notations of §4 are still in force.

**1.** *Preliminary lemmas.* For any element $z \in \mathrm{W}(k)$ of the Witt ring $\mathrm{W}(k)$, we will write $\bar{z}$ the class of $z$ in $\mathrm{W}/\mathrm{W}\omega = k$.

**Lemma 1.** *Let $a_i$ $(0 \leqslant i \leqslant n)$ be $n + 1$ elements of* W, *and let $\xi_0$ be an element of k verifying the equation*

$$\bar{a}_0 + \bar{a}_1\xi_0 + \bar{a}_2\xi_0^p + \cdots + \bar{a}_n\xi_0^{p^{n-1}} = 0. \tag{1}$$

*Then, if the $\bar{a}_i$ $(1 \leqslant i \leqslant n)$ are not all* 0, *there exists at least an element $x \in \mathrm{W}$ such that*

$$a_0 + a_1x + a_2x^\sigma + \cdots + a_nx^{\sigma^{n-1}} = 0 \tag{2}$$

*and that $\bar{x} = \xi_0$.*

It is enough to show that we may determine in W two sequences of elements $x_0, x_1, \ldots, x_n, \ldots$ and $b_1, b_2, \ldots, b_n, \ldots$ such that $\bar{x}_0 = \xi_0$ and that the following system of equations is satisfied:

$$(3)\qquad \begin{cases} a_0 + a_1x_0 + a_2x_0^\sigma + \cdots + a_nx_0^{\sigma^{n-1}} = -\omega b_1 \\ b_1 + a_1x_1 + a_2x_1^\sigma + \cdots + a_nx_1^{\sigma^{n-1}} = -\omega b_2 \\ \qquad\vdots \\ b_i + a_1x_i + a_2x_i^\sigma + \cdots + a_nx_i^{\sigma^{n-1}} = -\omega b_{i+1} \\ \qquad\vdots \end{cases}$$

To obtain $x_0$ satisfying the first equation (3), it is enough to take an arbitrary element $x_0$ such that $\bar{x}_0 = \xi_0$, and then the left-hand side of the the first equation (3) is a multiple of $\omega$ in W. Suppose the $x_h$ have been determined for $h \leqslant i - 1$ and the $b_h$ for $h \leqslant i$, satisfying the first $i$ equations (3). We then take in $k$ a root $\xi_i$ of the algebraic equation

$$\bar{b}_i + \bar{a}_1\xi_i + \bar{a}_2\xi_i^p + \cdots + \bar{a}_n\xi_i^{p^{n-1}} = 0 \tag{4}$$

which always exists since $k$ is algebraically closed and the coefficients $\bar{a}_i$ $(1 \leqslant i \leqslant n)$ are not all 0; then one takes for $x_i$ an element such that $\bar{x}_i = \xi_i$, and the left-hand side of the $(i + 1)$-st equation (3) is a multiple of $\omega$ in W, which determines $b_{i+1}$. Then the series

$$x = x_0 + x_1\omega + x_2\omega^2 + \cdots + x_i\omega^i + \cdots$$

is convergent in W, and $x$ is a solution of (2).

**Lemma 2.** *Let $r$ be an integer $\geqslant 1$.*

(i) *For any invertible element $a \in$ A, the right ideal $(\omega^r - at)$R of R is maximal* (in other words $\omega^r - at$ is irreducible).

(ii) *For all invertible elements $a \in$ A, the right (simple) R-modules R$/(\omega^r - at)$R are isomorphic.*

(i) As $c(\omega^r - at) = 1$, the relation $\omega^r - at = xy$ in R implies $c(x) + c(y) = 1$ for the costathm $c$, hence one of the two integers $c(x)$, $c(y)$ must be 0, and therefore $x$ or $y$ must be divisible by a power of $\omega$ if it is not invertible in R (§4, No. 1); but as $\omega^r - at$ is not divisible by a power of $\omega$, $x$ or $y$ must be invertible, and $\omega^r - at$ is irreducible.

(ii) We prove that there exists an invertible element

$$x = x_0 + x_1t + \cdots + x_nt^n + \cdots$$

in A (with coefficients $x_i \in$ W) such that

$$x(\omega^r - at) = (\omega^r - t)x, \tag{5}$$

which will obviously prove the assertion in (ii). Write

$$a = a_0 + a_1t + \cdots + a_nt^n + \cdots$$

with $a_i \in \mathrm{W}$ and $w(a_0) = 0$. Equation (5) is equivalent to the system of equations for the $x_i$

$$(6)\qquad \begin{cases} x_0^\sigma - a_0 x_0 = 0 \\ x_1^\sigma - a_0^\sigma x_1 = a_1 x_0 \\ \quad\vdots \\ x_n^\sigma - a_0^{\sigma^n} x_n = a_1^{\sigma^{n-1}} x_{n-1} + a_2^{\sigma^{n-2}} x_{n-2} + \cdots + a_n x_0 \\ \quad\vdots \end{cases}$$

and this system is solved by induction on $n$ and application of Lemma 1; since $\bar{a}_0 \neq 0$, we may suppose that $\bar{x}_0 \neq 0$ in $k$, and then $x_0$ is invertible in the ring W, and $x$ is invertible in A.

**Lemma 3.** *Let $r$ be an integer $\geqslant 1$, $a$ and $b$ two invertible elements in* A; *there exists an element $y \in \mathrm{R}$ such that*

$$(7)\qquad y(\omega^r - at) - (\omega^r - bt)y = 1.$$

Let

$$a = \sum_{i=0}^{\infty} a_i t^i, \quad b = \sum_{i=0}^{\infty} b_i t^i, \qquad \text{with} \quad \bar{a}_0 \neq 0 \quad \text{and} \quad \bar{b}_0 \neq 0,$$

and let us look for a solution of (7) of the form

$$y = \left(\sum_{i=0}^{\infty} y_i t^i\right) t^{-1} \qquad \text{with} \quad y_i \in \mathrm{W};$$

we have to solve a system of equations:

$$(8)\qquad \begin{aligned} b_0 y_0^\sigma - a_0^{\sigma^{-1}} y_0 &= 1 \\ b_0 y_1^\sigma - a_0 y_1 &= f_1(y_0) \\ &\ \vdots \\ b_0 y_n^{\sigma^n} - a_0^{\sigma^{n-1}} y_n &= f_n(y_0, \ldots, y_{n-1}) \\ &\ \vdots \end{aligned}$$

where the $f_i$ are linear in the $y_j$ of index $< i$ and their transforms by powers of $\sigma$. This is again solved by induction on $i$ and application of Lemma 1, since $\bar{b}_0 \neq 0$.

Let $x = x_0 + x_1 t + \cdots + x_n t^n + \cdots$ (with $x_i \in \mathrm{W}$) an element of A, such that $x_0 \neq 0$ and is not divisible in A by $\omega$, so that $h = c(x) > 0$. By definition of $c(x)$, we have $w(x_i) > 0$ for $0 \leqslant i < h$, and $x_h$ is invertible in W. Let $\gamma(x)$ be the smallest of the rational numbers $w(x_i)/(h - 1)$ for $0 \leqslant i < h$, and let us write $\gamma(x) = r(x)/s(x)$ where $r = r(x)$ and $s = s(x)$ are prime to each other; finally, let $j = j(x)$ be the smallest of the integers $l \leqslant h - 1$ such that

$$w(x_l)/(h - l) = \gamma(x);$$

we therefore have $h - j = qs$ and $w(x_j) = qs$ for an integer $q \geqslant 1$, and the preceding definitions imply the following relations:

$$(9) \qquad \begin{aligned} w(x_i) &> (h - i)r/s \qquad \text{for} \quad i < j; \\ w(x_j) &= (h - j)r/s \\ w(x_i) &\geqslant (h - i)r/s \qquad \text{for} \quad j < i < h \end{aligned}$$

("Newton's polygon").

**Lemma 4.** *Let $a$ be an element of* A *not divisible* (*in* A) *by $\omega$ nor by $t$, and such that $c(a) = h > 0$ and $s(a) = 1$. Then there exists in* A *two invertible elements $b$, $b'$ such that $a$ is divisible on the left by $\omega^r - bt$ and on the right by $\omega^r - b't$ (with $r = r(a)$).*

Let $j = j(a)$; we have

$$a = a_0 + a_1 t + \cdots + a_n t^n + \cdots \qquad \text{with} \quad a_i \in \mathrm{W} \text{ for all } i;$$

the assumptions are that, for $0 \leqslant i \leqslant h - 1$,

$$\begin{aligned} a_i &= \omega^{r(h-i)} u_i \qquad \text{with} \quad u_i \in \mathrm{W}, \\ w(u_j) &= 0 \qquad \text{and} \qquad w(a_h) = 0. \end{aligned}$$

We look for solutions of the equation

$$ax = y(\omega^r - t) \tag{10}$$

with $x = x_0 + x_1 t + \cdots + x_n t^n + \cdots$ and $w(x_0) = 0$, and

$$y = \omega^{r(h-1)} y_0 + \omega^{r(h-2)} y_1 t + \cdots + y_{h-1} t^{h-1} \tag{11}$$

with $y_i \in \mathrm{A}$. Equation (10) is then a consequence of the system

$$(12) \quad \begin{cases} u_0 x = y_0 \\ u_1 x^{\sigma} = y_1 - y_0 \\ \vdots \\ u_{h-1} x^{\sigma^{h-1}} = y_{h-1} - y_{h-2} \\ z x^{\sigma^h} = -y_{h-1} \end{cases}$$

where

$$z = a_h + a_{h+1} t + \cdots + a_{h+n} t^n + \cdots.$$

It is clear that this system has solutions if and only if $x$ satisfies the relation

$$u_0 x + u_1 x^{\sigma} + \cdots + u_{h-1} x^{\sigma^{h-1}} + z x^{\sigma^h} = 0 \tag{13}$$

which (since the $x_i$ are in W) is equivalent to the system of equations

$$(14) \quad \begin{cases} u_0 x_0 + u_1 x_0^{\sigma} + \cdots + u_{h-1} x_0^{\sigma^{h-1}} + a_h x_0^{\sigma^h} = 0 \\ u_0 x_i + u_1 x_i^{\sigma} + \cdots + u_{h-1} x_i^{\sigma^{h-1}} + a_h x_i^{\sigma^h} = -\sum_{l=1}^{i} a_{h+l} x_{i-l}^{\sigma^{h+l}} \\ \qquad \text{for} \quad i \geqslant 1. \end{cases}$$

Now, as $\bar{a}_h \neq 0$ and $\bar{u}_j \neq 0$, Lemma 1 proves that the first equation (14) has a solution $x_0$ such that $\bar{x}_0 \neq 0$, and then, by induction on $i$, that each of the other equations (14) has a solution $x_i$ in W, hence the existence of $x$ and $y$ satisfying (10) with $x$ *invertible* in A. We may therefore write

$$a = y x^{-1}(\omega^r - x x^{-\sigma} t)$$

which proves the divisibility on the right by an element of type $\omega^r - b't$ with $b'$ invertible in A; the divisibility on the left is proved in a similar way.

**Lemma 5.** *Let*

$$a = u(\omega^{r_1} - b_1 t)(\omega^{r_2} - b_2 t) \cdots (\omega^{r_q} - b_q t) \tag{15}$$

*where the $r_i$ are integers $\geqslant 1$, $u$ and the $b_i$ invertible elements in* A; *then the* R*-module* M $=$ R$/a$R *is isomorphic to the direct sum of the simple* R*-modules* R$/(\omega^{r_i} - t)$ $(1 \leqslant i \leqslant q)$.

We use induction on $q$; for $q = 1$, the lemma is Lemma 2(ii). Next consider the case $q = 2$; we may disregard the invertible element $u$, and therefore suppose that

$$a = (\omega^r - b_1 t)(\omega^s - b_2 t) = \omega^{r+s} - \omega^r b_2 t - \omega^s b_1 t + b_1 b_2^{\sigma} t^2.$$

Suppose first that $r \neq s$, and suppose first that $r > s$; then, with the notations introduced above, $c(a) = 2$, $\gamma(a) = s$, hence $r(a) = s$, $s(a) = 1$; it follows from Lemma 4 that we have $a = (\omega^s - b_2' t)a'$ for an invertible element $b'$ in A. But as we have $v(a) = s + v(a')$ and $c(a) = 1 + c(a')$, and $v(a) = r + s$, $c(a) = 2$, we have necessarily $a' = \omega^r - b_1' t$ with $b_1'$ inversible in A.

Now, the R-module M contains the submodules $\mathrm{M}_1$, generated by the class mod $a$R of $\omega^r - b_1 t$, and $\mathrm{M}_2$, generated by the class mod $a$R of $\omega^s - b_2' t$; we know (Lemma 2) that $\mathrm{M}_1$ (resp. $\mathrm{M}_2$) is isomorphic to R$/(\omega^r - t)$R (resp. to R$/(\omega^s - t)$R), and as $r \neq s$, $\mathrm{M}_1$ and $\mathrm{M}_2$ are *simple non isomorphic* R-modules (§3, No. 2). Furthermore, $\mathrm{M}/\mathrm{M}_1$ is isomorphic to $\mathrm{M}_2$ (Lemma 2), hence M is an R-module of length 2; as $\mathrm{M}_1 \cap \mathrm{M}_2 = 0$, M is the *direct sum* of $\mathrm{M}_1$ and $\mathrm{M}_2$. The argument is similar when $r < s$, writing this time

$$a = (\omega^r - b_1'' t)a''$$

by Lemma 2.

When $r = s$, we use Lemma 3 above, in conjunction with §4, No. 2, Lemma 1; the latter shows that M is the direct sum of two simple modules isomorphic to R$/(\omega^r - t)$R.

We finally consider the case in which $q$ is arbitrary; let us write

$$a' = (\omega^{r_1} - b_1 t)(\omega^{r_2} - b_2 t)$$

and let P be the submodule of M generated by the class mod $a$R of $a'$; the quotient module M/P is therefore isomorphic to R$/a'$R, hence the direct sum of two simple modules $M'_1$, $M'_2$; M is sum of the inverse images $M_1$, $M_2$ of $M'_1$, $M'_2$ in M. But $M_1$ and $M_2$, being submodules of a monogenic module $R/aR$, are themselves monogenic since R is a euclidean ring; as their length is $q - 1$, and their composition factors are isomorphic to modules of the form $R/(\omega^{r_i} - t)R$, we may apply to $M_1$ and $M_2$ the inductive hypothesis, which shows that they are sums of simple modules. Therefore M is sum of simple modules, hence *semisimple* ([6], §3, No. 3, Proposition 7); as its composition factors are isomorphic to the $R/(\omega^{r_i} - t)R$, M is isomorphic to the direct sum of these modules.

**Lemma 6.** *With the notations introduced above, if an element $a \in A$, not divisible (in A) by $\omega$ nor by $t$, is such that*

$$c(a) = h > 0, \qquad s(a) = 1, \qquad j(a) = 0,$$

*then $a$ is divisible (on the left or on the right) by a product of $h = c(a)$ factors of the form $\omega^r - b_i t$ $(1 \leqslant i \leqslant h)$, where the $b_i$ are invertible in A and $r = r(a)$.*

If we go back to the proof of Lemma 4, it is clear that $y_{h-1}$ is invertible in A as well as $y_0$ (since by assumption $w(u_0) = 0$); therefore

$$c(y) = h - 1, \quad s(y) = 1, \quad j(y) = 0, \quad r(y) = r.$$

We may therefore apply again Lemma 4 to $y$, and by induction on $h$ we obtain the lemma.

**2.** *Structure of finitely generated torsion* R*-modules.* We recall that, from the beginning of §4, the rings A and R have been constructed starting from a ring $W_e$. For any integer $d > 0$, let A′, R′ be the Hilbert-Witt ring and localized Hilbert-Witt ring defined similarly, but starting from the ring $W_{ed}$; then A′ (resp. R′) is obviously a free A-module (resp. a free R-module) on the left and on the right, with basis $1, \omega_{ed}, \ldots, \omega_{ed}^{d-1}$. For any right R-module M, the R′-module $M \otimes_R R'$ is therefore, when considered as an R-module, *isomorphic to the direct sum of d modules isomorphic to* M. In addition, if $M = R/aR$ for an $a \in R$, then $M' = R'/aR'$.

**Lemma 7.** *Let $a \in R$ be an element not divisible by $\omega = \omega_e$; there exists an integer $d > 0$ such that in the ring R′ corresponding to $W_{ed}$,*

$$a = u'(\omega'^{r_1} - b'_1 t) \cdots (\omega'^{r_q} - b'_q t) \tag{16}$$

*where* $\omega' = \omega_{ed}$, *the* $r_i$ *are integers* $\geqslant 1$, *the* $b'_i$ *are invertible in* $A'$ *and* $u'$ *is invertible in* $R'$.

We may assume that $h = c(a) > 0$, since otherwise $a$ is invertible in R. With the notations introduced in No. 1, let $d_1 = s(a)$; if $a$ is considered as an element *of the ring* $R_1$ corresponding to $W_{ed_1}$, then by definition $s(a) = 1$ and Lemma 4 shows that one may write

$$a = (\omega_{ed_1}^{m_1} - b_1 t)a_1$$

where $m_1 \geqslant 1$, $a_1 \in R_1$, and $b_1$ is invertible in $A_1$. As $c(a_1) = h - 1$ in $R_1$, we may use induction on $h$, passing at each step from a ring $R_i$ corresponding to $W_{e_i}$, to a ring $R_{i+1}$ corresponding to $W_{e_i d_i}$ for a suitable integer $d_i > 1$. This proves the lemma.

**Theorem 1.** *For any* $a \in R$ *not divisible by* $\omega$ *nor by* $t$, $R/aR$ *is a semisimple* R-*module.*

We have to prove that any nontrivial submodule N of $M = R/aR$ is a *direct summand* of M ([6], §3, No. 3, Proposition 7). As R is a euclidean ring, N is isomorphic to $R/a_2R$ for an element $a_2$ such that $a = a_1 a_2$, and we may assume that neither $a_1$ nor $a_2$ is invertible in R. Lemma 7 shows that, in a ring $R'$ corresponding to $W_{ed}$ for a suitable $d$, $a_1$ and $a_2$ are products of factors of type $\omega'^{r_i} - b'_i t$, and it then follows from Lemma 5 that $M' = R'/aR'$ is semisimple, and $N' = R'/a_2R'$ a direct summand of $M'$. We then conclude from §4, No. 2, Lemma 1 that there exist two elements $x'$, $y'$ in $R'$ such that $x'a_1 + a_2 y' = 1$. But if we write

$$x' = \sum_{i=0}^{d-1} x_i \omega_{ed}^i, \qquad y' = \sum_{i=0}^{d-1} y_i \omega_{ed}^i$$

with coefficients $x_i$, $y_i$ in R, the preceding relation yields in particular $x_0 a_1 + a_2 y_0 = 1$, and an application of §4, No. 2, Lemma 1 shows that N is a direct summand in M.

**Theorem 2.** *Let* $a \in R$ *be an element not divisible by* $\omega$; *in order that* $M = R/aR$ *be a simple* R-*module, a necessary and sufficient condition* *is that* M *be isomorphic to an* R-*module of type* $R/(\omega^m - t^n)R$, *where* $m$ *and* $n$ *are integers* $\geqslant 1$, *prime to each other.*

Suppose first that $a = \omega^m - t^n$, with $m$ and $n$ integers $\geqslant 1$ and prime to each other; let $d$ be an integer multiple of $n!$, and consider the ring R′ corresponding to $W_{ed}$; in that ring, it follows from the definitions of No. 1 that

$$c(a) = n, \quad r(a) = md/n, \quad s(a) = 1, \quad j(a) = 0;$$

applying Lemma 6, we see that, in R′, $a$ is a product of an invertible element and $n$ elements of type $\omega'^{md/n} - b'_i t$, where $\omega' = \omega_{ed}$ and the $b'_i$ are invertible in A′. Hence $R'/aR'$ is a direct sum of $n$ simple R′-modules, all isomorphic to $R'/(\omega'^{md/n} - t)R'$ by Lemma 5. Suppose $R/aR$ is not simple; we would then have

$$a = a_1 a_2 \qquad \text{with} \quad c(a_1) > 0 \quad \text{and} \quad c(a_2) > 0;$$

if we write $c(a_2) = n'$, we would then have $n' < n$; furthermore, the choice of $d$ is such that

$$s(a_2) = 1, \qquad r(a_2) = m'd/n',$$

and Lemma 4 shows that, in R′, $a_2$ is divisible by an element of type $\omega'^{m'd/n'} - b''t$ with $b''$ invertible in A′. But then

$$R'/(\omega'^{m'd/n'} - b''t)R'$$

would be isomorphic to a factor of a Jordan-Hölder series of $R'/aR'$, and by §4, No. 2 we should have $m'd/n' = md/n$, which is absurd since $n' < n$ and $m$ and $n$ are prime to each other.

Observe in addition that $R'/aR' = M \otimes_R R'$, as an R-*module*, is a direct sum of the $d$ modules $\omega'^i M$ $(0 \leqslant i < d - 1)$ which are all isomorphic to M, hence simple as we have just seen; the Krull-Schmidt theorem ([6], §2, No. 2, Theorem 1) shows that, as an R-*module*, M is isomorphic to $R'/(\omega'^{md/n} - t)R'$, which proves that the latter is a *simple* R-*module.*
Now suppose conversely that $M = R/aR$ is simple, and let

$$h = c(a) > 0, \quad r = r(a), \quad s = s(a), \quad j = j(a).$$

We take for $d$ a multiple of $s$, satisfying the conditions of Lemma 7 for $a$; then by Lemma 5, $M' = R'/aR'$ is direct sum of simple R'-modules, one of which is isomorphic to $R'/(\omega'^{rd/s} - t)R'$; but, as R-*module*, this last module is isomorphic to the *simple* R-module $R/(\omega^r - t^s)R$, as we have seen above. On the other hand, M', as an R-*module*, is isomorphic to the direct sum of the $d$ R-modules $\omega'^i M$ $(0 \leqslant i \leqslant d - 1)$ which are isomorphic to M, hence *simple*; the Jordan-Hölder theorem therefore shows that M is isomorphic to $R/(\omega^r - t^s)R$.

**Corollary.** *Every finitely generated torsion* R-*module is isomorphic to a direct sum of modules of type* $R/\omega^{r_i}R$ *and of type* $R/(\omega^{m_j} - t^{n_j})R$, *where the pairs* $(m_j, n_j)$ *consist of integers* $\geqslant 1$ *and prime to each other. Furthermore, the integers* $r_i$ *and the pairs* $(m_j, n_j)$ *are uniquely determined up to a permutation.*

The existence of the decomposition follows from Theorems 1 and 2 and from §4, Proposition 3; its uniqueness results from the Krull-Schmidt theorem and from the characterization of the rank and corank of a torsion R-module (§4, No. 2).

**3.** *Endomorphisms of indecomposable torsion* R-*modules.* As $\omega^m R$ is a *two-sided* ideal in R, it is clear that the ring of endomorphisms of the R-module $R/\omega^m R$ is isomorphic to the *ring* $R/\omega^m R$: the latter may be described as the ring of formal power series

$$a_h t^h + a_{h+1} t^{h+1} + \cdots,$$

with $h \in \mathbf{Z}$ and the coefficients $a_n$ *in the ring* $W/\omega^m W$, with the usual commutation law (§4, formula (1)).

As an R-module $R/(\omega^m - t^n)R$, with $m, n$ prime to each other and $\geqslant 1$, is simple, its ring of endomorphisms is a *sfield.* To determine that sfield, we introduce the following notations: $K_n$ is the field of fractions of the Witt ring $W(\mathbf{F}_{p^n})$, which is the unique unramified extension of degree $n$ of the $p$-adic field $\mathbf{Q}_p$ (which is the field of fractions of $W(\mathbf{F}_p)$); let $K_{e,n}$ be the completely ramified extension of degree $e$ of $K_n$, generated by a root $\omega_e$ of the polynomial $T^e - \omega$ of $K_n[T]$, and let $W_{e,n}$ be the ring of integers in $K_{e,n}$. It is clear that $K_{e,n}$ (resp. $W_{e,n}$) is the set of *invariant* elements under the automorphism $\sigma^n$ of $K_e$ (resp. $W_e$) (with the notations of §4, No. 1).

**Theorem 3.** *If m, n are two integers* $\geqslant 1$ *and prime to each other, the sfield of endomorphisms of the* R-*module* $R/(\omega^m - t^n)R$ *is isomorphic to the cyclic algebra* $(\omega^m, K_{e,n}, \sigma)$ [52]; *its center is the completely ramified extension* $V_e$ *of* $\mathbf{Q}_p$ *of degree e; the rank of the algebra* $(\omega^m, K_{e,n}, \sigma)$ *over* $V_e$ *is* $n^2$, *and its invariant is h/n, where h is the remainder of the euclidean division of m by n.*

Let $\varepsilon$ be the natural image of 1 in $M = R/(\omega^m - t^n)R$; for any endomorphism $u$ of M, we have

$$u(\varepsilon) = \varepsilon x \qquad \text{for an} \quad x \in R,$$

hence

$$u(\varepsilon t^n) = u(\varepsilon)t^n = \varepsilon x t^n \qquad \text{and} \qquad u(\varepsilon\omega^m) = u(\varepsilon)\omega^m = \varepsilon x \omega^m;$$

the element $x(\omega^m - t^n)$ must belong to the right ideal $(\omega^m - t^n)R$, and as

$$x(\omega^m - t^n) = (\omega^m - t^n)x + t^n(x - x^{\sigma^{-n}}),$$

this implies that

$$x - x^{\sigma^{-n}} = (\omega^m - t^n)a \qquad \text{for some} \quad a \in R.$$

From Lemma 1, it follows that there exists $y \in R$ such that $a = y - y^{\sigma^{-n}}$, and as $\sigma$ leaves $\omega$ and $t$ invariant, if we write

$$x' = x - (\omega^m - t^n)y,$$

we get $x' = x'^{\sigma^{-n}}$, or equivalently $x'^{\sigma^n} = x'$; furthermore $u(\varepsilon) = \varepsilon x'$. Conversely, for any $x' \in R$ such that $x'^{\sigma^n} = x'$, the homothetic mapping $z \mapsto x'z$ of R leaves the right ideal $(\omega^m - t^n)R$ stable, since $x'$ commutes with $t^n$; hence it gives in the quotient M an endomorphism $u_{x'}$ such that $u_{x'}(\varepsilon) = \varepsilon x'$. We have therefore defined a *surjective* mapping $x' \mapsto u_{x'}$ of the subring $R'$ of R consisting of the invariants of $\sigma^n$, onto the ring $\mathrm{End}_R(M)$; it is clear that

$$u_{x'y'} = u_{x'}u_{y'},$$

hence $x' \mapsto u_{x'}$ is a *ring* homomorphism. Finally the relation $u_{x'} = 0$

means that

$$x' \in (\omega^m - t^n)\mathrm{R} \cap \mathrm{R}' = (\omega^m - t^n)\mathrm{R}',$$

since $\omega^m - t^n$ belongs to R′.

The ring R′ consists of the formal power series

$$\sum_{i \geqslant h} x'_i t^i \qquad \text{with} \quad x'_i \in \mathrm{W}_{e,n}$$

and this implies that the right ideal $\mathfrak{J}$ in R′ generated by $\omega^m - t^n$ is in fact a two-sided ideal, since $\omega^m - t^n$ is in the center of R′. An element of R′ which is a polynomial in $t$ of degree $< n$,

$$a'_0 + a'_1 t + \cdots + a'_{n-1} t^{n-1} \qquad (\text{with } a'_i \in \mathrm{W}_{e,n})$$

and which belongs to $\mathfrak{J}$ is necessarily 0: indeed, if we write a relation

(17)

$$a'_0 + a'_1 t + \cdots + a'_{n-1} t^{n-1} = (\omega^m - t^n)(b'_0 + b'_1 t + \cdots + b'_i t^i + \cdots)$$

with $b'_i \in \mathrm{W}_{e,n}$, and identify the coefficients of terms in $t^{hn}$ for all $h \geqslant 0$, we obtain

$$a'_0 = \omega^m b'_0, \quad 0 = -b'_0 + \omega^m b'_n, \quad 0 = -b'_n + \omega^m b'_{2n}, \quad \ldots,$$

which gives

$$a'_0 = -\omega^{(h+1)m} b'_{hn} \qquad \text{for every} \quad h \geqslant 0,$$

a relation which is only possible in $\mathrm{W}_{e,n}$ if $a'_0 = 0$. But then $b'_0 = 0$, and as $t$ is invertible in R′, we may multiply both sides of (17) by $t^{-1}$ on the right, and we thus get a similar relation but with the left-hand side of degree $n - 2$; by induction, we conclude that

$$a'_0 = a'_1 = \cdots = a'_{n-1} = 0.$$

We therefore see that if $\tau$ is the image of $t$ in $\mathrm{R}'/\mathfrak{J}$, the elements $\tau^i$ for $0 \leqslant i \leqslant n - 1$ are linearly independent over the ring $\mathrm{W}_{e,n}$; furthermore, in $\mathrm{R}'/\mathfrak{J}$, the image of $\omega$ (which we may identify with $\omega$) is invertible, since $\omega^m = \tau^n$. The sfield $R'/\mathfrak{J}$, isomorphic to $\mathrm{End}_R(M)$, contains therefore the field of fractions $K_{e,n}$ of $W_{e,n}$, hence also the ring $K_{e,n}[\tau]$, which is a vector space over $K_{e,n}$ having as a basis the $\tau^i$ for $0 \leqslant i \leqslant n-1$. As we may write in R′

$$\sum_{i=0}^{\infty} x'_i t^i \equiv \sum_{j=0}^{n-1} \Big( \sum_{h=0}^{\infty} x_{j+hn} \omega^{hm} \Big) t^j \qquad \bmod \mathfrak{J}$$

we see that in fact we have $R'/\mathfrak{J} = K_{e,n}[\tau]$. The algebra $K_{e,n}[\tau]$ is isomorphic to the cyclic algebra $(\omega^m, K_{e,n}, \sigma)$ over the field $V_e$, since we have the relations $\tau x' = x'^{\sigma}\tau$ for $x' \in K_{e,n}$ and $\tau^n = \omega^m$ [52]. Q.E.D.

**4.** *Distinguished* A-*modules and torsion* R-*modules.* We still assume in this section that $k$ is algebraically closed. Let M be a distinguished right A-module of rank $n$ (§4, No. 3), and consider the corresponding R-module $M_{(R)} = M \otimes_A R$, which is a finitely generated torsion R-module, to which the preceding theory applies. The following theorem shows, however, that roughly only "half" of the finitely generated torsion R-modules may thus be derived from distinguished A-modules:

**Theorem 4.** *In order that a finitely generated torsion* R-*module, which is a direct sum of modules of type* $R/\omega^{r_i}R$ *and of type* $R/(\omega^{m_j} - t^{n_j})R$, *where the pairs* $(m_j, n_j)$ *consist of integers* $\geqslant 1$ *and prime to each other, be isomorphic to* $M_{(R)}$ *for a distinguished* A-*module* M, *a necessary and sufficient condition is that* $m_j \leqslant n_j$ *for every* $j$.

Indeed, if M is a distinguished A-module, it is naturally identified with a sub-A-module of $M_{(R)}$, the elements of $M_{(R)}$ being all of the form $xt^{-h}$ with $x \in M$ and $h \geqslant 0$ (§4, No. 3). If a sub-R-module N of $M_{(R)}$ is isomorphic to $R/(\omega^m - t^n)R$, we may therefore assume that it is generated by an element $x \in M$, whose annihilator in R is $(\omega^m - t^n)R$; hence

$$xt^{n-m} = x\omega^m t^{-m} = x\pi^m \in M$$

since M is distinguished (§4, No. 3); the same argument proves that, in fact, $xt^{h(n-m)} \in M$ for every integer $h > 0$. If we had $n < m$, this would imply that $x$ belongs to the intersection of the $Mt^q$ for all integers $q \geqslant 0$, which contradicts the assumption $x \neq 0$ (§4, No. 3, Corollary to Proposition 4).

Conversely, we are going to show that, for any integer $r > 0$ (resp. for any pair $(m, n)$ of integers prime to each other and such that $1 \leqslant m \leqslant n$) there is a distinguished A-module such that its tensor product with R is isomorphic to $R/\omega^r R$ (resp. $R/(\omega^m - t^n)R$).

In the construction which follows, we take for $m$ any integer $\geqslant 1$, and note $n$ any integer $\geqslant 0$ or the symbol $+\infty$. We write $M_{m,n}(k)$ the quotient of the right A-module $A^m$ by the sub-A-module, image of $A^m$ by the endomorphism $u$ whose matrix with respect to the canonical basis of $A^m$ is $I_m\omega - Vt$, where

$$V = \begin{pmatrix} 0 & 1 & 0 & \cdots & 0 \\ 0 & 0 & 1 & \cdots & 0 \\ \cdots & \cdots & \cdots & \cdots & \cdots \\ 0 & 0 & 0 & \cdots & 1 \\ t^n & 0 & 0 & \cdots & 0 \end{pmatrix} \tag{18}$$

with the conventions $t^\circ = 1$ and $t^{+\infty} = 0$; $M_{m,n} \otimes_A R$ is therefore isomorphic to the quotient of $R^m$ by the image $(u \otimes 1)(R^m)$, and $u \otimes 1$ has the same matrix as $u$ with respect to the canonical basis of $R^m$. However, as a matrix *with elements in* R, that matrix is seen to be *equivalent*, by "elementary transformations," successively to

$$\begin{pmatrix} -t & 0 & 0 & \cdots & 0 & \omega \\ \omega & -t & 0 & \cdots & 0 & 0 \\ 0 & \omega & -t & \cdots & 0 & 0 \\ \cdots & \cdots & \cdots & \cdots & \cdots & \cdots \\ 0 & 0 & 0 & \cdots & -t & 0 \\ 0 & 0 & 0 & \cdots & \omega & -t^{n+1} \end{pmatrix}$$

then to

$$(-t)^m \begin{pmatrix} 1 & 0 & 0 & \cdots & 0 & -\omega t^{-1} \\ 0 & 1 & 0 & \cdots & 0 & -\omega^2 t^{-2} \\ 0 & 0 & 1 & \cdots & 0 & -\omega^3 t^{-3} \\ \cdots & \cdots & \cdots & \cdots & \cdots & \cdots \\ 0 & 0 & 0 & \cdots & 1 & -\omega^{m-1} t^{-m+1} \\ 0 & 0 & 0 & \cdots & 0 & t^n - \omega^m t^{-m} \end{pmatrix}$$

and finally to

$$(-1)^{m-1}\mathrm{diag}(1, 1, \ldots, 1, \omega^m - t^{m+n})$$

where $t^{m+n}$ must be replaced by 0 if $n = +\infty$. Q.E.D.

**5.** *Descent to a perfect field.* We now return to the general assumptions of §4, no. 1, the field $k$ being only assumed to be perfect. We observe that the construction of the distinguished A-module $\mathrm{M}_{m,n}(k)$ in no. 4 carries over to that general situation. Furthermore:

**Proposition 1.** *The rank of* $\mathrm{M}_{m,n}(k)$ (§4, No. 3) *is equal to m and its corank* (§4, No. 3) *is equal to* $m + n$.

The first statement follows from §4, Proposition 7. Furthermore, let $\mathrm{N} = u(\mathrm{A}^m)$ with the notations of Theorem 4; $\mathrm{M}_{m,n}$ is isomorphic to $\mathrm{A}^m/\mathrm{N}$, hence $\mathrm{M}_{m,n}\omega$ to $(\mathrm{A}^m\omega + \mathrm{N})/\mathrm{N}$, and $\mathrm{M}_{m,n}/\mathrm{M}_{m,n}\omega$ to $\mathrm{A}^m/(\mathrm{A}^m\omega + \mathrm{N})$. But from the definition of $u$ and the form of the matrix $V$ in (18), it follows that, if $(e_j)_{1 \leqslant j \leqslant m}$ is the canonical basis of $\mathrm{A}^m$, $\mathrm{A}^m\omega + \mathrm{N}$ is the submodule of $\mathrm{A}^m$ generated by

$$e_1\omega, e_2\omega, \ldots, e_m\omega, e_1 t, e_2 t, \ldots, e_{m-1}t, e_m t^{n+1}.$$

Hence $\mathrm{A}^m/(\mathrm{A}^m\omega + \mathrm{N})$ is the direct product of $m - 1$ modules isomorphic to $\mathrm{A}/(\mathrm{A}\omega + \mathrm{A}t) = k$ and a module isomorphic to $\mathrm{A}/(\mathrm{A}\omega + \mathrm{A}t^{n+1})$; it is immediate from the definition of A (§4, No. 1) that this last module is an $(n + 1)$-dimensional $k$-vector space if $n$ is finite, an infinite-dimensional $k$-vector space if $n = +\infty$. Hence $\mathrm{M}_{m,n}/\mathrm{M}_{m,n}\omega$ has length $m + n$ if $n$ is finite, and infinite length if $n = +\infty$. QED.

**Proposition 2.** *The rank* (resp. *corank*) *of a distinguished* A-*module* M (*as defined in* §4, No. 3) *is equal to the rank of the torsion* R-*module* $\mathrm{M}_{(\mathrm{R})}$ (resp. *to the corank of* $\mathrm{M}_{(\mathrm{R})}$ *if* M *is equidimensional*) *as defined in* §4, No. 2.

When $k$ is algebraically closed, it follows from Theorem 4 and the corollary to Theorem 2 that $\mathrm{M}_{(\mathrm{R})}$ is isomorphic to $\mathrm{M}'_{(\mathrm{R})}$, where M′ is a direct sum of modules of type $\mathrm{M}_{m,n}(k)$. It follows from §4, Corollaries 1 and 2 to Proposition 8 that, to verify Proposition 2 for any distinguished A-module M, we only have to verify it for the modules $M_{m,n}(k)$, and then the conclusion follows at once from Proposition 1 and from the definition of the rank and corank of a torsion R-module $R/(\omega^m - t^{m+n})R$ (§4, No. 2).

To extend the result to the case in which $k$ is an arbitrary perfect field of characteristic $p$, let $k'$ be an algebraically closed extension of $k$, and A′ and R′ the Hilbert-Witt ring and the localized Hilbert-Witt ring corresponding to $k'$ and to the number $e$; it is clear that A and R are subrings of A′ and R′, respectively, and for any $a \neq 0$ in R, the stathm (resp. costathm) of *a in* R is the same as the stathm (resp. costathm) of *a in* R′. If M is a distinguished A-module, $M' = M \otimes_A A'$ is a distinguished A′-module having *same rank* (resp. *same corank*) as M (cf. §4, No. 3); furthermore

$$M'_{(R')} = M_{(R)} \otimes_R R',$$

hence, from the definition of the rank and corank of a torsion R-module (§4, No. 2) it follows that $M'_{(R')}$ has same rank (resp. same corank) as $M_{(R)}$; this ends the proof.

Still supposing that $k$ is an arbitrary perfect field of characteristic $p$, let us study in a little more detail the distinguished A-modules $M_{m,n}(k)$. By definition, $M_{m,n}(k)$ is a *monogenic* $A[\pi]$-*module*, with a generator $\bar{e}_1$; the images $\bar{e}_i$ $(1 \leqslant i \leqslant m)$ of the elements of the canonical basis $(e_i)$ of $A^m$ satisfy the conditions

$$(19) \qquad \bar{e}_2 = \bar{e}_1\pi, \quad \bar{e}_3 = \bar{e}_2\pi, \quad \ldots, \quad \bar{e}_m = \bar{e}_{m-1}\pi, \quad \bar{e}_m\pi = \bar{e}_1 t^n$$

and therefore the annihilator of $\bar{e}_1$ is the right ideal $(\pi^m - t^n)A[\pi]$ of $A[\pi]$.

**Proposition 3.** *The ring of endomorphisms of* $M_{r,\infty}(k)$ *is isomorphic to the quotient ring* $A[\pi]/\pi^r A[\pi]$.

Due to the relations $x\pi^r = \pi^r x^{\sigma^r}$ for $x \in R$, the ideal $\pi^r A[\pi]$ is a two-sided ideal in $A[\pi]$. Every endomorphism $u$ of a distinguished A-module M *commutes* with the semi-endomorphism $x \mapsto x\pi$, for the relation $y = x\pi$ is equivalent to $yt = x\omega$, and therefore $u(y)t = u(x)\omega$, i.e., $u(y) = u(x)\pi$; in other words, $u$ is also an $A[\pi]$-*endomorphism*. The proposition immediately follows from these remarks, since any endomorphism of the ring $B = A[\pi]/\pi^r A[\pi]$, considered as a right B-module, is obtained by multiplication on the left by an element of B.

**Proposition 4.** *For $m \geqslant 1$, $n > 0$, m and $m + n$ prime to each other, the ring of endomorphisms of the* A*-module* $M_{m,n}(k)$ *is isomorphic to the subring of the sfield* $K_{e,m+n}[\tau]$ *(Theorem 3) generated by the ring of integers* $W_{e,m+n}$ *of the valued field* $K_{e,m+n}$ *and by the elements $\tau$ and $\varpi = \omega\tau^{-1}$ (linked by the relation $\varpi^m = \tau^n$). This subring is an order of* $K_{e,m+n}[\tau]$*; it is a maximal order if and only if $m = 1$ or $n = 1$.*

Any endomorphism of $M_{m,n}(k)$ is obtained by multiplication on the left in $A[\pi]$ by an element $z \in A[\pi]$, and passage to the quotient modulo the ideal $(\pi^m - t^n)A[\pi]$; such an endomorphism exists if and only if it extends to an endomorphism of $R/(\omega^m - t^{m+n})R$, hence (Theorem 3) if and only if $z$ also belongs to the subring R′ of R consisting of formal power series in which all the coefficients belong to $W_{e,m+n}$; hence the description of the ring

$$B = \mathrm{End}(M_{m,n}(k)).$$

It is known [52] that the normalized valuation $w$ of the sfield $K_{e,m+n}[\tau]$ is such that

$$w(\omega) = m + n \qquad \text{and} \qquad w(\tau) = m;$$

let $\alpha$, $\beta$ be the smallest integers $> 0$ such that

$$\alpha(m + n) = \beta m + 1;$$

then $w(\omega^\alpha\tau^{-\beta}) = 1$, hence (*loc.cit.*) the maximal order C of $K_{e,m+n}[\tau]$ is the subring $W_{e,m+n}[\omega^\alpha\tau^{-\beta}]$. We now consider separately three cases:

1° $m = 1$, which implies $\alpha = 1$, $\beta = n$; then

$$C = W_{e,m+n}[\omega\tau^{-n}] = W_{e,m+n}[\tau]$$

since $\omega = \tau^{n+1}$; therefore $C \subset B$, and as obviously $B \subset C$, we have $B = C$.

2° $n = 1$, hence $\alpha = \beta = 1$; then

$$C = W_{e,m+n}[\omega\tau^{-1}] = W_{e,m+n}[\varpi],$$

and the same argument shows that $B = C$.

3° $m > 1$ and $n > 1$; then necessarily $\alpha > \beta$, otherwise we would have $\alpha(m + n) \leqslant \alpha m + 1$, which is absurd. But then $\omega^\alpha t^{-\beta} = \pi^\alpha t^{-(\beta-\alpha)}$ cannot belong to $A[\pi]$, for it is readily verified that if an element

$$\sum_k a_k t^k \in R' \qquad \text{with} \quad a_k \in W_{e,m+n}$$

belongs to $A[\pi]$, then necessarily $w(a_k) + k \geqslant 0$ for all exponents $k$; this shows that $B \neq C$ in that case.

We have excluded the case $n = 0$ in Proposition 4; in that case we must have $m = 1$ and $K_{e,m+n}[\tau]$ is reduced to its center $\mathbf{Q}_p(\omega^{1/e})$; furthermore $\omega = 1$, hence the ring of endomorphisms of $M_{m,n}(k)$ is then isomorphic to the ring of integers of the valued field $\mathbf{Q}_p(\omega^{1/e})$.

## §6. Applications to commutative reduced infinitesimal groups

**1.** *The classification "up to isogeny."* It follows, from §3, Theorem 1 and Proposition 7, §4, that to every reduced $n$-dimensional commutative bigebra $\mathfrak{G}$ over a perfect field $k$ is associated a *distinguished* module $\mathscr{E}(\mathfrak{G})$ of rank $n$ over the Hilbert-Witt ring $\mathscr{A} \subset \mathrm{End}(\mathfrak{G}_\infty(k))$ defined in §3, No. 2 (which corresponds to the choice of $e = 1$ in the definition of §4, No. 1), and conversely to any distinguished $\mathscr{A}$-module M of rank $n$ corresponds an $n$-dimensional commutative bigebra $\mathfrak{G}$ over $k$ such that $\mathscr{E}(\mathfrak{G})$ is isomorphic to M; furthermore that correspondence between reduced finite-dimensional commutative bigebras and distinguished $\mathscr{A}$-modules of finite rank sends isomorphic bigebras onto isomorphic modules and vice versa (§3, Theorem 2); finally, $\mathfrak{G}$ and $\mathfrak{G}'$ are *isogenous* if and only if the modules $\mathscr{E}(\mathfrak{G})_{(\mathscr{R})}$ and $\mathscr{E}(\mathfrak{G}')_{(\mathscr{R})}$ over the localized Hilbert-Witt ring $\mathscr{R}$ are isomorphic.

When $k$ is *algebraically closed*, we have constructed (§5, Theorem 4) distinguished $\mathscr{A}$-modules $M_{r,\infty}(k)$ and $M_{m,n}(k)$ of respective ranks $r$ and $m$, such that every distinguished $\mathscr{A}$-module M of finite rank is such that $M_{(\mathscr{R})}$ is isomorphic to a direct sum of $\mathscr{R}$-modules $(M_j)_{(\mathscr{R})}$, where $M_j$ is of one of the types $M_{r,\infty}(k)$ or $M_{m,n}(k)$. Denote by $\mathfrak{G}_{m,n}(k)$ the bigebra corresponding to $\mathrm{M}_{m,n}(k)$ and by $\mathbf{G}_{m,n}(k)$ the corresponding formal group; we may therefore say (§3, No. 3, Proposition 3) that *any* finite-dimensional commutative reduced formal group over $k$ is *isogenous to a direct product of groups* $\mathbf{G}_{m,n}(k)$. Furthermore, the groups $\mathbf{G}_{m,n}(k)$ with finite $n$ are *simple* (§5, No. 2, Theorem 2) and the groups $\mathbf{G}_{r,\infty}(k)$ are "indecomposable" in the sense that they are not isogenous to a direct product of two nontrivial reduced subgroups (§4, No. 2, Proposition 3).

The structure of the bigebras $\mathfrak{G}_{r,\infty}(k)$ may be described more explicitly. From the definition of the corresponding matrix $V$ (§5, No. 4, formula (18)) and the interpretation of the matrix $V$ in §3, No. 4, formula (10), it follows that the privileged canonical basis $(Z'_\alpha)$ of $\mathfrak{G}_{r,\infty}(k)$ is such that it satisfies the relations

$$(1) \qquad X'^p_{h0} = 0, \qquad X'^p_{h1} = X'_{h0}, \qquad \ldots, \qquad X'^p_{h,r-1} = X'_{h,r-2}.$$

However, up to reindexing the $X'_{hi}$ by replacing $X'_{hi}$ with $X'_{h,r-1-i}$, we may easily exhibit a bigebra having a privileged canonical basis with the same multiplication table. Namely, we consider, in the Witt bigebra $\mathfrak{G}_\infty(k)$ (§2, No. 1) the subalgebra $\mathfrak{H}_r$ generated by the $X_{hi}$ with $i \geqslant r$; it is a bigebra and has as privileged canonical basis the $Z_{v^r(\alpha)}$ it is therefore isomorphic to $\mathfrak{G}_\infty(k)$. If $\mathrm{J}_r$ is the biideal generated by $\mathfrak{H}_r^+$ in $\mathfrak{G}_\infty(k)$ the quotient bigebra $\mathfrak{G}_\infty(k)/\mathrm{J}_r$ is obviously isomorphic to $\mathfrak{G}_{r,\infty}(k)$. We say that $\mathfrak{G}_{r,\infty}(k)$ (resp. $\mathbf{G}_{r,\infty}(k)$) is the *truncated Witt bigebra* (resp. the *truncated Witt formal group*) *of dimension r*. This description, together with the determination of the sub-$\mathscr{R}$-modules of $\mathscr{R}/\omega^r\mathscr{R}$ (§4, No. 2, Proposition 3) show that, up to isogeny, the only reduced subbigebras of $\mathfrak{G}_{r,\infty}(k)$ are the quotients $\mathfrak{H}_s/\mathrm{J}'_{r,s}$, where $\mathrm{J}'_{r,s}$ is the ideal in $\mathfrak{H}_s$ generated by $\mathfrak{H}_r^+$, for $s < r$; that quotient is isomorphic to $\mathfrak{G}_{r-s,\infty}(k)$. As a subbigebra of $\mathfrak{G}_{r,\infty}(k)$, its augmentation ideal generates an ideal $\mathrm{J}''_{r,s}$ of $\mathfrak{G}_{r,\infty}(k)$, which is the image of $\mathrm{J}_s$ in $\mathfrak{G}_{r,\infty}(k)$, and the quotient bigebra $\mathfrak{G}_{r,\infty}(k)/\mathrm{J}''_{r,s}$ is isomorphic to $\mathfrak{G}_{s,\infty}(k)$.

We observe that the bigebras $\mathfrak{G}_{m,n}(k)$ may be defined for *any* perfect field $k$, since the definition of $\mathrm{M}_{m,n}(k)$ is valid in that general case (§5, No. 5). It is clear that, when we extend $k$ to its algebraic closure $\bar{k}$, the bigebra $\mathfrak{G}_{m,n}(k) \otimes_k \bar{k}$ is identical to $\mathfrak{G}_{m,n}(\bar{k})$. This immediately implies that the bigebras $\mathfrak{G}_{r,\infty}(k)$ are still *indecomposable* and the bigebras $\mathfrak{G}_{m,n}(k)$ with $n$ finite ($m$ and $n$ prime to each other) are still *simple*.

**2.** *Reduced infinitesimal groups of dimension* 1. It follows from §4, No. 3, Proposition 7 that any distinguished $\mathscr{A}$-module of rank 1 is isomorphic to a quotient $\mathscr{A}/(\omega - a\mathbf{t})\mathscr{A}$ for some $a \in \mathrm{A}$. If $a = 0$, we obtain a *unique* infinitesimal group (up to isomorphism), namely $\mathbf{G}_{1,\infty}(k)$, described above, characterized by the fact that $X_h'^p = 0$ for every $h \geqslant 0$ (we suppress of course here the index $i \in \mathrm{I}$, since the Lie algebra is one-dimensional). It is easy to see that $\mathbf{G}_{1,\infty}(k)$ is isomorphic to the *formal additive group* $\mathbf{G}_a(k)$ (II, §2, No. 6, *Example*); we have seen that the privileged basis $(Z_r)_{r \geqslant 0}$ for the bigebra $\mathfrak{G}_a(k)$ has the multiplication law

$$Z_r Z_s = \binom{r+s}{s} Z_{r+s} \tag{2}$$

and that if

$$r = r_0 + r_1 p + \cdots + r_n p^n + \cdots$$

is the $p$-adic expression of the integer $r$, we have

$$Z_r = \prod_{h=0}^{\infty} (X_h^{r_h}/r_h!). \tag{3}$$

From (2) we get $X_h^p = 0$, proving the isomorphism asserted above.

Also note that from §5, Proposition 3, it follows that the ring of endomorphisms of $\mathfrak{G}_{1,\infty}(k)$ is isomorphic to the ring $\mathscr{A}/\omega\mathscr{A}$, i.e., the *Hilbert ring* of formal power series $\sum_{i=0}^{\infty} \xi_i t^i$ with coefficients *in the field* $k$, with the commutation law $t\xi = \xi^p t$ for $\xi \in k$.

For $a \neq 0$, we may here, for any perfect field $k$, write explicitly the conditions for two distinguished $\mathscr{A}$-modules $\mathscr{A}/(\omega - a\mathbf{t})\mathscr{A}$, $\mathscr{A}/(\omega - b\mathbf{t})\mathscr{A}$ to be isomorphic, and completely determine the additive group of homomorphisms of one into the other. In the first place, the corresponding $\mathscr{R}$-modules may be isomorphic only if $a$, $b$, written as formal power series in $\mathbf{t}$, have the same order (§4, Corollary 1 to Proposition 8); we may therefore consider only the $\mathscr{A}$-modules $\mathscr{A}/(\omega - a\mathbf{t}^h)\mathscr{A}$, $\mathscr{A}/(\omega - b\mathbf{t}^h)\mathscr{A}$, where

$$a = a_0 + a_1\mathbf{t} + \cdots + a_i\mathbf{t}^i + \cdots, \qquad b = b_0 + b_1\mathbf{t} + \cdots + b_i\mathbf{t}^i + \cdots,$$

with $a_0 \neq 0$ and $b_0 \neq 0$. Using §3, No. 4, formula (15), the homomorphisms of $\mathscr{A}/(\omega - a\mathbf{t}^h)\mathscr{A}$ into $\mathscr{A}/(\omega - b\mathbf{t}^h)\mathscr{A}$ will correspond to elements

$$x = x_0 + x_1\mathbf{t} + \cdots + x_i\mathbf{t}^i + \cdots \quad \text{in } \mathscr{A} \qquad (\text{the } x_i \in \mathrm{W}(k))$$

for which there exists another element

$$y = y_0 + y_1\mathbf{t} + \cdots + y_i\mathbf{t}^i + \cdots \qquad (\text{the } y_i \in \mathrm{W}(k))$$

satisfying the relation

$$x(\omega - a\mathbf{t}^h) = (\omega - b\mathbf{t}^h)y \tag{4}$$

(several different values of $x$ may give the same homomorphism). The relation (4) is equivalent to the infinite system of equations

$$\begin{aligned} x_i &= y_i \qquad \text{for} \quad 0 \leqslant i \leqslant h-1 \\ x_h\omega - a_0x_0 &= y_h\omega - b_0y^{\sigma^h} \\ &\vdots \\ x_{h+n}\omega - a_0^{\sigma^n}x_n - a_1^{\sigma^{n-1}}x_{n-1} - \cdots - a_nx_0 & \\ &= y_{h+n}\omega - b_0y_n^{\sigma^h} - b_1y_{n-1}^{\sigma^{h+1}} - \cdots - b_ny_0^{\sigma^{h+n}} \end{aligned}$$

which may also be written

$$\begin{aligned} b_0x_0^{\sigma^h} - a_0x_0 &= (y_h - x_h)\omega \\ b_0x_1^{\sigma^h} - a_0^{\sigma}x_1 &= (y_{h+1} - x_{h+1})\omega + a_1x_0 - b_1x_0^{\sigma^h} \\ &\vdots \\ b_0x_n^{\sigma^h} - a_0^{\sigma^n}x_n &= (y_{h+n} - x_{h+n})\omega + b_0(x_n - y_n)^{\sigma^h} \\ &\qquad + \Phi_n(x_0, \ldots, x_{n-1}, y_0, \ldots, y_{n-1}) \\ &\vdots \end{aligned} \tag{5}$$

As $h \geqslant 1$, the solution of this system leads to the inductive process consisting in the following operations, where $\bar{x}$ is the natural image in $k$ of an element $x \in \mathrm{W}(k)$: the first equation implies for $\xi_0 = \bar{x}_0$ the algebraic equation

$$\bar{b}_0\xi_0^{p^h} - \bar{a}_0\xi_0 = 0 \tag{6}$$

in $k$; the next $h-1$ equations yield by induction equations in $k$ for $\xi_i = \bar{x}_i$:

$$\bar{b}_0\xi_i^{p^h} - \bar{a}_0^{p^i}\xi_i = \Psi_i(\xi_0, \xi_1, \ldots, \xi_{i-1}) \qquad (1 \leqslant i \leqslant h-1). \tag{7}$$

*Suppose* these equations have been solved; then we pick arbitrarily $x_i$

in the class $\xi_i$ for $0 \leqslant i \leqslant h - 1$, and the first $h$ equations (5) determine uniquely the differences $y_h - x_h, \ldots, y_{2h-1} - x_{2h-1}$. With the help of these expressions, we may eliminate the elements $y_h, y_{h+1}, \ldots, y_{2h-1}$ from the equations (5) corresponding to $n \geqslant h$. Consider then those equations for which $h \leqslant n \leqslant 2h - 1$; they imply by reduction mod $\omega$ the algebraic equations in $k$:

$$(8)\quad \bar{b}_0 \xi_{h+i}^{p^h} - \bar{a}_0^{p^{h+i}} \xi_{h+i} = \Psi_{h+i}(\xi_0, \xi_1, \ldots, \xi_{h+i-1}), \qquad 0 \leqslant i \leqslant h - 1.$$

Suppose again these equations have been solved; then we take arbitrarily $x_i$ in $\xi_i$ for $h \leqslant i \leqslant 2h - 1$; the equations (5) for $h \leqslant n \leqslant 2h - 1$ determine the differences

$$y_{2h} - x_{2h}, \quad \ldots, \quad y_{3h-1} - x_{3h-1},$$

which enables one to eliminate the elements $y_{2h}, \ldots, y_{3h-1}$ from Eq. (5). This process may be continued inductively, and we see that the only "obstructions" to the existence of a homomorphism come from the possibility of solving *in* $k$, Eq. (6), (7), and (8) and their analogs through the induction.

We will limit ourselves to the two extreme cases (for a more general discussion, see [26]):

I) $k$ is *algebraically closed.* As $h \geqslant 1$, there are *no obstructions* if at least one of the two elements $\bar{a}_0$, $\bar{b}_0$ is not 0. In particular, we see that *all distinguished* $\mathscr{A}$*-modules* $\mathscr{A}/(\omega - \mathbf{a}t^k)\mathscr{A}$ *with* $\bar{a}_0 \neq 0$ (i.e., $a$ invertible in $\mathscr{A}$) *are isomorphic,* and not only isogenous. If we return to the interpretation of the matrix $V$ in §3, No. 4, Eq. (10), we see that *all reduced bigebras of dimension 1 for which*

$$(9)\quad X_n^p = 0 \quad \textit{for} \quad n \leqslant h - 1, \quad X_{h-1}^p \neq 0, \quad \textit{and} \quad X_{h-1}^p \in \mathfrak{g}_0$$

*are isomorphic to* $\mathfrak{G}_{1,h-1}(k)$; the number $h$ is called the *height* of the bigebra (or of the corresponding formal group). One may also say that $h$ is the smallest number $n$ for which $[p]X_n \neq 0$ (§1, No. 2, Proposition 1), which also amounts to saying that in the algebra $\mathfrak{G}^* = k[[t]]$ the endomorphism corresponding by transposition to the endomorphism $[p]$ of $\mathfrak{G}$ (or to the "formal $p$-th power" in the corresponding formal group) transforms $t$ into a power series of the form $\lambda t^{p^h} + \cdots$ with $\lambda \neq 0$ in $k$ and the unwritten terms of degree $> p^h$.

The bigebra $\mathfrak{G}_{1,h-1}(k)$ may be described explicitly by the construction of §3, No. 4: it is the quotient of $\mathfrak{G}_\infty(k)$ by the ideal generated by the $T^p$ for $n < h - 1$, and the $T^p_{n+h-1} - T_n$ for $n \geqslant 0$, so that in that bigebra we have

$$(10)\quad X_n^p = 0 \quad \text{for} \quad n < h - 1, \qquad X^p_{n+h-1} = X_n \quad \text{for} \quad n \geqslant 0$$

and the corresponding privileged basis is given by

$$Z_r = \mathrm{E}_r(X_0, X_1, \ldots) \quad \text{for} \quad r > 0,$$

where the $\mathrm{E}_r$ are the hyperexponential polynomials (§3, No. 1).

II) $k$ is the *prime field* $\mathbf{F}_p$; as $\xi \mapsto \xi^p$ is then the identity, hence also $\sigma$, the compatibility condition (4) with $\xi_0 \neq 0$ yields the infinite system of conditions

$$(11)\qquad \bar{a}_i = \bar{b}_i \quad \text{in } k \qquad \text{for every} \quad i \geqslant 0$$

which are necessary and sufflcient for the isomorphism of the $\mathscr{A}$-modules $\mathscr{A}/(\omega - a\mathbf{t}^h)\mathscr{A}$ and $\mathscr{A}/(\omega - b\mathbf{t}^h)\mathscr{A}$. There is here no distinction between isomorphism and isogeny, and we see therefore that the set of isomorphism classes of bigebras, corresponding to the same integer $h$, has the *power of the continuum.*

**3.** *The multiplicative group.* The bigebra $\mathfrak{G}_{1,0}(k)$ corresponding to the $\mathscr{A}$-module $\mathscr{A}/(\omega - \mathbf{t})\mathscr{A}$ is characterized among all bigebras $\mathfrak{G}_{1,n}(k)$ as the only one for which the Frobenius homomorphism is *bijective.* It is interesting to have a criterion enabling one to check this property when one knows only the "group law" of a reduced formal group $\mathbf{G}$ of dimension 1, i.e., the comultiplication $c$ in its contravariant bigebra $\mathfrak{G}^* = k[[t]]$:

$$(12)\qquad c(t) = t + v_1(t)x + v_2(t)x^2 + \cdots$$

where the $v_j$ are formal power series in $k[[t]]$:

**Proposition 1.** *In order that $\mathbf{G}$ be isomorphic to $\mathbf{G}_{1,0}(k)$, it is necessary and sufficient that, in the formal power series $1/v_1(t)$, the coefficient of $t^{p-1}$ be $\neq 0$.*

If we consider $X_0$ as a linear operator on $\mathfrak{G}^*$ (II, §3, No. 10), we have

$X_0 = v_1 D$ where $D = d/dt$ (II, §3, No. 10, formula (34)). Hence we have

$$X_0^p = ((v_1 D)^{p-1} \cdot v_1)D + \cdots,$$

where the unwritten terms are linear combinations with coefficients in $k[[t]]$ of the operators $D^2, D^3, \ldots, D^{p-1}$ (since $D^p = 0$); as we know *a priori* that $X_0^p = \gamma X_0$ for some $\gamma \in k$ (II, §2, No. 2), all these terms in fact vanish and $\gamma$ is just the *constant term* of the formal power series $(v_1 D)^{p-1} \cdot v_1$. The proposition will result from the following lemma:

**Lemma 1.** *In a commutative k-algebra* A *over a field k of characteristic* $p > 0$, *for every k-derivation* D *of* A *and every element* $z \in \mathrm{A}$,

$$(13) \qquad (z\mathrm{D})^{p-1} \cdot z = -z\mathrm{D}^{p-1}(z^{p-1}).$$

Indeed, if this lemma is proved, we replace A by $\mathfrak{G}^*$, $z$ by $v_1$, and D by $d/dt$; $\gamma$ is therefore the constant term in $-\mathrm{D}^{p-1}(v_1^{p-1})$, and as $(p-1)! = -1$ in $k$, that term is finally the coefficient of $t^{p-1}$ in $v_1^{p-1}$; but as

$$v_1(t) = 1 + a_1 t + \cdots, \qquad v_1^p(t) = 1 + a_1^p t^p + \cdots;$$

the coefficient of $t^{p-1}$ in $v_1^{p-1}$ is also the coefficient of $t^{p-1}$ in $v_1^{-1}$, proving Proposition 1.

Lemma 1 is itself a consequence of a general formula of Hochschild which applies to every commutative algebra R over the prime field $\mathbf{F}_p$ and every derivation $\delta$ of R: for any $r \in \mathrm{R}$, we have

$$(14) \qquad (r\delta)^p = r^p \delta^p + ((r\delta)^{p-1}) \cdot r.\delta.$$

This formula is first applied to the field of fractions R of the algebra of polynomials $\mathbf{F}_p[\mathrm{T}, \mathrm{U}_0, \mathrm{U}_1, \ldots]$ in infinitely many indeterminates, and to the derivation $\delta$ of R defined by

$$\delta(\mathrm{U}_i) = \mathrm{U}_{i+1} \qquad \text{and} \qquad \delta(\mathrm{T}) = \mathrm{U}_0^{-1};$$

taking, in (14), $r = \mathrm{U}_0$ and taking the values of both sides at $\mathrm{T} \in \mathrm{R}$, we get

$$0 = \mathrm{U}_0^p \delta^{p-1}(\mathrm{U}_0^{-1}) + ((\mathrm{U}_0\delta)^{p-1}(\mathrm{U}_0))\mathrm{U}_0^{-1},$$

hence

$$(U_0\delta)^{p-1}(U_0) = -U_0^{p+1}\delta^{p-1}(U_0^{-1})$$

and as $\delta(U_0) = 0$, this formula may also be written

$$(U_0\delta)^{p-1}(U_0) = -U_0\delta^{p-1}(U_0^{p-1}). \tag{15}$$

But there is a homomorphism $\varphi$ of the ring of polynomials $\mathbf{F}_p[U_0, U_1, \ldots]$ in A such that

$$\varphi(U_0) = z, \qquad \varphi(U_i) = D^i(z), \qquad \varphi \circ \delta = D \circ \varphi.$$

Applying $\varphi$ to both sides of (15), we obtain (13), proving completely Proposition 1.

The most interesting case of application of Proposition 1 is the *formal multiplicative group* $\mathbf{G}_m(k)$, which is associated to the algebraic multiplicative group $\mathbf{G}_m$ (I, §2, No. 12), and therefore is a reduced group of dimension 1 for which the coproduct in the contravariant bigebra is given by

$$c(x) = x \otimes 1 + x \otimes x + 1 \otimes x$$

or, when $\mathfrak{G}^*(k)$ is identified with $k[[t, x]]$,

$$c(t) = t + x + tx;$$

as here $v_1(t) = 1 + t$, the condition of Proposition 1 is trivially satisfied, and $\mathfrak{G}_m(k)$ is isomorphic to the bigebra $\mathfrak{G}_{1,0}(k)$ when $k$ is algebraically closed. But, in fact, this isomorphism holds for *any* perfect field $k$, for we have only to prove it exists for $k = \mathbf{F}_p$, and it is enough to show that in $\mathfrak{G}_m(k)$ we have then $X_n^p = X_n$ for every $n \geqslant 0$. But it is clear that the endomorphism of the algebra $\mathfrak{G}_m^*(k)$ corresponding to $[p]$ in $\mathfrak{G}_m(k)$ (or to the "formal $p$-th power in $\mathbf{G}_m(k)$") is identical to the Frobenius homomorphism; in $\mathfrak{G}_m(k)$ we have therefore $V = [p]$ and it follows from §1, No. 4, Proposition 1 that F is the identity in $\mathfrak{G}_m(k)$.

## 4. *Commutativity of reduced infinitesimal groups of dimension 1.*

**Theorem 1** (M. Lazard). *A reduced infinitesimal group* $\mathbf{G}$ *of dimension 1 is commutative.*

The following proof is due to J. P. Serre [44] and applies to formal groups over any reduced ring. We identify $\mathfrak{G}^*$ with $k[[t]]$, $\mathfrak{G}^* \hat{\otimes} \mathfrak{G}^*$ with $k[[x, y]]$, and we write

$$c(t) = f(x, y) = x + y + \cdots$$

for the coproduct. The three canonical generic points X, Y, X′ of $\mathbf{G}$ with values in

$$A = \mathfrak{G}^* \hat{\otimes} \mathfrak{G}^* \hat{\otimes} \mathfrak{G}^*,$$

identified with $k[[x, y, x']]$ are the homomorphisms of $\mathfrak{G}^*$ into A sending $t$ to $x$, $y$, $x'$, respectively. It is clear that the point $u = XYX^{-1}$ takes its values in $\mathfrak{G}^* \hat{\otimes} \mathfrak{G}^*$, hence $u(t) = h(x, y)$. Specializing X to the unit element of $\mathbf{G}_A$ (sending $t$ to 0) we see that $h(0, y) = y$, hence we may write

$$h(x, y) = y + \sum_{n=1}^{\infty} r_n(x) y^n, \tag{16}$$

where the $r_n$ are formal power series in $k[[x]]$ and $r_n(0) = 0$ for all $n$.

We prove the theorem by contradiction, assuming that $\mathbf{G}$ is *not commutative*. Then $XYX^{-1} \neq Y$, so that one of the $r_n$ must be $\neq 0$; let $m$ be the smallest $n$ having that property, and distinguish two cases:

I) $m = 1$, so that

$$h(x, y) \equiv y(1 + r_1(x)) \qquad \text{mod } y^2.$$

We have the relation

$$(XX')Y(XX')^{-1} = X(X'YX'^{-1})X^{-1}$$

in the group $\mathbf{G}_A$, which is equivalent to

$$h(f(x, x'), y) = h(x, h(x', y)), \tag{17}$$

and therefore yields the relation

$$y(1 + r_1(f(x, x'))) \equiv y(1 + r_1(x)(1 + r_1(x')) \qquad \text{mod } y^2$$

or equivalently

$$r_1(f(x, x')) = r_1(x) + r_1(x') + r_1(x)r_1(x').$$

But by definition of the multiplicative group, this means that the mapping ${}^t v$ of $k[[x]]$ into $k[[t]]$, such that ${}^t v(x) = r_1(t)$, is a bigebra homomorphism of $\mathfrak{G}_m^*$ into $\mathfrak{G}^*$, corresponding to a formal homomorphism $\mathbf{v}: \mathbf{G} \to \mathbf{G}_m$ which is not trivial.

II) $m \geqslant 2$, so that we now have

$$h(x, y) \equiv y + r_m(x)y^m \qquad \text{mod } y^{m+1}.$$

From (17) we now get

$$y + r_m(f(x, x'))y^m \equiv y + r_m(x')y^m + r_m(x)(y + r_m(x')y^m)^m \qquad \text{mod } y^{m+1}$$
$$\equiv y + (r_m(x) + r_m(x'))y^m \qquad \text{mod } y^{m+1}$$

since $m \geqslant 2$, or equivalently

$$r_m(f(x, x')) = r_m(x) + r_m(y').$$

The same argument shows this time that, if ${}^t v(x) = r_m(t)$, ${}^t v$ is a bigebra homomorphism of $\mathfrak{G}_a^*$ into $\mathfrak{G}^*$, corresponding to a nontrivial formal homomorphism $\mathbf{v}$ of $\mathbf{G}$ into the additive group $\mathbf{G}_a$.

The contradiction will appear from the fact that, if we have a nontrivial homomorphism $\mathbf{v}: \mathbf{G} \to \mathbf{G}'$ of $\mathbf{G}$ into any *commutative* reduced formal group $\mathbf{G}'$, $\mathbf{G}$ itself must be *commutative*. Indeed, replacing $\mathbf{G}'$ by $\mathbf{v}(\mathbf{G})$, we may assume that $\mathbf{G}'$ itself is one-dimensional, hence

$$\mathfrak{G}'^* = k[[x]] \qquad \text{and} \qquad {}^t v(x) = g(t)$$

where $g(t) = \sum_{r=1}^{\infty} a_r t^r$ is a nonzero formal power series without constant term; let $h \geqslant 1$ be the smallest integer such that $a_r \neq 0$. We have to prove that the point

$$w = \mathrm{XYX}^{-1}\mathrm{Y}^{-1}$$

in the group $\mathbf{G}_\mathrm{A}$ is the identity. Suppose not, so that that point would be such that

$$w(t) = \mathrm{C}(x, y) \in k[[x, y]] \qquad \text{with} \quad \mathrm{C}(x, y) = \mathrm{C}_m(x, y) + \cdots$$

where $C_m(x, y) \neq 0$ is a homogeneous polynomial of degree $m$ and the unwritten terms have total degree $\geqslant m + 1$. Now by assumption $\mathbf{v}_A(XYX^{-1}Y^{-1})$ is the neutral element of $\mathbf{G}_A$; however, this element sends $x$ to

$$g(C(x, y)) = a_h(C_m(x, y))^h + \cdots,$$

the unwritten terms having total degree $\geqslant mh + 1$; as $g(C(x, y))$ should be 0, we have reached a contradiction. QED.

It should be observed that Theorem 1 does not generalize to *nonreduced* infinitesimal groups of dimension 1. Indeed, take

$$\mathfrak{G}^* = k[[t]]/(t^{p^{h+k}}) \qquad \text{with} \quad h \geqslant 1, \quad k \geqslant 1, \quad h \neq k;$$

then if we define

$$c(t) = x + y + x^{p^h} y^{p^k},$$

it is readily verified that $c$ is a coproduct $\mathfrak{G}^* \to \mathfrak{G}^* \otimes \mathfrak{G}^*$ which is coassociative but not cocommutative and defines therefore a nonreduced noncommutative infinitesimal group $\mathbf{G}$ of dimension 1.

CHAPTER IV

# Representable Reduced Infinitesimal Groups

**1.** *Conventions and notations.* In this chapter, "formal group" will mean "reduced infinitesimal formal group of finite dimension"; no other kind of formal group will ever be considered. The basic field $k$ will always be assumed to be *algebraically closed*, but its characteristic will be *arbitrary* (unless the contrary is explicitly stated).

It will be suggestive to abuse language so as to have statements on formal groups closer in appearance to the "naive" point of view of "formal Lie groups" (I, §2, No. 12). If $\mathbf{G}$ is a formal group of dimension $n$, $\mathfrak{G}$ its covariant bigebra, and $\mathfrak{g}_0 \subset \mathfrak{G}$ its Lie algebra, we know that, to each basis $(X_{0i})$ of $\mathfrak{g}_0$ over $k$ (with $1 \leqslant i \leqslant n$), we may associate a system $(x_i)_{1 \leqslant i \leqslant n}$ of elements of $\mathfrak{G}^*$ such that $\mathfrak{G}^*$ is isomorphic to the algebra of formal power series $k[[x_1, \ldots, x_n]]$ (the $x_i$ being considered as *indeterminates*); any other system $(x'_1, \ldots, x'_n)$ having that property is given by $n$ formal power series without constant term

$$x'_i = f_i(x_1, \ldots, x_n)$$

whose Jacobian

$$\partial(f_1, \ldots, f_n)/\partial(x_1, \ldots, x_n)$$

has a constant term $\neq 0$ ("change of variables"). The completed tensor product

$$\mathrm{A} = \mathfrak{G}^* \mathbin{\hat{\otimes}} \mathfrak{G}^* \mathbin{\hat{\otimes}} \cdots \qquad \text{(a finite number of times)}$$

is identified to an algebra of formal power series

$$k[[x_1, \ldots, x_n; y_1, \ldots, y_n; \ldots]]$$

in indeterminates the number of which is a multiple of $n$. A point U of the group $\mathbf{G}_{\mathrm{A}}$ (I, §3, No. 1) is entirely determined by its values $\mathrm{U}(x_i)$ for $1 \leqslant i \leqslant n$, which are formal power series without constant term

$$u_i(x_1, \ldots, x_n; y_1, \ldots, y_n; \ldots) \qquad (1 \leqslant i \leqslant n),$$

and may be given arbitrarily; we will identify U with this system of power series. In particular, the independent canonical generic points of $\mathbf{G}_{\mathrm{A}}$ are the systems

$$\mathrm{X} = (x_1, \ldots, x_n), \qquad \mathrm{Y} = (y_1, \ldots, y_n)$$

of $n$ indeterminates. If

$$c : \mathfrak{G}^* \to \mathfrak{G}^* \mathbin{\hat{\otimes}} \mathfrak{G}^*$$

is the coproduct of $\mathfrak{G}^*$, it is therefore identified with the point

$$\mathrm{XY} = (\varphi_1(x_1, \ldots, x_n; y_1, \ldots, y_n), \ldots, \varphi_n(x_1, \ldots, x_n; y_1, \ldots, y_n))$$

in A; for any two points U, V of $\mathbf{G}_{\mathrm{A}}$, UV will be the system of formal power series obtained by substituting $u_i$ and $v_i$ to $x_i$ and $y_i$ $(1 \leqslant i \leqslant n)$ in each $\varphi_i(\mathrm{X}, \mathrm{Y})$. Similarly, if $a\colon \mathfrak{G}^* \to \mathfrak{G}^*$ is the antipodism of $\mathfrak{G}^*$, and

$$a(x_i) = \theta_i(x_1, \ldots, x_n),$$

where $\theta_i$ is a formal power series without constant term $(1 \leqslant i \leqslant n)$, $\mathrm{X}^{-1}$ is the point

$$(\theta_1(x_1, \ldots, x_n), \ldots, \theta_n(x_1, \ldots, x_n))$$

in A, and $\mathrm{U}^{-1}$ the system of formal power series obtained by substituting $u_i$ to $x_i$ $(1 \leqslant i \leqslant n)$ in each $\theta_i(\mathrm{X})$.

If $\mathbf{H}$ is a (reduced!) formal subgroup of $\mathbf{G}$ of dimension $m$, we know that the ideal $\mathfrak{H}^0$ in $\mathfrak{G}^*$ may be taken as generated by $x_{m+1}, \ldots, x_n$ for a suitable choice of the elements $x_i$; we say that **H** is

$$\text{“\textit{defined by the equations}} \qquad x_{m+1} = \cdots = x_n = 0\text{”}.$$

If $j\colon \mathfrak{H} \to \mathfrak{G}$ is the natural injection of bigebras, we have therefore

$$^t j(x_i) = 0 \qquad \text{for} \quad m+1 \leqslant i \leqslant n,$$

and the

$$x_i' = {}^t j(x_i) \qquad \text{for} \quad 1 \leqslant i \leqslant m$$

are elements of $\mathfrak{H}^*$ for which $\mathfrak{H}^*$ is identified with $k[[x_1', \ldots, x_m']]$; if

$$\mathrm{X}' = (x_1', \ldots, x_m'), \qquad \mathrm{Y}' = (y_1', \ldots, y_m')$$

are two independent generic points of **H**, we have therefore

$$\mathrm{X}'\mathrm{Y}' = (\varphi_1'(x_1', \ldots, x_m'; y_1', \ldots, y_m'), \ldots, \varphi_m'(x_1', \ldots, x_m'; y_1', \ldots, y_m'))$$

with

$$\varphi_i'(x_1', \ldots, x_m'; y_1', \ldots, y_m') = \varphi_i(x_1', \ldots, x_m', 0, \ldots, 0; y_1', \ldots, y_m', 0, \ldots, 0) \qquad \text{for} \quad 1 \leqslant i \leqslant m$$

and

$$\varphi_i(x_1, \ldots, x_m, 0, \ldots, 0; y_1, \ldots, y_m, 0, \ldots, 0) = 0 \qquad \text{for} \quad m+1 \leqslant i \leqslant n,$$

and similarly for the inverse $\mathrm{X}'^{-1}$.

Finally, let **G**, **G**′ be two formal groups of dimensions $n$, $n'$, and let $\mathbf{u}\colon \mathbf{G} \to \mathbf{G}'$ be a formal homomorphism. If $\mathfrak{G}^*$ and $\mathfrak{G}'^*$ are identified with

$$k[[x_1, \ldots, x_n]], \qquad k[[x_1', \ldots, x_{n'}']],$$

we may write respectively,

$$^t u(x_j') = u_j(x_1, \ldots, x_n)$$

where the $u_j$ are formal power series without constant term, for $1 \leqslant j \leqslant n'$;

the point $\mathbf{u}_A(X)$ of $\mathbf{G}'_A$ is therefore identified with the system of $n'$ power series $u_j$ $(1 \leqslant j \leqslant n')$, which we will simply write $\mathbf{u}(X)$. Similarly, for any point W in $\mathbf{G}_A$, identified with $n$ formal power series $w_i$ $(1 \leqslant i \leqslant n)$ without constant term, $\mathbf{u}_A(W)$, which we simply write $\mathbf{u}(W)$, is the system of $n'$ formal power series obtained by substituting $w_i$ to $x_i$ $(1 \leqslant i \leqslant n)$ in each $u_j$ $(1 \leqslant j \leqslant n')$. The conditions for the formal power series $u_j$ to define a formal homomorphism are then simply expressed by

$$\mathbf{u}(XY) = \mathbf{u}(X)\mathbf{u}(Y)$$

for the canonical generic points.

If $\mathbf{G}''$ is a third formal group of dimension $n''$, $\mathbf{v}\colon \mathbf{G}' \to \mathbf{G}''$ a formal homomorphism, and if $\mathfrak{G}''^*$ is identified with $k[[x''_1, \ldots, x''_{n''}]]$, so that, for a generic point $X' = (x'_1, \ldots, x'_{n'})$ of $\mathbf{G}'$, $\mathbf{v}(X')$ is a system of power series without constant term

$$v_k(x'_1, \ldots, x'_{n'}) \qquad \text{for} \quad 1 \leqslant k \leqslant n'',$$

then, if we write $\mathbf{w} = \mathbf{v} \circ \mathbf{u}$, $\mathbf{w}(X)$ is identified with the system of $n''$ power series obtained by substituting $u_j$ to $x'_j$ $(1 \leqslant j \leqslant n')$ in each $v_k$ $(1 \leqslant k \leqslant n'')$.

In particular, if $\mathbf{H}'$ is a formal subgroup of $\mathbf{G}'$, the condition $\mathbf{u}(\mathbf{G}) \subset \mathbf{H}'$ amounts to saying that $\mathbf{u} = \mathbf{j} \circ \mathbf{v}$, where $\mathbf{j}\colon \mathbf{H}' \to \mathbf{G}'$ is the natural injection and $\mathbf{v}\colon \mathbf{G} \to \mathbf{H}'$ a formal homomorphism. If the $x'_j$ have been chosen so that $\mathbf{H}'$ is defined by

$$x'_{m+1} = \cdots = x'_{n'} = 0,$$

the condition $\mathbf{u}(\mathbf{G}) \subset \mathbf{H}'$ is then simply expressed by writing that the formal power series $u_{m+1}, \ldots, u_n$ all vanish.

If K is an algebraically closed extension of $k$, we recall that the formal group $\mathbf{G}_{(K)}$ obtained by extending the scalars to K corresponds, after a choice of indeterminates $x_1, \ldots, x_n$ for $\mathbf{G}$, to the *same* formal power series

$$\varphi_i(x_1, \ldots, x_n; y_1, \ldots, y_n)$$

as $\mathbf{G}$, but these series are considered as elements of

$$K[[x_1, \ldots, x_n; y_1, \ldots, y_n]].$$

**2.** *Formalization of algebraic groups.* We will limit ourselves to *affine algebraic groups* in the sense of Borel [4]: such groups form a category dual to the category of *reduced finitely generated commutative k-bigebras*, and the affine group G associated to such a bigebra A is not the affine group scheme Spec(A), but the "Serre" variety Spm(A), "maximal spectrum" of A, consisting of the closed points of Spec(A) (see [28]).* It is known that such a group can always be considered (in infinitely many ways) as a closed subgroup (for the Zariski topology) of a general linear group GL(N) for some integer N; the algebra A corresponding to G can then be identified with the algebra of the restrictions to G of the "regular functions" on GL(N), i.e., rational functions, quotient of a polynomial in the $N^2$ coordinates $x_{ij}$ ($1 \leqslant i, j \leqslant N$) by a power of the determinant $\det(x_{ij})$; we write $A = R[G]$ for that algebra (which is independent of the embedding of G into a GL(N)).

We recall a number of elementary facts about Serre affine varieties (remember they are always defined over an *algebraically closed* field $k$), and simple points in these varieties. Let $V = Spm(A)$ be such a variety and $a$ a simple point of V. Then $a$ has an affine neighborhood contained in only one irreducible component of V, and as we are only interested in what happens locally around $a$, we may suppose that V is *irreducible*, hence A an *integral domain*. If $\mathfrak{m}$ is the maximal ideal of A consisting of all $f \in A$ vanishing at $a$, the local ring $A_{\mathfrak{m}}$ of V at $a$ is also an integral domain contained in the field of fractions of A and containing A; furthermore, it is a *regular* ring. Its completion with respect to the $\mathfrak{m}A_{\mathfrak{m}}$-adic topology is identified with the completion of A for the $\mathfrak{m}$-adic topology; it is again an integral domain in which $A_{\mathfrak{m}}$ is embedded. More precisely, if V has dimension $n$, the $k$-vector space $\mathfrak{m}/\mathfrak{m}^2$ has dimension $n$; if $s_1, \ldots, s_n$ are $n$ elements of $A \subset A_{\mathfrak{m}}$ whose classes mod $\mathfrak{m}^2$ form a basis of that subspace, the continuous algebra homomorphism of $k[[x_1, \ldots, x_n]]$ ($x_j$ indeterminates) into the completion $\hat{A}_{\mathfrak{m}}$ sending $x_j$ to $s_j$ for $1 \leqslant j \leqslant n$ is an *isomorphism*; a system of elements $s_j$ ($1 \leqslant j \leqslant n$) having the preceding property is called a *regular system of parameters* of $A_{\mathfrak{m}}$ (or of $\hat{A}_{\mathfrak{m}}$).

Let $W = Spm(B)$ be another affine irreducible variety, $b$ a simple point of W, and $\mathfrak{n}$ the maximal ideal of B to which it corresponds. A *morphism* $u: V \to W$ corresponds to a uniquely determined *comorphism* $v: B \to A$, which is a homomorphism of $k$-algebras defined by

$$(v(f))(x) = f(u(x)) \qquad \text{for} \quad f \in B,$$

when

* All results on affine groups which we will use are found in [4] or [15].

$$A = R[V] \qquad \text{and} \qquad B = R[W]$$

are considered as rings of "regular functions" on V and W. To say that $u(a) = b$ means that $\mathfrak{n} = v^{-1}(\mathfrak{m})$, and therefore $v$ is continuous for the $\mathfrak{m}$-adic and $\mathfrak{n}$-adic topologies, and may be continuously extended in a unique way to a homomorphism $\hat{v}: \hat{B}_{\mathfrak{n}} \to \hat{A}_{\mathfrak{m}}$. If $(s_1, \ldots, s_n)$ is a regular system of parameters for $A_{\mathfrak{m}}$, $(t_1, \ldots, t_m)$ a regular system of parameters for $\mathbf{B}_{\mathfrak{n}}$, then $\hat{v}$ is entirely determined by the $m$ power series without constant term

$$\hat{v}(t_j) = u_j(s_1, \ldots, s_n) \qquad (1 \leqslant j \leqslant m),$$

when the $s_i$ and $t_j$ are identified to indeterminates.

If $u$ is *open* at the point $a$, i.e., maps every neighborhood of $a$ onto a neighborhood of $b$, then $\hat{v}$ is *injective*, and the $m$ power series $\hat{v}(t_j)$ can be completed to a regular system of parameters of $\hat{A}_{\mathfrak{m}}$.

If V is a *closed subvariety* of W and $u$ the natural injection, A is identified to $B/\mathfrak{J}$, where $\mathfrak{J}$ is the ideal defining V; there is then a regular system of parameters $(t_1, \ldots, t_m)$ of $B_{\mathfrak{n}}$ such that $t_{n+1}, \ldots, t_m$ belong to $\mathfrak{J}$ and generate $\mathfrak{J}_{\mathfrak{n}}$ and the images $v(t_1), \ldots, v(t_n)$ form a regular system of parameters of $A_{\mathfrak{m}}$.

Finally, if we consider the product $V \times W$, with the same notations, we have

$$V \times W = \operatorname{Spm}(A \otimes_k B),$$

and $\mathfrak{m} \otimes B + A \otimes \mathfrak{n}$ is the maximal ideal corresponding to the point $(a, b)$, which is simple on $V \times W$; the projections $\mathrm{pr}_1$ and $\mathrm{pr}_2$ correspond to the natural injective homomorphisms

$$A \to A \otimes B, \qquad B \to A \otimes B;$$

the $m + n$ elements

$$s_j \otimes 1 \qquad (1 \leqslant j \leqslant n), \qquad 1 \otimes t_k \qquad (1 \leqslant k \leqslant m)$$

form a regular system of parameters at $(a, b)$, and the completed local ring at that point is naturally isomorphic to the completed tensor product $\hat{A}_{\mathfrak{m}} \hat{\otimes} \hat{B}_{\mathfrak{n}}$.

Suppose now that $G = \operatorname{Spm}(A)$ is an *affine group*. Then, to the morphism $(x, y) \mapsto xy$ of $G \times G$ into G corresponds a comorphism

$$c : A \to A \otimes_k A$$

which is coassociative; in addition, if $\mathfrak{m}$ is the maximal ideal of A corresponding to the neutral element $e$ of G, we have a counit

$$\gamma : A \to k = A/\mathfrak{m},$$

and to the morphism $x \mapsto x^{-1}$ corresponds an antipodism $a: A \to A$; in other words, A is a *commutative bigebra* over $k$, with an antipodism. Furthermore, as multiplication is a morphism sending the simple point $(e, e)$ of $G \times G$ to the simple point $e$ of G, $c$ can be extended by continuity to a homomorphism

$$\hat{c} : \hat{A}_{\mathfrak{m}} \to \hat{A}_{\mathfrak{m}} \hat{\otimes} \hat{A}_{\mathfrak{m}},$$

and this defines on $\hat{A}_{\mathfrak{m}}$, isomorphic to $k[[x_1, \ldots, x_n]]$, a structure of *contravariant bigebra* $\mathfrak{G}^*$ of a reduced infinitesimal group **G**. We will say that this group is the *formalization* of the affine algebraic group G, and write it also $\mathbf{\Phi}(G)$. When one takes as indeterminates $x_i$ in $\mathfrak{G}^*$ a regular system of parameters of $A_{\mathfrak{m}}$ belonging to A, the formal power series $\hat{c}(x_i)$ in

$$k[[x_1, \ldots, x_n, y_1, \ldots, y_n]]$$

are the elements $c(x_i)$ $(1 \leqslant i \leqslant n)$.

It is clear by definition that the formalization of an affine group and of its neutral component are identical.

Now let $G' = \mathrm{Spm}(B)$ be a second affine group, and $u: G \to G'$ a homomorphism of affine groups; if $v: B \to A$ is the corresponding comorphism, it is a bigebra homomorphism, hence its continuous extension $\hat{v}: \mathfrak{G}'^* \to \mathfrak{G}^*$ is also a bigebra homomorphism. There is therefore a unique formal homomorphism $\mathbf{u}: \mathbf{G} \to \mathbf{G}'$ such that, if we denote by $u_*: \mathfrak{G} \to \mathfrak{G}'$ the corresponding homomorphism of covariant bigebras, we have ${}^t u_* = \hat{v}$; we will say that $\mathbf{u}$ (also written $\mathbf{\Phi}(u)$) is the *formalization* of the homomorphism $u$ of affine groups. It is immediate that, if $u': G' \to G''$ is a second homomorphism of affine groups, then

$$\mathbf{\Phi}(u' \circ u) = \mathbf{\Phi}(u') \circ \mathbf{\Phi}(u).$$

If we take as indeterminates $x'_j$ ($1 \leqslant j \leqslant n'$) in $\mathfrak{G}'^*$ a regular system of parameters for the local ring of B corresponding to the neutral element $e'$, then the elements $v(x'_j)$ are formal power series $u_j(x_1, \ldots, x_n)$ in the indeterminates $x_i$ chosen in $\mathfrak{G}^*$, and for this choice of indeterminates, the formal homomorphism $\mathbf{u}$ is defined by the formal power series $u_j$ ($1 \leqslant j \leqslant n'$). In particular, if $u(\mathrm{G}) = \{e'\}$, all the formal power series $u_j$ are 0, hence $\mathbf{u}(\mathbf{G}) = \mathbf{e}'$.

**3.** *Formalization of closed subgroups and quotient groups of affine algebraic groups.* Let G be an affine group of dimension $n$, H a closed subgroup of G of dimension $m < n$, $j$: H → G the natural injection. We have seen in No. 2 that if G = Spm(A), H = Spm(B) with $\mathrm{B} = \mathrm{A}/\mathfrak{J}$, and $v\colon \mathrm{A} \to \mathrm{A}/\mathfrak{J}$ is the comorphism of $j$, it is possible to choose a regular system of parameters of the local ring $\mathrm{A}_{\mathfrak{m}}$ of A at the point $e$ consisting of elements $s_1, \ldots, s_n$ such that $s_{m+1}, \ldots, s_n$ belong to $\mathfrak{J}$ and $v(s_1), \ldots, v(s_m)$ form a regular system of parameters of the local ring $\mathrm{B}_{\mathfrak{n}}$ of B at $e$. It is then clear, from the definition of the formal homomorphism $\mathbf{j}\colon \mathbf{H} \to \mathbf{G}$, that $\mathbf{j}$ is an isomorphism of $\mathbf{H}$ onto the formal subgroup of $\mathbf{G}$ defined by

$$x_{m+1} = \cdots = x_n = 0,$$

when the $s_j$ are identified with indeterminates $x_j$ ($1 \leqslant j \leqslant n$). We will identify $\mathbf{H}$ with that formal subgroup; it is clear that if $\mathrm{H}_1$ is a closed subgroup of H, then $\mathbf{H}_1 \subset \mathbf{H}$ in the sense of I, §3, No. 7.

**Proposition 1.** *Let $u$ be a homomorphism of an affine group* G *into an affine group* G′, *having kernel* N, *and let $j$*: N → G, *$j'$*: $u$(G) → G′ *be the natural injections. Then* $\mathbf{j}(\mathbf{N})$ *is the reduced kernel of* $\mathbf{u}\colon \mathbf{G} \to \mathbf{G}'$,

$$\mathbf{j}'(\Phi(u(\mathrm{G}))) = \mathbf{u}(\mathbf{G})$$

*and the ranks of $u$ and* $\mathbf{u}$ *are equal.*

It is known that $u$(G) is a *closed* subgroup of G′, hence its formalization $\Phi(u(\mathrm{G}))$ and the formal homomorphism

$$\mathbf{j}'\colon \Phi(u(\mathrm{G})) \to \mathbf{G}'$$

are defined, $\mathbf{j}'$ being injective as seen above, and we may consider $u$ as composed of $j'$ and a *surjective* homomorphism of affine groups. We may therefore restrict ourselves to the case in which $u$ is *surjective*. From the relation $u(j(\mathrm{N})) = e'$ it follows that $\mathbf{u}(\mathbf{j}(\mathbf{N})) = \mathbf{e}'$, and as $\mathbf{j}(\mathbf{N})$ is reduced, we have

$$\mathbf{j}(\mathbf{N}) \subset \mathbf{N}' = {}^r\mathbf{u}^{-1}(\mathbf{e}').$$

Suppose that $\mathbf{N}' \neq \mathbf{j}(\mathbf{N})$; if

$$\dim(\mathrm{G}) = n, \qquad \dim(\mathrm{G}') = m,$$

we would then have

$$\dim(\mathbf{N}') > n - m.$$

If $\mathbf{H} = \mathbf{u}(\mathbf{G})$, we would then have $\dim(\mathbf{H}) < m$; however, if $v$ is the comorphism corresponding to $u$, we have seen in No. 2 that $\hat{v} = {}^t u_*$ is an *injective* homomorphism of $\mathfrak{G}'^*$ into $\mathfrak{G}^*$, whose image is isomorphic to $\mathfrak{H}^*$, and is therefore an algebra of power series in $m$ indeterminates, which contradicts the relation $\dim(\mathbf{H}) < m$.

**Corollary 1.** *If* N *is a closed invariant subgroup of an affine group* G, $\Phi(\mathrm{N})$ *is an invariant formal subgroup of* $\Phi(\mathrm{G})$, *and* $\Phi(\mathrm{G})/\Phi(\mathrm{N})$ *is isomorphic to* $\Phi(\mathrm{G}/\mathrm{N})$.

It is known that G/N has a natural structure of affine group for which the natural mapping $u\colon \mathrm{G} \to \mathrm{G}/\mathrm{N}$ is a homomorphism of affine groups; we apply Proposition 1 to that homomorphism.

**Corollary 2.** *For any closed subgroup* F *of* G,

$$\Phi(u(\mathrm{F})) = \mathbf{u}(\Phi(\mathrm{F})),$$

*and for any closed subgroup* F′ *of* $u(\mathrm{G})$,

$$\Phi(u^{-1}(\mathrm{F}')) = {}^r\mathbf{u}^{-1}(\Phi(\mathrm{F}')).$$

The first statement is an obvious consequence of Proposition 1; on the other hand, Proposition 1 shows that the dimensions of $u^{-1}(\mathrm{F}')$ and of $^r\mathbf{u}^{-1}(\mathbf{\Phi}(\mathrm{F}'))$ are the same, and the second statement then follows from the fact that $\mathbf{\Phi}(u^{-1}(\mathrm{F}'))$ is a reduced group whose image by $\mathbf{u}$ is contained in $\mathbf{\Phi}(\mathrm{F}')$, hence

$$\mathbf{\Phi}(u^{-1}(\mathrm{F}')) \subset {}^r\mathbf{u}^{-1}(\mathbf{\Phi}(\mathrm{F}')).$$

**Proposition 2.** *If* $\mathrm{H}_1$, $\mathrm{H}_2$ *are two closed subgroups of* G *which centralize each other, then the formal subgroups* $\mathbf{H}_1$ *and* $\mathbf{H}_2$ *of* $\mathbf{G}$ *centralize each other.*

Let

$$j_1 : \mathrm{H}_1 \to \mathrm{G}, \qquad j_2 : \mathrm{H}_2 \to \mathrm{G}$$

be the natural injections. By definition, the morphisms

$$(z_1, z_2) \mapsto j_1(z_1)j_2(z_2) \qquad \text{and} \qquad (z_1, z_2) \mapsto j_2(z_2)j_1(z_1)$$

of $\mathrm{H}_1 \times \mathrm{H}_2$ into G are the same, hence they have the same comorphisms; but if $\mathrm{X}_1$, $\mathrm{X}_2$ are generic points of $\mathbf{H}_1$ and $\mathbf{H}_2$, respectively, these two comorphisms map a regular system of parameters of $\mathfrak{G}^*$, respectively, on the system of power series $\mathbf{j}_1(\mathrm{X}_1)\mathbf{j}_2(\mathrm{X}_2)$ and $\mathbf{j}_2(\mathrm{X}_2)\mathbf{j}_1(\mathrm{X}_1)$, hence the conclusion (I, §3, No. 10).

**Proposition 3.** *Let* $\mathrm{N}_1$, $\mathrm{N}_2$ *be two connected closed invariant subgroups of* G. *Then*

$$\mathbf{\Phi}((\mathrm{N}_1, \mathrm{N}_2)) = [\mathbf{\Phi}(\mathrm{N}_1), \mathbf{\Phi}(\mathrm{N}_2)].$$

Let $u$ be the natural surjective homomorphism $\mathrm{G} \to \mathrm{G}/(\mathrm{N}_1, \mathrm{N}_2)$ (it is known that $(\mathrm{N}_1, \mathrm{N}_2)$ is a closed connected invariant subgroup of G); as the subgroups $u(\mathrm{N}_1)$ and $u(\mathrm{N}_2)$ commute, the formal invariant subgroups $\mathbf{u}(\mathbf{N}_1)$ and $\mathbf{u}(\mathbf{N}_2)$ of $\mathbf{u}(\mathbf{G})$ also commute (Proposition 2), and as $\mathbf{\Phi}((\mathrm{N}_1, \mathrm{N}_2))$ is the reduced kernel of $\mathbf{u}$ (Proposition 1) and $[\mathbf{N}_1, \mathbf{N}_2]$ is a reduced subgroup of $\mathbf{G}$ (II, §3, No. 9), we have by definition (I, §3, No. 16)

$$[\mathbf{N}_1, \mathbf{N}_2] \subset \mathbf{\Phi}((\mathrm{N}_1, \mathrm{N}_2)).$$

Let

$$\dim[\mathbf{N}_1, \mathbf{N}_2] = m, \qquad \dim\Phi((\mathrm{N}_1, \mathrm{N}_2)) = q$$

and put $\mathrm{N} = (\mathrm{N}_1, \mathrm{N}_2)$ for convenience. Let $j\colon \mathrm{N} \to \mathrm{G}$ be the natural injection, and ${}^t j_*\colon \mathfrak{G}^* \to \mathfrak{N}^*$ the corresponding comorphism extended by continuity. We may assume that a regular system of parameters $x_1, \ldots, x_n$ of the local ring of G at $e$, identified with $n$ indeterminates in $\mathfrak{G}^*$, has been chosen such that the

$$y_i = {}^t j_*(x_i) \qquad \text{for} \quad 1 \leqslant i \leqslant q$$

constitute a regular system of parameters for the local ring of N at $e$, identified to $q$ indeterminates in $\mathfrak{N}^*$. On the other hand, there is a system of $n$ formal power series

$$x_i' = \psi_i(x_1, \ldots, x_n) \qquad (1 \leqslant i \leqslant n)$$

without constant term and with a jacobian having nonvanishing constant term, such that the formal subgroup $[\mathbf{N}_1, \mathbf{N}_2]$ of $\mathbf{G}$ is defined by

$$x_i' = 0 \qquad \text{for} \quad m + 1 \leqslant i \leqslant n.$$

Now it is known that there exists an integer $r$ such that the morphism

$$(1) \qquad (z_1', \ldots, z_r', z_1'', \ldots, z_r'') \mapsto z_1' z_1'' z_1'^{-1} z_1''^{-1} \cdots z_r' z_r'' z_r'^{-1} z_r''^{-1}$$

of the variety $\mathrm{N}_1^r \times \mathrm{N}_2^r$ into the variety $(\mathrm{N}_1, \mathrm{N}_2)$ is *surjective.* Let

$$j_1\colon \mathrm{N}_1 \to \mathrm{G}, \qquad j_2\colon \mathrm{N}_2 \to \mathrm{G}$$

be the natural injections, and let

$$\mathrm{X}_j' \quad (1 \leqslant j \leqslant r), \qquad \mathrm{X}_j'' \quad (1 \leqslant j \leqslant r)$$

be $r$ independent generic points of $\mathbf{N}_1$ and $\mathbf{N}_2$, respectively. Then if $v$ is the comorphism of the morphism (1), $(v(y_i))$, for $1 \leqslant i \leqslant q$, is the system of $q$ power series denoted by

$$\mathbf{j}_1(\mathrm{X}_1')\mathbf{j}_2(\mathrm{X}_1'')\mathbf{j}_1(\mathrm{X}_1')^{-1}\mathbf{j}_2(\mathrm{X}_1'')^{-1} \cdots \mathbf{j}_1(\mathrm{X}_r')\mathbf{j}_2(\mathrm{X}_r'')\mathbf{j}_1(\mathrm{X}_r')^{-1}\mathbf{j}_2(\mathrm{X}_r'')^{-1}$$

and from the definition of $[\mathbf{N}_1, \mathbf{N}_2]$ it follows that we have

$$\psi_i(v(y_1), \ldots, v(y_q), 0, \ldots, 0) = 0 \qquad \text{for} \quad m + 1 \leqslant i \leqslant n. \tag{2}$$

But as the morphism (1) is surjective, its comorphism $v$ is injective and the series (2) are the images by $v$ of the elements of $\mathfrak{N}^*$ identified to the power series

$$\psi_i(y_1, \ldots, y_q, 0, \ldots, 0).$$

These power series are then all 0 for $m + 1 \leqslant i \leqslant n$, which means that $\mathbf{j}(\mathbf{N}) \subset [\mathbf{N}_1, \mathbf{N}_2]$. QED.

**Corollary.** *In order that a connected affine group* G *be solvable* (resp. *nilpotent*, resp. *commutative*), *it is necessary and sufficient that its formalization* $\mathbf{G}$ *be solvable* (resp. *nilpotent*, resp. *commutative*).

**Proposition 4.** *Let* $H_1$, $H_2$ *be two closed connected subgroups of an affine group* G. *If we denote by* $\sup(H_1, H_2)$ *the smallest closed subgroup of* G *containing* $H_1$ *and* $H_2$, *then*

$$\Phi(\sup(H_1, H_2)) = \Phi(H_1) \vee \Phi(H_2).$$

As

$$\Phi(H_1) \subset \Phi(\sup(H_1, H_2)) \qquad \text{and} \qquad \Phi(H_2) \subset \Phi(\sup(H_1, H_2)),$$

we obviously have

$$\Phi(H_1) \vee \Phi(H_2) \subset \Phi(\sup(H_1, H_2)).$$

On the other hand, it is known that there is an integer $r > 0$ such that the morphism

$$(z'_1, \ldots, z'_r, z''_1, \ldots, z''_r) \mapsto z'_1 z''_1 z'_2 z''_2 \cdots z'_r z''_r \tag{3}$$

of the variety $H_1^r \times H_2^r$ into the variety $\sup(H_1, H_2)$ is *surjective*. The same argument as in Proposition 3 then shows that

$$\mathbf{\Phi}(\sup(H_1, H_2)) \subset \mathbf{\Phi}(H_1) \vee \mathbf{\Phi}(H_2),$$

since with the same notations

$$\mathbf{j}_1(X_1')\mathbf{j}_2(X_1'')\mathbf{j}_1(X_2')\mathbf{j}_2(X_2'') \cdots \mathbf{j}_1(X_r')\mathbf{j}_2(X_r'')$$

is a point of the formal subgroup $\mathbf{\Phi}(H_1) \vee \mathbf{\Phi}(H_2)$ of $\mathbf{G}$.

**4.** *Representable formal groups.* We will say that a formal group $\mathbf{G}$ is *representable* if there exists an *isogeny* of $\mathbf{G}$ into a formal group $\mathbf{\Phi}(G')$, where $G'$ is an affine algebraic group. If $\mathbf{GL}(N)$ is the formal group $\mathbf{\Phi}(GL(N))$, it is equivalent to say that there is an isogeny of $\mathbf{G}$ into some $\mathbf{GL}(N)$, since any affine algebraic group is a closed subgroup of some $GL(N)$.

We will see later examples of formal groups which are *not representable.* However, we have the general result:

**Theorem 1.** *For any formal group* $\mathbf{G}$, *the quotient* $\mathbf{G}/{}^r\mathscr{Z}(\mathbf{G})$ *of* $\mathbf{G}$ *by its reduced center* (II, §3, No. 5) *is representable.*

Let

$$X = (x_1, \ldots, x_n), \qquad Z = (z_1, \ldots, z_n)$$

be independent generic points of $\mathbf{G}$ (No. 1), and let K be an algebraically closed field containing the integral domain $k[[z_1, \ldots, z_n]]$. Let $\mathfrak{m}$ be the maximal ideal of the ring of formal power series $K[[x_1, \ldots, x_n]]$; for any $h > 0$, $\mathfrak{m}^h$ is the ideal of formal power series having only terms of degree $\geqslant h$. To each $f \in \mathfrak{m}$, the formal power series $f(ZXZ^{-1})$ is defined, since $ZXZ^{-1}$ is a system of $n$ formal power series in $x_1, \ldots, x_n$ without constant term and whose coefficients are in K; write this formal power series, considered as element of $\mathfrak{m}$, as $f_Z$. It is clear that the mapping $f \mapsto f_Z$ is a linear mapping of each $\mathfrak{m}^h$ into itself, hence defines for each $h$ a linear mapping $j_Z$ of the K-vector space $\mathfrak{m}/\mathfrak{m}^h$ onto itself; furthermore, as $(f_Z)_{Z^{-1}} = f$, we have

$$j_Z \circ j_{Z^{-1}} = 1$$

and $j_Z$ is an *automorphism* of the K-vector space $\mathfrak{m}/\mathfrak{m}^h$. Now the classes

$\bar{x}^\alpha$ of the monomials $x^\alpha$ for $0 < |\alpha| < h$ form a *basis* of the vector space $\mathfrak{m}/\mathfrak{m}^h$, and we may therefore write

$$j_Z(\bar{x}^\alpha) = \sum_{0<|\beta|<h} c_{\alpha\beta}(Z)\bar{x}^\beta,$$

where the $c_{\alpha\beta}(Z)$ are formal power series belonging to $k[[z_1, \ldots, z_n]]$, with no constant term if $\alpha \neq \beta$ and constant term 1 if $\alpha = \beta$. Finally if Z′ is a third independent generic point of **G**, we have $(f_Z)_{Z'} = f_{ZZ'}$, hence

$$j_{ZZ'} = j_Z \circ j_{Z'},$$

in other words, if

$$\dim_K(\mathfrak{m}/\mathfrak{m}^h) = n_h,$$

the $n_h^2$ formal power series $c_{\alpha\beta}(Z) - \delta_{\alpha\beta}$ can be written $\mathbf{u}_h(Z)$, where $\mathbf{u}_h$ is a *formal homomorphism* of **G** into $\mathbf{GL}(n_h)$. Let $\mathbf{N}_h$ be the reduced kernel of $\mathbf{u}_h$; it is clear that $\mathbf{N}_{h+1} \subset \mathbf{N}_h$, hence there is a smallest $h$ such that

$$\mathbf{N}_k = \mathbf{N}_h \qquad \text{for all} \quad k \geqslant h.$$

If $\mathbf{v}: \mathbf{N}_h \to \mathbf{G}$ is the natural injection and Y a generic point of $\mathbf{N}_h$, we therefore have

$$f - f_{\mathbf{v}(Y)} \in \mathfrak{m}^k \qquad \text{for every} \quad k \geqslant h,$$

which of course implies

$$f = f_{\mathbf{v}(Y)} \qquad \text{for all} \quad f \in \mathfrak{m},$$

in particular for all $f \in \mathfrak{G}^{*+}$, which means that

$$\mathbf{v}(Y)X\mathbf{v}(Y)^{-1} = \mathbf{e},$$

hence $\mathbf{N}_h$ is contained in the center of **G**, and as it is reduced, it is contained in the reduced center ${}^r\mathscr{Z}(\mathbf{G})$. On the other hand, if

$$\mathbf{w} : {}^r\mathscr{Z}(\mathbf{G}) \to \mathbf{G}$$

is the natural injection and T a generic point of ${}^r\mathscr{Z}(\mathbf{G})$, T may also be considered as a generic point of ${}^r(\mathscr{L}(\mathbf{G}_{(K)}))$, hence we have $f_{w(T)} = f$ for every $f \in \mathfrak{m}$, in other words ${}^r\mathscr{L}(\mathbf{G}) \subset \mathbf{N}_k$ for any $k$; we therefore have ${}^r\mathscr{L}(\mathbf{G}) = \mathbf{N}_h$, and $\mathbf{G}/{}^r\mathscr{L}(\mathbf{G})$ is isogenous to a subgroup of $\mathbf{GL}(n_h)$ (II, §3, No. 7).

**5.** *Algebraic hull of a formal subgroup.* Let G be an affine group, $\mathbf{H}$ a formal subgroup of the formalized group $\mathbf{G}$. If A = R[G] and M is a closed *subvariety* of G containing the neutral element $e$, it is defined by an ideal $\mathfrak{J}$ of A, and R[M] = A/$\mathfrak{J}$. If $\mathfrak{m}$ is the maximal ideal of A corresponding to $e$, we have $\mathfrak{J} \subset \mathfrak{m}$, and the natural homomorphism $A \to A/\mathfrak{J}$ extends to a continuous homomorphism $\mathfrak{G}^* = \hat{A} \to \hat{A}/\bar{\mathfrak{J}}$. On the other hand, as $\mathbf{H}$ is a formal subgroup of $\mathbf{G}$, we have a continuous homomorphism $\mathfrak{G}^* \to \mathfrak{H}^*$. We will say that $\mathbf{H}$ is *contained in* M if the homomorphism $\mathfrak{G}^* \to \mathfrak{H}^*$ *factorizes* through the homomorphism $\mathfrak{G}^* \to \hat{A}/\bar{\mathfrak{J}}$, or equivalently if $\mathfrak{H}^\circ \supset \bar{\mathfrak{J}}$. Another equivalent way of expressing this relation is to say that if X is a generic point of $\mathbf{H}$, $\mathbf{j}: \mathbf{H} \to \mathbf{G}$ the natural injection, and $\mathbf{j}(X)$ the corresponding system of $n$ formal power series in $\mathfrak{G}^*$, then we must have $f(\mathbf{j}(X)) = 0$ for any $f \in \mathfrak{J}$.

We may always assume that G is a closed subvariety of some affine space $k^N$, hence M is also a closed subvariety of $k^N$, and $\mathfrak{J}$ is the image in A of the ideal $\mathfrak{I}$ of all polynomials $P \in k[t_1, \ldots, t_N]$ which vanish in M (as $k$ is infinite, we freely identify polynomials with functions in $k^N$); we may also assume that the point $e$ corresponds to the maximal ideal $(t_1, \ldots, t_N)$ of $k[t_1, \ldots, t_N]$; if ${}^t v_*$ is the natural homomorphism $k[[t_1, \ldots, t_N]] \to \mathfrak{H}^*$, corresponding to a system

$$\mathbf{v}(X) = (v_1(x_1, \ldots, x_n), \ldots, v_N(x_1, \ldots, x_n))$$

of N formal power series, the condition $\mathbf{H} \subset M$ amounts to saying that, for every $P \in \mathfrak{I}$, the formal power series

$$P(v_1(x_1, \ldots, x_n), \ldots, v_N(x_1, \ldots, x_n)) = 0. \tag{4}$$

For any given formal subgroup $\mathbf{H}$ of $\mathbf{G}$, there is therefore a *smallest* closed subvariety of G containing $\mathbf{H}$: its ideal in $k[t_1, \ldots, t_N]$ consists of *all* polynomials P satisfying (4).

**Proposition 5.** *For any formal subgroup* $\mathbf{H}$ *of the formalized group* $\mathbf{G} = \Phi(G)$ *of an affine group* G, *the smallest closed subvariety* $A(\mathbf{H})$ *of* G *containing* **H** *is a closed connected subgroup of* G, *and* **H** *is contained in the formal group* $\Phi(\mathrm{A}(\mathbf{H}))$.

With the preceding notations, if $\mathfrak{I}$ is the ideal of A(**H**), we will prove that, if $\mathrm{P} \in \mathfrak{I}$ and $y$, $z$ are two points of A(**H**), $\mathrm{P}(y \cdot z) = 0$ (the product being taken in G) which will show that $y \cdot z \in \mathrm{A}(\mathbf{H})$ and prove that A(**H**) is a group. Denote by ${}_y\mathrm{P}$ any of the polynomials whose restriction to G is the function $s \mapsto \mathrm{P}(y \cdot s)$ of R[G]; it is a polynomial in $t_1, \ldots, t_\mathrm{N}$ whose coefficients are polynomials in the coordinates of $y \in k^\mathrm{N}$ (since the coordinates of $y \cdot z$ are functions of $\mathrm{R}[\mathrm{G} \times \mathrm{G}]$, hence restrictions to $\mathrm{G} \times \mathrm{G}$ of polynomials in the coordinates of $y$ and $z$). We may therefore write

$$(5) \qquad {}_y\mathrm{P}(v_1(\mathrm{X}), \ldots, v_\mathrm{N}(\mathrm{X})) = \sum_\alpha \mathrm{Q}_\alpha(y_1, \ldots, y_\mathrm{N}) x^\alpha$$

where the $\mathrm{Q}_\alpha$ are polynomials of $k[t_1, \ldots, t_\mathrm{N}]$ (since a term in $x^\alpha$ can only come from terms of total degree $\leqslant |\alpha|$ in the power series $v_j(\mathrm{X})$), and the $y_i$ the coordinates of $y$. As $\mathrm{P} \in \mathfrak{I}$, we have, for a second independent generic point X′ of **H**,

$$\mathrm{P}(v_1(\mathrm{XX'}), \ldots, v_\mathrm{N}(\mathrm{XX'})) = 0,$$

but as $\mathbf{v}: \mathbf{H} \to \mathbf{G}$ is a formal homomorphism, the power series $v_j(\mathrm{XX'})$ is obtained by substituting the $v_i(\mathrm{X})$ and $v_i(\mathrm{X'})$ $(1 \leqslant i \leqslant \mathrm{N})$ to the coordinates of $y$ and $z$ in the $j$-th coordinate of $y \cdot z$; hence from (5) we derive the relation

$$\sum_\alpha \mathrm{Q}_\alpha(v_1(\mathrm{X}), \ldots, v_\mathrm{N}(\mathrm{X})) x'^\alpha = 0$$

and that relation implies

$$\mathrm{Q}_\alpha(v_1(\mathrm{X}), \ldots, v_\mathrm{N}(\mathrm{X})) = 0 \qquad \text{for each } \alpha.$$

By definition, this means that $\mathrm{Q}_\alpha \in \mathfrak{I}$ for each $\alpha$, hence, by definition of A(**H**), $\mathrm{Q}_\alpha(y_1, \ldots, y_\mathrm{N}) = 0$ for each $\alpha$, and therefore ${}_y\mathrm{P} \in \mathfrak{I}$; but then $\mathrm{P}(y \cdot z) = 0$ for every $z \in \mathrm{A}(\mathbf{H})$, proving that A(**H**) is a closed subgroup of G. To prove that A(**H**) is connected, observe that if $\mathfrak{I}_0$ is the ideal defining its neutral component, then, for any polynomial $\mathrm{P} \in \mathfrak{I}_0$, there is a polynomial $\mathrm{R} \in k[t_1, \ldots, t_\mathrm{N}]$ with constant term $\neq 0$ such that $\mathrm{PR} \in \mathfrak{I}$; as the power series $\mathrm{R}(v_1(\mathrm{X}), \ldots, v_\mathrm{N}(\mathrm{X}))$ has a nonvanishing constant term, we must have $P(v_1(X), \ldots, v_N(X)) = 0$, which shows that $\mathfrak{I}_0 = \mathfrak{I}$.

Finally, if we write $L = A(\mathbf{H})$, $\mathfrak{L}^*$ is identified to $\hat{A}/\mathfrak{I}$, and as by definition $\mathfrak{G}^* \to \mathfrak{H}^*$ factorizes into $\mathfrak{G}^* \to \mathfrak{L}^* \to \mathfrak{H}^*$, this means that

$$\mathbf{H} \subset \mathbf{L} = \Phi(L).$$

We will say that the affine group $A(\mathbf{H})$ is the *algebraic hull* of the formal subgroup $\mathbf{H}$ of $\mathbf{G}$. For reasons of dimensions, it is clear that, for any connected closed subgroup H of G, $A(\Phi(H)) = H$. A formal subgroup $\mathbf{H}$ of $\mathbf{G}$ is called an *algebraic formal subgroup* of $\mathbf{G}$ if $\Phi(A(\mathbf{H})) = \mathbf{H}$.

Well-known examples of nonalgebraic formal subgroups of algebraic groups exist for fields $k$ of characteristic 0 (namely, the formal groups corresponding to nonalgebraic Lie subgroups of an algebraic Lie group, for instance complexifications of nonclosed Lie subgroups of compact real Lie groups). It is also easy to give examples of commutative nonalgebraic formal subgroups of the formalization of a torus when $k$ has characteristic $p > 0$. One takes the torus $G = k^{*2}$, and for H the image by a formal homomorphism $u$ of the multiplicative formal group $\mathbf{G}_m(k)$, where the two formal power series defining $u$ are

$$u_1(x) = x \qquad \text{and} \qquad u_2(x) = (1 + x)^\zeta - 1,$$

$\zeta$ being a $p$-adic integer which is not a rational number.

Observe that from the relation $A(\Phi(L)) = L$ for any closed subgroup L of G, it follows that for any such subgroup for which $\mathbf{H} \subset \Phi(L)$ we have $A(\mathbf{H}) \subset L$; $A(\mathbf{H})$ is therefore the *smallest* closed subgroup such that $\mathbf{H} \subset \Phi(L)$.

**Corollary 1.** *If* $\mathbf{H}_1$, $\mathbf{H}_2$ *are two formal subgroups of* $\mathbf{G} = \Phi(G)$,

$$A(\mathbf{H}_1 \vee \mathbf{H}_2) = \sup(A(\mathbf{H}_1), A(\mathbf{H}_2)).$$

We obviously have

$$\sup(A(\mathbf{H}_1), A(\mathbf{H}_2)) \subset A(\mathbf{H}_1 \vee \mathbf{H}_2).$$

On the other hand,

$$\Phi(\sup(A(\mathbf{H}_1), A(\mathbf{H}_2))) = \Phi(A(\mathbf{H}_1)) \vee \Phi(A(\mathbf{H}_2))$$

by Proposition 4, and *a fortiori*

$$\Phi(\sup(A(\mathbf{H}_1), A(\mathbf{H}_2))) \supset \mathbf{H}_1 \vee \mathbf{H}_2,$$

whence

$$A(\mathbf{H}_1 \vee \mathbf{H}_2) \subset A(\Phi(\sup(A(\mathbf{H}_1), A(\mathbf{H}_2)))) = \sup(A(\mathbf{H}_1), A(\mathbf{H}_2)).$$

It is possible to give examples for which

$$A(\mathbf{H}_1 \wedge H_2) \neq A(\mathbf{H}_1) \cap A(\mathbf{H}_2).$$

**Corollary 2.** *If* $H_1$, $H_2$ *are two closed subgroups of an affine group* G, *then*

$$\Phi(H_1 \cap H_2) = \Phi(H_1) \wedge \Phi(H_2).$$

It is clear that $\Phi(H_1 \cap H_2)$ is contained in both $\Phi(H_1)$ and $\Phi(H_2)$, and being reduced it is contained in $\Phi(H_1) \wedge \Phi(H_2)$. On the other hand, $A(\Phi(H_1) \wedge \Phi(H_2))$ is contained in both $A(\Phi(H_1)) = H_1$ and $A(\Phi(H_2)) = H_2$, hence in the neutral component of $H_1 \cap H_2$, hence

$$\Phi(H_1) \wedge \Phi(H_2) \subset \Phi(A(\Phi(H_1) \wedge \Phi(H_2))) \subset \Phi(H_1 \cap H_2).$$

**Proposition 6.** *Let* G *be an affine group,* $\mathbf{H}$ *a formal subgroup of* $\mathbf{G} = \Phi(G)$, $\mathbf{N}$ *an invariant formal subgroup of* $\mathbf{H}$. *Then*:

(i) $A(\mathbf{N})$ *is an invariant closed subgroup of* $A(\mathbf{H})$.

(ii) $\mathbf{N}$ *is an invariant formal subgroup of* $\Phi(A(\mathbf{H}))$.

We keep the same notations as in the beginning of this section.

(i) If $\mathfrak{L}$ is the ideal of $A(\mathbf{N})$ in $k[t_1, \ldots, t_N]$, we have to prove that if $P \in \mathfrak{L}$, $y \in A(\mathbf{N})$, and $z \in A(\mathbf{H})$, then $P(zyz^{-1}) = 0$. Denote by ${}^zP$ any of the polynomials whose restriction to G is the function $s \mapsto P(zsz^{-1})$; it is again a polynomial in $t_1, \ldots, t_N$ whose coefficients are polynomials in the N coordinates of $z$ (here we use the fact that the coordinates of $z^{-1}$ are polynomials in those of $z$, since $s \mapsto s^{-1}$ is a morphism of the closed variety G onto itself). Let ${}^t w_*$ be the natural homomorphism $k[[t_1, \ldots, t_N]] \to \mathfrak{N}^*$, corresponding to a system

$$\mathbf{w}(U) = (w_1(u_1, \ldots, u_m), \ldots, w_N(u_1, \ldots, u_m))$$

of N formal power series. We may write

$$(6)\qquad {}^{z}\mathrm{P}(w_1(\mathrm{U}), \ldots, w_{\mathrm{N}}(\mathrm{U})) = \sum_{\alpha} \mathrm{R}_\alpha(z_1, \ldots, z_{\mathrm{N}})u^\alpha$$

where the $\mathrm{R}_\alpha$ are polynomials of $k[t_1, \ldots, t_{\mathrm{N}}]$ and the $z_i$ coordinates of $z$. As $\mathbf{v}(\mathrm{X})\mathbf{w}(\mathrm{U})\mathbf{v}(\mathrm{X})^{-1}$ is a point of $\mathbf{N}$, we have, since $\mathrm{P} \in \mathfrak{L}$,

$$\sum_{\alpha} \mathrm{R}_\alpha(v_1(\mathrm{X}), \ldots, v_{\mathrm{N}}(\mathrm{X}))u^\alpha = 0$$

hence

$$\mathrm{R}_\alpha(v_1(\mathrm{X}), \ldots, v_{\mathrm{N}}(\mathrm{X})) = 0 \qquad \text{for all } \alpha.$$

In other words, the polynomials $\mathrm{R}_\alpha$ belong to the ideal $\mathfrak{I}$ of $\mathrm{A}(\mathbf{H})$, hence $\mathrm{R}_\alpha(z_1, \ldots, z_{\mathrm{N}}) = 0$ for all $\alpha$, and from (6) it follows that ${}^{z}\mathrm{P} \in \mathfrak{L}$, which by definition implies $\mathrm{P}(zyz^{-1}) = 0$.

(ii) By II, §3, No. 2, Corollary 2 to Proposition 3, it is possible to find a regular system of parameters $s_1, \ldots, s_q$ of $\mathfrak{G}^*$, identified with indeterminates, such that the formal subgroups $\Phi(\mathrm{A}(\mathbf{H}))$, $\mathbf{H}$ and $\mathbf{N}$ of $\mathbf{G}$ are, respectively, defined by the equations $s_k = 0$ for $k > l$, $s_k = 0$ for $k > n$ and $s_k = 0$ for $k > m$ $(m \leqslant n \leqslant l)$. For any point $y \in \mathrm{G}$, the $k$-th component of the system $y\mathbf{w}(\mathrm{U})y^{-1}$ of N power series can be written

$$(7)\qquad (y\mathbf{w}(\mathrm{U})y^{-1})_k = \sum_{\alpha} \mathrm{Q}_{k\alpha}(y_1, \ldots, y_{\mathrm{N}})u^\alpha$$

where the $\mathrm{Q}_{k\alpha}$ are polynomials. From the assumption that $\mathbf{N}$ is invariant in $\mathbf{H}$, it follows that, for $k > m$, the power series

$$(\mathbf{v}(\mathrm{X})\mathbf{w}(\mathrm{U})\mathbf{v}(\mathrm{X})^{-1})_k = \sum_{\alpha} \mathrm{Q}_{k\alpha}(v_1(\mathrm{X}), \ldots, v_{\mathrm{N}}(\mathrm{X}))u^\alpha = 0,$$

in other words,

$$\mathrm{Q}_{k\alpha}(v_1(\mathrm{X}), \ldots, v_{\mathrm{N}}(\mathrm{X})) = 0$$

for these $k$ and all $\alpha$; but this means that the $\mathrm{Q}_{k\alpha}$ belong to the ideal $\mathfrak{I}$ of $\mathrm{A}(\mathbf{H})$, hence $\mathrm{Q}_{k\alpha}(y_1, \ldots, y_{\mathrm{N}}) = 0$ for any point $y \in \mathrm{A}(\mathbf{H})$. But if we write $\mathrm{H}' = \mathrm{A}(\mathbf{H})$ and denote by ${}^{t}r_*$ the natural homomorphism

$$k[[t_1, \ldots, t_{\mathrm{N}}]] \to \mathfrak{H}'^*,$$

corresponding to a system

$$\mathbf{r}(V) = (r_1(v_1, \ldots, v_l), \ldots, r_N(v_1, \ldots, v_l))$$

of N formal power series, we conclude from the preceding result that we also have

$$Q_{k\alpha}(r_1(V), \ldots, r_N(V)) = 0 \qquad \text{for} \quad k > m;$$

but this means that the components

$$(\mathbf{r}(V)\mathbf{w}(U)\mathbf{r}(V)^{-1})_k = \sum_a Q_{k\alpha}(r_1(V), \ldots, r_N(V))u^\alpha = 0 \qquad \text{for all} \quad k > m,$$

and this implies that $\mathbf{N}$ is an *invariant formal subgroup* of $\Phi(A(\mathbf{H}))$.

It should be observed that, even if $\mathbf{H}/\mathbf{N}$ is an algebraic formal group, $A(\mathbf{H})/A(\mathbf{N})$ may have a dimension strictly greater than $\dim(\mathbf{H}/\mathbf{N})$.

**Lemma 1.** *Let* G, G′ *be two affine algebraic groups, which are closed subvarieties of some affine space* $k^N$, *and let f be a morphism of the variety* G *into the variety* G′. *Let* $\mathbf{H}$ *be a formal subgroup of* $\mathbf{G} = \Phi(G)$, M *a closed subvariety of* G′. *If, for any polynomial* P *in the ideal of* M *in* $k[t_1, \ldots, t_N]$, $P \circ f$ *belongs to the ideal of* $A(\mathbf{H})$, *then* $A(\mathbf{H}) \subset f^{-1}(M)$.

By definition of a morphism of varieties, $P \circ f$, restricted to G, is a function of R[G], and these functions generate the ideal of R[G] defining $f^{-1}(M)$; the result then follows from the definitions.

**Proposition 7.** *Let* G, G′ *be two affine groups, which are, respectively, closed subvarieties of* $k^N$ *and* $k^{N'}$. *For any formal subgroup* $\mathbf{H}$ *of* $\mathbf{G} = \Phi(G)$ *and any formal subgroup* $\mathbf{H}'$ *of* $\mathbf{G}' = \Phi(G')$,

$$A(\mathbf{H} \times \mathbf{H}') = A(\mathbf{H}) \times A(\mathbf{H}').$$

Let $\mathbf{v}(X)$ the system of N power series corresponding to $\mathbf{H}$, $\mathbf{v}'(X')$ the system of N′ power series corresponding to $\mathbf{H}'$. If P′ is a polynomial of $k[t'_1, \ldots, t'_{N'}]$, belonging to the ideal of $A(\mathbf{H}')$, then, considered as a polynomial of $k[t_1, \ldots, t_N, t'_1, \ldots, t'_{N'}]$, it satisfies the condition

$$P'(v_1(X), \ldots, v_N(X), v'_1(X'), \ldots, v'_{N'}(X')) = 0,$$

hence, applying Lemma 1 to the projection $pr_2 : G \times G' \to G'$, we see that

$$A(\mathbf{H} \times \mathbf{H}') \subset pr_2^{-1}(A(\mathbf{H}'));$$

a similar argument for $pr_1$ proves that

$$A(\mathbf{H} \times \mathbf{H}') \subset A(\mathbf{H}) \times A(\mathbf{H}').$$

Conversely, for any polynomial P of $k[t_1, \ldots, t_N, t'_1, \ldots, t'_{N'}]$, we may write

$$P(t_1, \ldots, t_N, v'_1(X'), \ldots, v'_{N'}(X')) = \sum_{\alpha} Q_\alpha(t_1, \ldots, t_N) x'^\alpha$$

and if P belongs to the ideal of $A(\mathbf{H} \times \mathbf{H}')$, we must have

$$P(v_1(X), \ldots, v_N(X), v'_1(X'), \ldots, v'_{N'}(X')) = \sum_{\alpha} Q_\alpha(v_1(X), \ldots, v_N(X)) x'^\alpha = 0$$

hence $Q_\alpha$ must belong to the ideal of $A(\mathbf{H})$ for every $\alpha$. For any $s \in A(\mathbf{H})$ we have therefore

$$P(s_1, \ldots, s_N, v'_1(X'), \ldots, v'_{N'}(X')) = 0;$$

if we apply Lemma 1 to the morphism $f : y \mapsto (s, y)$ of G′ into $G \times G'$, we see that

$$A(\mathbf{H}') \subset f^{-1}(A(\mathbf{H} \times \mathbf{H}')),$$

which shows that

$$\{s\} \times A(\mathbf{H}') \subset A(\mathbf{H} \times \mathbf{H}'),$$

and finally

$$A(\mathbf{H}) \times A(\mathbf{H}') \subset A(\mathbf{H} \times \mathbf{H}').$$

## 6. *Commutator subgroups of representable formal groups.*

**Proposition 8.** *Let* G *be an affine group,* $\mathbf{H}$ *a formal subgroup of* $\mathbf{G} = \Phi(G)$.

(i) *If* $\mathbf{M}$, $\mathbf{N}$ *are two invariant formal subgroups of* $\mathbf{H}$, *then*

$$A([\mathbf{M}, \mathbf{N}]) = (A(\mathbf{M}), A(\mathbf{N})).$$

(ii) *For any integer* $r \geqslant 1$,

$$A(\mathscr{D}^r(\mathbf{H})) = \mathscr{D}^r(A(\mathbf{H})) \quad \text{and} \quad A(\mathscr{C}^r(\mathbf{H})) = \mathscr{C}^r(A(\mathbf{H})).$$

(iii) *In order that* $\mathbf{H}$ *be commutative* (resp. *nilpotent*, resp. *solvable*), *it is necessary and sufficient that* $A(\mathbf{H})$ *be commutative* (resp. *nilpotent*, resp. *solvable*).

(iv) *Let* $\mathbf{N}$ *be a formal invariant subgroup of* $\mathbf{H}$. *If* $\mathbf{H}/\mathbf{N}$ *is commutative* (resp. *nilpotent*, resp. *solvable*), *then* $A(\mathbf{H})/A(\mathbf{N})$ *is commutative* (resp. *nilpotent*, resp. *solvable*).

It is clear that (ii) and (iii) follow from (i); to say that $\mathbf{H}/\mathbf{N}$ is commutative (resp. nilpotent, resp. solvable) means that $\mathscr{D}(\mathbf{H}) \subset \mathbf{N}$ (resp. $\mathscr{C}^r(\mathbf{H}) \subset \mathbf{N}$ for some $r$, resp. $\mathscr{D}^r(\mathbf{H}) \subset \mathbf{N}$ for some $r$), and therefore (iv) follows from (ii). We are therefore reduced to proving (i). As

$$\mathbf{M} \subset \Phi(A(\mathbf{M})) \quad \text{and} \quad \mathbf{N} \subset \Phi(A(\mathbf{N})),$$

we have

$$[\mathbf{M}, \mathbf{N}] \subset [\Phi(A(\mathbf{M})), \Phi(A(\mathbf{N}))] = \Phi((A(\mathbf{M}), A(\mathbf{N})))$$

by Proposition 3, noting that $A(\mathbf{M})$ and $A(\mathbf{N})$ are connected closed invariant subgroups of $A(\mathbf{H})$ by Proposition 6. Hence

$$A([\mathbf{M}, \mathbf{N}]) \subset A(\Phi((A(\mathbf{M}), A(\mathbf{N})))) = (A(\mathbf{M}), A(\mathbf{N})).$$

Conversely, let $\mathbf{v}(X)$ and $\mathbf{v}'(X')$ be the systems of power series corresponding to $\mathbf{M}$ and $\mathbf{N}$; for any polynomial $P \in k[t_1, \ldots, t_N]$, and any point $y \in G$, denote by ${}^yP$ any of the polynomials whose restriction to $G$ is the function

$$s \mapsto P(ysy^{-1}s^{-1});$$

its coefficients (as a polynomial in $t_1, \ldots, t_N$) are polynomials in the coordinates of $y$. We may write

$$^{y}\mathrm{P}(v'_1(\mathrm{X}'), \ldots, v'_{\mathrm{N}}(\mathrm{X}')) = \sum_{\alpha} \mathrm{Q}_{\alpha}(y_1, \ldots, y_{\mathrm{N}})x'^{\alpha}$$

where the $\mathrm{Q}_\alpha$ are polynomials and the $y_i$ coordinates of $y$. By definition, if P belongs to the ideal of $\mathrm{A}([\mathbf{M}, \mathbf{N}])$, we have (I, §3, No. 16)

$$\sum_{\alpha} \mathrm{Q}_{\alpha}(v_1(\mathrm{X}), \ldots, v_{\mathrm{N}}(\mathrm{X}))x'^{\alpha} = 0,$$

hence all formal power series $\mathrm{Q}_\alpha(v_1(\mathrm{X}), \ldots, v_{\mathrm{N}}(\mathrm{X})) = 0$, which means that the $\mathrm{Q}_\alpha$ belong to the ideal of $\mathrm{A}(\mathbf{M})$. Now take $y \in \mathrm{A}(\mathbf{M})$; we have therefore

$$^{y}\mathrm{P}(v'_1(\mathrm{X}'), \ldots, v'_{\mathrm{N}}(\mathrm{X}')) = 0,$$

in other words $^{y}\mathrm{P}$ belongs to the ideal of $\mathrm{A}(\mathbf{N})$; Lemma 1 applied to the morphism $s \mapsto ysy^{-1}s^{-1}$ of the variety G into itself proves that, for any $z \in \mathrm{A}(\mathbf{N})$, we have

$$yzy^{-1}z^{-1} \in \mathrm{A}([\mathbf{M}, \mathbf{N}]),$$

hence

$$(\mathrm{A}(\mathbf{M}), \mathrm{A}(\mathbf{N})) \subset \mathrm{A}([\mathbf{M}, \mathbf{N}]).$$

**Theorem 2.** *Let* G *be an affine group. For any formal subgroup* $\mathbf{H}$ *of* $\mathbf{G} = \Phi(\mathrm{G})$ *we have*

$$\mathscr{D}(\mathbf{H}) = \Phi(\mathscr{D}(\mathrm{A}(\mathbf{H}))) = \mathscr{D}(\Phi(\mathrm{A}(\mathbf{H}))) = \Phi(\mathrm{A}(\mathscr{D}(\mathbf{H}))),$$

*in other words the formal subgroup* $\mathscr{D}(\mathbf{H})$ *is algebraic.*

The theorem, for the case of complex Lie groups, was first stated by E. Cartan [8], who did not publish any proof; the first proof for $k$ of characteristic 0 was given by Chevalley [14]; the following proof is an adaptation of his. We take as in Proposition 6 a system of indeterminates $s_1, \ldots, s_q$ of $\mathfrak{G}^*$ such that the formal subgroups $\Phi(\mathrm{A}(\mathbf{H}))$, $\mathbf{H}$, and $\mathscr{D}(\mathbf{H})$ are, respectively, defined by $s_k = 0$ for $k > l$, $s_k = 0$ for $k > n$, and $s_k = 0$ for $k > m$ ($m \leqslant n \leqslant l$). With the notations of the proof of Proposition 6, let X, X′ be two independent generic points of $\mathbf{H}$; for any point $y \in \mathrm{G}$, the $k$-th component of the system

$$y\mathbf{v}(X)y^{-1}\mathbf{v}(X)^{-1}$$

may be written

$$(y\mathbf{v}(X)y^{-1}\mathbf{v}(X)^{-1})_k = \sum_\alpha R_{k\alpha}(y_1, \ldots, y_N)x^\alpha$$

where the $R_{k\alpha}$ are polynomials. By assumption (I, §3, No. 16), we have

$$(\mathbf{v}(X')\mathbf{v}(X)\mathbf{v}(X')^{-1}\mathbf{v}(X)^{-1})_k = 0 \qquad \text{for} \quad k > n$$

hence

$$R_{k\alpha}(v_1(X), \ldots, v_N(X)) = 0 \qquad \text{for} \quad k > n \quad \text{and all } \alpha.$$

This implies that, for any $y \in A(\mathbf{H})$, we have

$$(y\mathbf{v}(X)y^{-1}\mathbf{v}(X)^{-1})_k = 0 \qquad \text{for any} \quad k > n,$$

hence also

(8) $$(\mathbf{r}(V)\mathbf{v}(X)\mathbf{r}(V)^{-1}\mathbf{v}(X)^{-1})_k = 0 \qquad \text{for} \quad k > n,$$

where V is a generic point of $\mathbf{\Phi}(A(\mathbf{H}))$. Now, for any point $y \in G$, write similarly

$$(\mathbf{r}(V)y\mathbf{r}(V)^{-1}y^{-1})_k = \sum_\alpha S_{k\alpha}(y_1, \ldots, y_N)v^\alpha$$

where the $S_{k\alpha}$ are polynomials. From (8) we deduce that

$$S_{k\alpha}(v_1(X), \ldots, v_N(X)) = 0 \qquad \text{for} \quad k > n \quad \text{and all } \alpha.$$

We then have similarly

$$(\mathbf{r}(V)y\mathbf{r}(V)^{-1}y^{-1})_k = 0 \qquad \text{for} \quad k > n$$

and every $y \in A(\mathbf{H})$, hence finally

$$(\mathbf{r}(V)\mathbf{r}(V')\mathbf{r}(V)^{-1}\mathbf{r}(V')^{-1})_k = 0 \qquad \text{for} \quad k > n$$

where V and V′ are independent generic points of $\mathbf{\Phi}(A(\mathbf{H}))$. As we know from Proposition 6 that $\mathscr{D}(\mathbf{H})$ is an invariant formal subgroup of $\mathbf{\Phi}(\mathrm{A}(\mathbf{H}))$, we conclude (I, §3, No. 16) that

$$\mathscr{D}(\mathbf{\Phi}(\mathrm{A}(\mathbf{H}))) \subset \mathscr{D}(\mathbf{H});$$

as the converse inclusion is trivial,

$$\mathscr{D}(\mathbf{H}) = \mathscr{D}(\mathbf{\Phi}(\mathrm{A}(\mathbf{H}))).$$

But from Proposition 3 we have

$$\mathscr{D}(\mathbf{\Phi}(\mathrm{A}(\mathbf{H}))) = \mathbf{\Phi}(\mathscr{D}(\mathrm{A}(\mathbf{H}))). \quad \text{QED.}$$

**Corollary.** *The formal group* $\mathbf{\Phi}(\mathrm{A}(\mathbf{H}))/\mathbf{H}$ *is commutative.*

Indeed, we have

$$\mathscr{D}(\mathbf{\Phi}(\mathrm{A}(\mathbf{H}))) = \mathscr{D}(\mathbf{H}) \subset \mathbf{H},$$

hence $\mathbf{\Phi}(\mathrm{A}(\mathbf{H}))/\mathbf{H}$ is isomorphic to a quotient group of the commutative formal group $\mathbf{\Phi}(\mathrm{A}(\mathbf{H}))/\mathscr{D}(\mathbf{\Phi}(\mathrm{A}(\mathbf{H})))$.

**7.** *Normalizers and centralizers of representable groups.* Let G be an affine group, and for any $s \in \mathrm{G}$, let $\mathrm{Int}(s)$ be the inner automorphism $y \mapsto sys^{-1}$ of G; by formalization it gives rise to a formal automorphism $\mathbf{Int}(s)$ of the formal group $\mathbf{G} = \mathbf{\Phi}(\mathrm{G})$ (No. 2).

**Proposition 9.** *Let* G *be an affine group,* $\mathbf{H}$ *a formal subgroup of* $\mathbf{G} = \mathbf{\Phi}(\mathrm{G})$. *The set of elements* $s \in \mathrm{G}$ *such that* $\mathbf{Int}(s)(\mathbf{H}) \subset \mathbf{H}$ (resp. *such that the restriction of* $\mathbf{Int}(s)$ *to* $\mathbf{H}$ *is the natural injection of* $\mathbf{H}$ *into* $\mathbf{G}$) *is a closed subgroup* ${}^a\mathscr{N}(\mathbf{H})$ (resp. ${}^a\mathscr{Z}(\mathbf{H})$) *of* G, *called the algebraic normalizer* (resp. *the algebraic centralizer*) *of* $\mathbf{H}$ *in* G; *these groups are such that*

$$\mathbf{\Phi}({}^a\mathscr{N}(\mathbf{H})) = \mathscr{N}(\mathbf{H}), \qquad \mathbf{\Phi}({}^a\mathscr{Z}(\mathbf{H})) = {}^r\mathscr{Z}(\mathbf{H}) \tag{9}$$

(*in other words,* $\mathscr{N}(\mathbf{H})$ *and* ${}^r\mathscr{Z}(\mathbf{H})$ *are algebraic formal groups*) *and*

$${}^a\mathscr{N}(\mathbf{H}) \subset \mathscr{N}(\mathrm{A}(\mathbf{H})), \qquad {}^a\mathscr{Z}(\mathbf{H}) = \mathscr{Z}(\mathrm{A}(\mathbf{H})), \tag{10}$$

*where* $\mathscr{Z}(\mathrm{A}(\mathbf{H}))$ *and* $\mathscr{N}(\mathrm{A}(\mathbf{H}))$ *are the centralizer and the normalizer of* $\mathrm{A}(\mathbf{H})$ *in* G.

For reasons of dimension, if $\mathbf{Int}(s)(\mathbf{H}) \subset \mathbf{H}$, both formal subgroups must be equal, from which it follows at once that ${}^a\mathscr{N}(\mathbf{H})$ is a subgroup of G. Take as in Proposition 6 a system of indeterminates $s_1, \ldots, s_q$ of $\mathfrak{G}^*$ such that $\mathbf{H}$ is defined by $s_k = 0$ for $k > n$, and write, for $y \in \mathrm{G}$,

$$(y\mathbf{v}(\mathrm{X})y^{-1})_k = \sum_\alpha \mathrm{Q}_{k\alpha}(y_1, \ldots, y_\mathrm{N})x^\alpha$$

where the $\mathrm{Q}_{k\alpha}$ are polynomials. To say that $y \in {}^a\mathscr{N}(\mathbf{H})$ means that

$$\mathrm{Q}_{k\alpha}(y_1, \ldots, y_\mathrm{N}) = 0 \qquad \text{for all } \alpha \quad \text{and for} \quad k > n,$$

hence ${}^a\mathscr{N}(\mathbf{H})$ is a closed subgroup of G. Furthermore, if $\mathbf{L}$ normalizes $\mathbf{H}$, $\mathbf{j}\colon \mathbf{L} \to \mathbf{G}$ is the natural injection and Z a generic point of $\mathbf{L}$, we must have

$$(\mathbf{j}(\mathrm{Z})\mathbf{v}(\mathrm{X})\mathbf{j}(\mathrm{Z})^{-1})_k = 0 \qquad \text{for all} \quad k > n$$

which is equivalent to

$$\mathrm{Q}_{k\alpha}(j_1(\mathrm{Z}), \ldots, j_\mathrm{N}(\mathrm{Z})) = 0 \qquad \text{for all } \alpha \quad \text{and for} \quad k > n,$$

and therefore means that $\mathbf{L} \subset \Phi({}^a\mathscr{N}(\mathbf{H}))$, and this proves the first relation (9). In addition, for any $y \in \mathrm{G}$, $\mathrm{Int}(y)$ is an automorphism of G, hence for any formal subgroup $\mathbf{L}$ of $\mathbf{G}$,

$$\mathrm{Int}(y)(\mathrm{A}(\mathbf{L})) = \mathrm{A}(\mathbf{Int}(y)(\mathbf{L})),$$

which proves the first relation (10).

Similarly, if we write

$$(y\mathbf{v}(\mathrm{X})y^{-1}\mathbf{v}(\mathrm{X})^{-1})_k = \sum_\alpha \mathrm{R}_{k\alpha}(y_1, \ldots, y_\mathrm{N})x^\alpha$$

to say that $y \in {}^a\mathscr{Z}(\mathbf{H})$ means that

$$\mathrm{R}_{k\alpha}(y_1, \ldots, y_\mathrm{N}) = 0 \qquad \text{for } \textit{all } k \text{ and all } \alpha,$$

and to say that $\mathbf{L}$ centralizes $\mathbf{H}$ means that

$$R_{k\alpha}(j_1(Z), \ldots, j_N(Z)) = 0 \qquad \text{for all } k \text{ and all } \alpha,$$

hence the second relation (9). Furthermore, if $y \in \mathscr{Z}(A(\mathbf{H}))$, the restriction of $\mathrm{Int}(y)$ to $A(\mathbf{H})$ is the identity, hence also the restriction of $\mathbf{Int}(y)$ to $\Phi(A(\mathbf{H}))$, and *a fortiori* its restriction to $\mathbf{H}$, which shows that

$$\mathscr{Z}(A(\mathbf{H})) \subset {}^{a}\mathscr{Z}(\mathbf{H}).$$

On the other hand, in the formal group $\Phi({}^{a}\mathscr{N}(\mathbf{H})) = \mathscr{N}(\mathbf{H})$, $\mathbf{H}$ and ${}^{r}\mathscr{Z}(\mathbf{H})$ are invariant formal subgroups, hence, by Proposition 8,

$$({}^{a}\mathscr{Z}(\mathbf{H}), A(\mathbf{H})) = A([\Phi({}^{a}\mathscr{Z}(\mathbf{H})), \mathbf{H}]) = A([{}^{r}\mathscr{Z}(\mathbf{H}), \mathbf{H}]) = e$$

which proves that ${}^{a}\mathscr{Z}(\mathbf{H}) \subset \mathscr{Z}(A(\mathbf{H}))$.

**8.** *Semisimple and simple formal groups.* We will say that a formal group $\mathbf{G}$ is *semisimple* if it does not contain an invariant commutative formal subgroup distinct from $\mathbf{e}$. A *simple* formal group is obviously semisimple if it is not commutative.

**Theorem 3.** (i) *A simple formal group is either commutative (and then isogenous to a group* $\mathbf{G}_{m,n}(k)$, cf. Chapter III) *or isogenous to the formalization* $\Phi(G)$ *of a simple (noncommutative) affine group* G (i.e., *a Chevalley group* [15]). *A semisimple formal group is isogenous to a direct product of simple (noncommutative) affine groups.*

(ii) *Conversely, for any semisimple* (resp. *noncommutative simple*) *affine group* G, *the formal group* $\Phi(G)$ *is semisimple* (resp. *simple*).

If $\mathbf{G}$ is a semisimple formal group, its reduced center must be $\mathbf{e}$ by definition, hence $\mathbf{G}$ is representable (Theorem 1). Among all isogenies $\mathbf{u}$ of $\mathbf{G}$ into formalizations of affine groups, consider one for which the dimension of $A(\mathbf{u}(\mathbf{G}))$ is *minimum*. Then $A(\mathbf{u}(\mathbf{G}))$ cannot have a noncommutative invariant closed subgroup N, for then there would exist a homomorphism $f$ of $A(\mathbf{u}(\mathbf{G}))$ into an affine group $G'$, with kernel N; as $\mathbf{u}(\mathbf{G}) \wedge \Phi(N)$ is an invariant formal subgroup of $\mathbf{u}(\mathbf{G})$, it must be reduced to $\mathbf{e}$, hence the restriction of $\mathbf{f} = \Phi(f)$ to $\mathbf{u}(\mathbf{G})$ is an isogeny, and $\mathbf{f} \circ \mathbf{u}$ would be an isogeny of $\mathbf{G}$ into $\Phi(G')$; furthermore, the dimension of $A(\mathbf{f}(\mathbf{u}(\mathbf{G})))$ would be at most $\dim(A(\mathbf{u}(\mathbf{G}))) - \dim(N)$, contradicting our choice of $\mathbf{u}$. Therefore $A(\mathbf{u}(\mathbf{G}))$ is connected and semisimple, hence equal to its commutator subgroup $\mathscr{D}(\mathrm{A}(\mathbf{u}(\mathbf{G})))$; but then, by Theorem 2,

$$\mathbf{u}(\mathbf{G}) \subset \Phi(\mathrm{A}(\mathbf{u}(\mathbf{G}))) = \mathscr{D}(\mathbf{u}(\mathbf{G})) \subset \mathbf{u}(\mathbf{G}),$$

hence $\mathbf{u}(\mathbf{G}) = \Phi(\mathrm{A}(\mathbf{u}(\mathbf{G})))$ is algebraic. The same argument proves that if $\mathbf{G}$ is simple and noncommutative, $\mathrm{A}(\mathbf{u}(\mathbf{G}))$ has no invariant closed subgroup of dimension $> 0$ other than itself.

Suppose conversely G is a semisimple connected affine group; if $\mathbf{G} = \Phi(\mathrm{G})$ contained a commutative invariant formal subgroup $\mathbf{N}$ of dimension $> 0$, $\mathrm{A}(\mathbf{N})$ would be a commutative invariant closed subgroup of G (Propositions 6 and 8); hence we must have $\mathbf{N} = \mathbf{e}$ and $\mathbf{G}$ is semisimple. Similarly, if G is simple and noncommutative, $\mathbf{G}$ cannot contain an invariant formal subgroup $\mathbf{N}$ of dimension $> 0$ other than itself: otherwise $\mathrm{A}(\mathbf{N})$ would be an invariant connected closed subgroup of G of dimension $> 0$, hence we would have $\mathrm{A}(\mathbf{N}) = \mathrm{G}$; but then, by Theorem 2,

$$\mathscr{D}(\mathbf{N}) = \Phi(\mathscr{D}(\mathrm{A}(\mathbf{N}))) = \Phi(\mathscr{D}(\mathrm{G})) = \Phi(\mathrm{G}) = \mathbf{G},$$

which proves that $\mathbf{N} = \mathbf{G}$, and $\mathbf{G}$ is simple.

**Corollary.** *Any invariant formal subgroup and any quotient of a semisimple formal group is semisimple.*

If $\mathbf{H}$ is semisimple and $\mathbf{N}$ an invariant formal subgroup of $\mathbf{H}$, $\mathrm{A}(\mathbf{N})$ is an invariant closed subgroup of the semisimple group $\mathrm{A}(\mathbf{H})$, hence is semisimple; therefore $\Phi(\mathrm{A}(\mathbf{N})) \supset \mathbf{N}$ is a formal semisimple group, hence

$$\mathscr{D}(\Phi(\mathrm{A}(\mathbf{N}))) = \Phi(\mathrm{A}(\mathbf{N})) \supset \mathbf{N};$$

but as

$$\mathscr{D}(\Phi(\mathrm{A}(\mathbf{N}))) = \mathscr{D}(\mathbf{N}) \subset \mathbf{N},$$

we have $\Phi(\mathrm{A}(\mathbf{N})) = \mathbf{N}$ and $\mathbf{N}$ is semisimple; furthermore

$$\mathbf{H}/\mathbf{N} = \Phi(\mathrm{A}(\mathbf{H}))/\Phi(\mathrm{A}(\mathbf{N}))$$

is isomorphic to $\Phi(\mathrm{A}(\mathbf{H})/\mathrm{A}(\mathbf{N}))$ (Corollary 1 to Proposition 1), hence is semisimple.

**9.** *Formal solvable groups.* We will denote as usual by T(N) the closed subgroup of GL(N) consisting of all upper triangular matrices (matrices $(s_{ij})$ with $s_{ij} = 0$ for $i > j$), and by U(N) the invariant closed subgroup of T(N) consisting of the *unipotent* upper triangular matrices (i.e., those with $s_{ii} = 1$ for each $i$).

**Proposition 10.** *For any solvable formal subgroup* $\mathbf{H}$ *of* $\mathbf{GL}(\mathrm{N})$, *there exists* $s \in \mathrm{GL}(\mathrm{N})$ *such that*

$$\mathrm{Int}(s)(\mathbf{H}) \subset \Phi(\mathrm{T}(\mathrm{N})).$$

Indeed, $\mathrm{A}(\mathbf{H})$ is a closed solvable subgroup of GL(N) (Proposition 8), hence, by the Lie-Kolchin theorem, there is an $s \in \mathrm{GL}(\mathrm{N})$ such that $\mathrm{Int}(s)(\mathrm{A}(\mathbf{H}))$ is contained in T(N), hence the result.

**Proposition 11.** *For any solvable formal group* $\mathbf{G}$, *the derived group* $\mathscr{D}(\mathbf{G})$ *is nilpotent.*

If $\mathbf{G}$ is representable, this follows at once from Proposition 10, for we may then assume that $\mathbf{G} \subset \Phi(\mathrm{T}(\mathrm{N}))$, and then

$$\mathscr{D}(\mathbf{G}) \subset \mathscr{D}(\Phi(\mathrm{T}(\mathrm{N}))) = \Phi(\mathscr{D}(\mathrm{T}(\mathrm{N}))) = \Phi(\mathrm{U}(\mathrm{N}))$$

(Proposition 3), and $\Phi(\mathrm{U}(\mathrm{N}))$ is nilpotent (Corollary to Proposition 3). Take now an arbitrary solvable formal group $\mathbf{G}$, and let ${}^r\mathbf{Z}$ be its reduced center; then $\mathbf{G}/{}^r\mathbf{Z}$ is representable (Theorem 1), hence $\mathscr{D}(\mathbf{G}/{}^r\mathbf{Z})$ is nilpotent. But $\mathscr{D}(\mathbf{G}/{}^r\mathbf{Z})$ is isogenous to $\mathscr{D}(\mathbf{G})/({}^r\mathbf{Z} \wedge \mathscr{D}(\mathbf{G}))$; as the reduced centralizer ${}^r\mathscr{Z}(\mathscr{D}(\mathbf{G}))$ of $\mathscr{D}(\mathbf{G})$ in $\mathbf{G}$ obviously contains ${}^r\mathbf{Z} \wedge \mathscr{D}(\mathbf{G})$, so does the reduced center

$${}^r\mathbf{Z}' = {}^r\mathscr{Z}(\mathscr{D}(\mathbf{G})) \wedge \mathscr{D}(\mathbf{G})$$

of $\mathscr{D}(\mathbf{G})$; hence $\mathscr{D}(\mathbf{G})/{}^r\mathbf{Z}'$, which is isogenous to a quotient group of $\mathscr{D}(\mathbf{G})/({}^r\mathbf{Z} \wedge \mathscr{D}(\mathbf{G}))$, is nilpotent. But then

$$\mathscr{C}^r(\mathscr{D}(\mathbf{G})) \subset {}^r\mathbf{Z}' \qquad \text{for some} \quad r > 0;$$

hence, by definition, $\mathscr{C}^{r+1}(\mathscr{D}(\mathbf{G})) = \mathbf{e}$, and therefore $\mathscr{D}(\mathbf{G})$ is nilpotent.

We now define the *radical* $\mathbf{Rad}(\mathbf{G})$ of a formal group $\mathbf{G}$ in the same way as the radical Rad(G) of an algebraic group G: if $\mathbf{N}_1$, $\mathbf{N}_2$ are two invariant solvable formal subgroups of $\mathbf{G}$, so is $\mathbf{N}_1 \vee \mathbf{N}_2$, for $(\mathbf{N}_1 \vee \mathbf{N}_2)/\mathbf{N}_1$ is isogenous to $\mathbf{N}_2/(\mathbf{N}_1 \wedge \mathbf{N}_2)$, hence solvable, and as $\mathbf{N}_1$ is solvable, so is $\mathbf{N}_1 \vee \mathbf{N}_2$; there is therefore a *largest* invariant solvable formal subgroup $\mathbf{Rad}(\mathbf{G})$ of $\mathbf{G}$. It is clear that $\mathbf{G}/\mathbf{Rad}(\mathbf{G})$ cannot contain a nontrivial invariant commutative formal subgroup $\mathbf{H}'$, for its reduced inverse image $\mathbf{H}$ in $\mathbf{G}$ would be an extension of $\mathbf{H}'$ by $\mathbf{Rad}(\mathbf{G})$, hence would be invariant and solvable, contrary to the definition of $\mathbf{Rad}(\mathbf{G})$; therefore $\mathbf{G}/\mathbf{Rad}(\mathbf{G})$ is *semisimple.*

Conversely, if $\mathbf{N}$ is an invariant solvable formal subgroup of $\mathbf{G}$ such that $\mathbf{G}/\mathbf{N}$ is semisimple, then $\mathbf{N} = \mathbf{Rad}(\mathbf{G})$, for $\mathbf{Rad}(\mathbf{G})/\mathbf{N}$ is an invariant solvable subgroup of $\mathbf{G}/\mathbf{N}$, hence must be reduced to $\mathbf{e}$, otherwise its last derived subgroup not equal to $\mathbf{e}$ would be a commutative nontrivial invariant subgroup of $\mathbf{G}/\mathbf{N}$.

**Proposition 12.** *Let* G *be an affine group. For any formal subgroup* $\mathbf{H}$ *of* $\mathbf{G} = \Phi(G)$,

$$A(\mathbf{Rad}(\mathbf{H})) = \mathrm{Rad}(A(\mathbf{H})) \qquad \Phi(A(\mathbf{H})) = \mathscr{D}(\mathbf{H}) \vee \Phi(A(\mathbf{Rad}(\mathbf{H})))$$
$$\textit{and} \quad \mathbf{H} \wedge \Phi(A(\mathbf{Rad}(\mathbf{H}))) = \mathbf{Rad}(\mathbf{H}).$$

It is clear that $A(\mathbf{Rad}(\mathbf{H}))$ is a connected solvable invariant closed subgroup of $A(\mathbf{H})$ (Propositions 6 and 8). All we have to show is that $A(\mathbf{H})/A(\mathbf{Rad}(\mathbf{H}))$ is semisimple. Let us first observe that we have

$$\mathbf{H} = \mathscr{D}(\mathbf{H}) \vee \mathbf{Rad}(\mathbf{H}). \tag{11}$$

Indeed, as $\mathbf{H}/\mathbf{Rad}(\mathbf{H})$ is semisimple, we have

$$\mathscr{D}(\mathbf{H}/\mathbf{Rad}(\mathbf{H})) = \mathbf{H}/(\mathbf{Rad}(\mathbf{H});$$

but on the other hand

$$\mathscr{D}(\mathbf{H}/\mathbf{Rad}(\mathbf{H})) = (\mathscr{D}(\mathbf{H}) \vee \mathbf{Rad}(\mathbf{H}))/\mathbf{Rad}(\mathbf{H}))$$

hence our assertion.

From this result we deduce (Corollary 1 to Proposition 5)

$$A(\mathbf{H}) = \sup(A(\mathscr{D}(\mathbf{H})), A(\mathbf{Rad}(\mathbf{H}))) \tag{12}$$

and as $\mathbf{\Phi}(\mathrm{A}(\mathscr{D}(\mathbf{H}))) = \mathscr{D}(\mathbf{H})$ (Theorem 2), (12) and Proposition 4 imply

$$\mathbf{\Phi}(\mathrm{A}(\mathbf{H})) = \mathscr{D}(\mathbf{H}) \vee \mathbf{\Phi}(\mathrm{A}(\mathbf{Rad}(\mathbf{H}))) \tag{13}$$

and *a fortiori*

$$\mathbf{\Phi}(\mathrm{A}(\mathbf{H})) = \mathbf{H} \vee \mathbf{\Phi}(\mathrm{A}(\mathbf{Rad}(\mathbf{H}))). \tag{14}$$

On the other hand, the reduced intersection $\mathbf{H} \wedge \mathbf{\Phi}(\mathrm{A}(\mathbf{Rad}(\mathbf{H})))$ is an invariant solvable formal subgroup of $\mathbf{\Phi}(\mathrm{A}(\mathbf{H}))$, hence of $\mathbf{H}$, which contains $\mathbf{Rad}(\mathbf{H})$, hence is equal to $\mathbf{Rad}(\mathbf{H})$. From that remark and (14), we conclude that $\mathbf{\Phi}(\mathrm{A}(\mathbf{H}))/\mathbf{\Phi}(\mathrm{A}(\mathbf{Rad}(\mathbf{H})))$ is isogenous to $\mathbf{H}/\mathbf{Rad}(\mathbf{H})$, hence semisimple; but as it is isomorphic to $\mathbf{\Phi}(\mathrm{A}(\mathbf{H})/\mathrm{A}(\mathbf{Rad}(\mathbf{H})))$ (Corollary 1 to Proposition 1), $\mathrm{A}(\mathbf{H})/\mathrm{A}(\mathbf{Rad}(\mathbf{H}))$ is semisimple by Theorem 3.

**Proposition 13.** *Let* G *be an affine group,* $\mathbf{H}$ *a formal subgroup of* $\mathbf{G} = \mathbf{\Phi}(\mathrm{G})$.

(i) *For any maximal solvable formal subgroup* $\mathbf{S}$ *of* $\mathbf{H}$, $\mathrm{A}(\mathbf{S})$ *is the unique Borel subgroup* B *of* $\mathrm{A}(\mathbf{H})$ *such that* $\mathbf{S} \subset \mathbf{\Phi}(\mathrm{B})$. *Conversely, for any Borel subgroup* B *of* $\mathrm{A}(\mathbf{H})$, $\mathbf{H} \wedge \mathbf{\Phi}(\mathrm{B})$ *is a maximal solvable formal subgroup of* $\mathbf{H}$.

(ii) *If* $\mathbf{S}_1$, $\mathbf{S}_2$ *are two maximal solvable subgroups of* $\mathbf{H}$, *there exists* $s \in \mathrm{A}(\mathbf{H})$ *such that* $\mathbf{Int}(s)(\mathbf{S}_1) = \mathbf{S}_2$.

Any maximal solvable formal subgroup $\mathbf{S}$ of a formal group $\mathbf{H}$ must contain $\mathbf{Rad}(\mathbf{H})$, for otherwise $\mathbf{S} \vee \mathbf{Rad}(\mathbf{H})$, being isogenous to an extension of the solvable group $\mathbf{S}/(\mathbf{S} \wedge \mathbf{Rad}(\mathbf{H}))$ by $\mathbf{Rad}(\mathbf{H})$, would be distinct from $\mathbf{S}$ and contain it. The maximal solvable subgroups of $\mathbf{H}$ are therefore the inverse images of the maximal solvable subgroups of the formal semisimple group $\mathbf{H}/\mathbf{Rad}(\mathbf{H})$. If $\mathbf{H} \subset \mathbf{G}$ is semisimple and B is a Borel subgroup of $\mathrm{A}(\mathbf{H})$, then $\mathbf{\Phi}(\mathrm{B})$ is solvable (Corollary to Proposition 3), and if $\mathbf{S}$ is a solvable formal subgroup containing $\mathbf{\Phi}(\mathrm{B})$, then $\mathrm{A}(\mathbf{S})$ is a connected solvable affine group (Proposition 8) containing $\mathrm{A}(\mathbf{\Phi}(\mathrm{B})) = \mathrm{B}$, hence equal to B; in other words $\mathbf{\Phi}(\mathrm{B})$ is maximal solvable.

Now suppose $\mathbf{H} \subset \mathbf{G}$ is arbitrary; is $\mathbf{S}$ is a maximal solvable subgroup of $\mathbf{H}$,

$$\mathbf{S}' = \mathbf{S} \vee \mathbf{\Phi}(\mathrm{A}(\mathbf{Rad}(\mathbf{H})))$$

is maximal solvable in $\mathbf{\Phi}(\mathrm{A}(\mathbf{H}))$, for it follows from Proposition 12 that

$\Phi(\mathrm{A}(\mathbf{H}))/\mathbf{Rad}(\mathbf{H})$ is isogenous to the direct product of $\mathbf{H}/\mathbf{Rad}(\mathbf{H})$ and $\Phi(\mathrm{A}(\mathbf{Rad}(\mathbf{H})))/\mathbf{Rad}(\mathbf{H})$; if $\mathbf{S}'' \supset \mathbf{S}'$, $\mathbf{S}''/\Phi(\mathrm{A}(\mathbf{Rad}(\mathbf{H})))$ is isogenous to $(\mathbf{S}'' \wedge \mathbf{H})/\mathbf{Rad}(\mathbf{H})$, and if $\mathbf{S}''$ is solvable, so is $\mathbf{S}'' \wedge \mathbf{H}$, which contains $\mathbf{S}$, and therefore

$$\mathbf{S}'' \wedge \mathbf{H} = \mathbf{S} \qquad \text{and} \qquad \mathbf{S}'' = (\mathbf{S}'' \wedge \mathbf{H}) \vee \Phi(\mathrm{A}(\mathbf{Rad}(\mathbf{H}))) = \mathbf{S}';$$

furthermore we have $\mathbf{S} = \mathbf{S}' \wedge \mathbf{H}$. Now we have seen above that $\mathbf{S}' = \Phi(\mathrm{B})$, where B is a Borel subgroup of $\mathrm{A}(\mathbf{H})$, hence $\mathbf{S} = \Phi(\mathrm{B}) \wedge \mathbf{H}$. On the other hand,

$$\mathrm{A}(\mathbf{S}') = \mathrm{B} = \sup(\mathrm{A}(\mathbf{S}), \mathrm{A}(\mathbf{Rad}(\mathbf{H}))) = \mathrm{A}(\mathbf{S})$$

since $\mathbf{S} \supset \mathbf{Rad}(\mathbf{H})$. In addition, we see that

$$(15) \qquad \dim(\mathrm{A}(\mathbf{S})) = \dim \mathbf{S} + \dim(\mathrm{Rad}(\mathrm{A}(\mathbf{H}))) - \dim(\mathbf{Rad}(\mathbf{H})).$$

If $\mathrm{B}_1$ is a Borel subgroup of $\mathrm{A}(\mathbf{H})$ such that $\mathbf{S} \subset \Phi(\mathrm{B}_1)$, then

$$\mathrm{A}(\mathbf{S}) \subset \mathrm{A}(\Phi(\mathrm{B}_1)) = \mathrm{B}_1,$$

and as $\mathrm{A}(\mathbf{S})$ is a Borel subgroup, $\mathrm{B}_1 = \mathrm{A}(\mathbf{S})$.

Conversely, let $\mathrm{B}'$ be any Borel subgroup of $\mathrm{A}(\mathbf{H})$; then there is an $s \in \mathrm{A}(\mathbf{H})$ such that $\mathrm{Int}(s)(\mathrm{B}) = \mathrm{B}'$, and as $\mathbf{H}$ is an invariant subgroup of $\Phi(\mathrm{A}(\mathbf{H}))$, we have

$$\mathbf{Int}(s)(\mathbf{H} \wedge \Phi(\mathrm{B})) = \mathbf{H} \wedge \Phi(\mathrm{B}'),$$

hence $\mathbf{H} \wedge \Phi(\mathrm{B}')$ is maximal solvable in $\mathbf{H}$.

Finally, if $\mathbf{S}_1$, $\mathbf{S}_2$ are any two maximal solvable subgroups of $\mathbf{H}$, there is an $s \in \mathrm{A}(\mathbf{H})$ such that

$$\mathrm{Int}(s)(\mathrm{A}(\mathbf{S}_1)) = \mathrm{A}(\mathbf{S}_2),$$

from which we conclude similarly that

$$\mathbf{Int}(s)(\mathbf{S}_1) = \mathbf{Int}(s)(\mathbf{H} \wedge \Phi(\mathrm{A}(\mathbf{S}_1))) = \mathbf{H} \wedge \Phi(\mathrm{A}(\mathbf{S}_2)) = \mathbf{S}_2.$$

**Corollary 1.** *In any formal group* $\mathbf{G}$, *the smallest subgroup containing all maximal solvable subgroups is* $\mathbf{G}$, *and the largest subgroup contained in all maximal solvable subgroups is* $\mathbf{Rad}(\mathbf{G})$.

As all maximal solvable subgroups of **G** contain **Rad(G)**, one is immediately reduced to the case **Rad(G)** = **e**, i.e., **G** semisimple, hence **G** = Φ(G), where G is an affine connected semisimple group. But it is known that the intersection of the Borel subgroups of G is reduced to *e*, and the union of the Borel subgroups of G is G itself; as the maximal solvable subgroups of **G** are the Φ(B), where B is a Borel subgroup of G, this proves the corollary.

**Corollary 2.** *In any formal group* **G**, *the normalizer of a maximal solvable subgroup* **S** *is equal to* **S**.

Let **N** be the normalizer of **S** in **G**; if **N** ≠ **S**, it follows from Corollary 1 that there are solvable subgroups **S**′ ≠ **e** in **N**/**S**; but then the reduced inverse image $\mathbf{S}_1$ of **S**′ in **N** is solvable, since $\mathbf{S}_1/\mathbf{S}$ is solvable (as isogenous to **S**′). But this contradicts the maximality of **S**, hence **N** = **S**.

**Corollary 3.** *In any representable formal group* **G**, *the reduced center of* **G** *is equal to the reduced center of any maximal solvable subgroup* **S** *of* **G**.

As A(**S**) is a Borel subgroup of A(**G**), it is known that the center of A(**S**) is equal to the center of A(**G**); but if Z is the center of A(**G**), the reduced center of **G** is equal to Φ(Z) by Proposition 9, and similarly for **S**, hence the corollary.

I do not know if Corollary 3 extends to arbitrary formal groups, nor if two maximal solvable subgroups of an arbitrary formal group are always isomorphic.

**10.** *Formal tori.* Up to now, all the results of this chapter were independent of the characteristic of the field *k*. When that characteristic is 0, the natural correspondence between formal groups and their Lie algebras (II, §2, No. 5) enables one to translate all these results into results concerning "algebraic" Lie algebras, their "algebraic hulls," etc. (see [14]), taking into account the additional fact that, due to Ado's theorem, *every* formal Lie group is representable when *k* has characteristic 0.

From now on until the end of the chapter, we again assume on the contrary, that *k is an algebraically closed field of characteristic* $p > 0$.

We will say that a formal Lie group $\mathbf{T}$ over $k$ is a *formal torus* if it is isomorphic to a direct product $(\mathbf{G}_m(k))^n$ of multiplicative formal groups. Observe that this definition could still be given when $k$ has characteristic 0, but then $\mathbf{G}_m(k)$ is isomorphic to the additive formal group $\mathbf{G}_a(k)$, and due to this "degeneracy," most of the results we are going to prove about formal tori are *false* for fields of characteristic 0.

The results of Chapter III show that any formal commutative group is *isogenous* to the direct product of a formal torus and a subgroup which does not contain any torus $\neq \mathbf{e}$ (III, §6, No. 1). But it is possible for formal tori to obtain more precise results involving *isomorphisms* and not only isogenies.

**Proposition 14.** (i) *Any formal commutative group* $\mathbf{G}$ *is the direct product of the largest formal torus* $\mathbf{C}$ *in* $\mathbf{G}$ *and of the largest subgroup* $\mathbf{R}$ *of* $\mathbf{G}$ *containing no formal torus* $\neq \mathbf{e}$. *The covariant bigebras* $\mathfrak{C}$ *and* $\mathfrak{R}$ *of* $\mathfrak{G}$ *are characterized by the following properties*: $\mathfrak{C}$ *is the largest subbigebra* $\mathfrak{H}$ *of* $\mathfrak{G}$ *such that the restriction of the Frobenius homomorphism* F *to* $\mathfrak{H}$ *is a bijection of* $\mathfrak{H}$ *onto itself, and the biideal generated by* $\mathfrak{R}^+$ *is the set of all* $Z \in \mathfrak{G}$ *such that* $\mathrm{F}^h(Z) = 0$ *for some integer* $h \geqslant 1$.

(ii) *Any formal subgroup* $\mathbf{T}$ *of* $\mathbf{C}$ *is a formal torus and a direct factor of* $\mathbf{C}$ *and therefore there is at least one formal torus* $\mathbf{T}'$ *such that* $\mathbf{C} = \mathbf{T} \times \mathbf{T}'$.

(i) The Frobenius homomorphism F is a semilinear mapping of each of the finite-dimensional vector spaces $\mathfrak{g}_r$ and $\mathfrak{s}_r$ into itself. But for such a vector space E over $k$, the Fitting theory ([31], [21]) implies that E is the direct sum of two uniquely determined subspaces $\mathrm{E}_0 \oplus \mathrm{E}_1$, such that:

1° F is a bijection of $\mathrm{E}_0$ onto itself, and there is a basis $(e_\lambda)$ of $\mathrm{E}_0$ over $k$ such that $\mathrm{F} \cdot e_\lambda = e_\lambda$ for every $\lambda$.

2° For any element $z \in \mathrm{E}_1$, there is an integer $h \geqslant 1$ such that $\mathrm{F}^h \cdot z = 0$.

It immediately follows from these properties that the only elements $z \in \mathrm{E}$ such that $\mathrm{F} \cdot z = z$ are the linear combinations of the $e_\lambda$ with coefficients such that $\xi^p = \xi$, i.e., belonging to the prime field $\mathbf{F}_p$.

If we apply these results to $\mathfrak{g}_r$ or $\mathfrak{s}_r$, we have decompositions

$$\mathfrak{g}_r = \mathfrak{g}_r^{(0)} \oplus \mathfrak{g}_r^{(1)},$$

where $\mathfrak{g}_r^{(0)}$ and $\mathfrak{g}_r^{(1)}$ are *p-Lie algebras*, and

$$\mathfrak{s}_r = \mathfrak{s}_r^{(0)} \oplus \mathfrak{s}_r^{(1)},$$

where $\mathfrak{s}_r^{(0)}$ is a *subbigebra* and $\mathfrak{s}_r^{(1)}$ a *biideal.* We are going to show that the $\mathfrak{g}_r^{(0)}$ are the higher Lie algebras and the $\mathfrak{s}_r^{(0)}$ the Frobenius subbigebras of the bigebra **C** of a formal torus.

First of all, we may find a basis $(X_{0i})_{1 \leqslant i \leqslant n}$ of the Lie algebra $\mathfrak{g}_0$ of $\mathfrak{G}$ such that the $X_{0i}$ $(1 \leqslant i \leqslant m)$ form a basis of $\mathfrak{g}_0^{(0)}$ with $X_{0i}^p = X_{0i}$, and the $X_{0i}$ for $m + 1 \leqslant i \leqslant n$ a basis of $\mathfrak{g}_0^{(1)}$. We first construct a privileged basis $(Z_\alpha^{(0)})$ of $\mathfrak{G}$ corresponding to that basis $(X_{0i})_{1 \leqslant i \leqslant n}$. We will now modify by induction that privileged basis in such a way that, at the $r$-th stage:

1° the $Z_\alpha^{(r)}$ with $\alpha(i) = 0$ for $i > m$ and $\mathrm{ht}(\alpha) \leqslant r$ form a privileged basis of $\mathfrak{s}_r^{(0)}$;

2° the $Z_\alpha^{(r)}$ with $\alpha(i) \neq 0$ for some $i > m$ and $\mathrm{ht}(\alpha) \leqslant r$ form a basis of $\mathfrak{s}_r^{(1)}$;

3° we have

$$(X_{hi}^{(r)})^p = X_{hi}^{(r)} \qquad \text{for} \quad 1 \leqslant i \leqslant m \quad \text{and} \quad h \leqslant r,$$

and there is an integer $l \geqslant 1$ such that

$$(X_{hi}^{(r)})^{p^l} = 0 \qquad \text{for} \quad i > m \quad \text{and} \quad h \leqslant r;$$

4° with the notations of III, §2, No. 2, Theorem 2, there is a homomorphism $u_r$ of $(\mathfrak{G}_\infty(k))^{\otimes n}$ onto $\mathfrak{G}$ such that

$$u_r(Z_{\lambda(\alpha)}) = Z_\alpha^{(r)} \qquad \text{for} \quad \mathrm{ht}(\alpha) \leqslant r.$$

Suppose we have obtained such a basis; as the multiplication law of the basis $(Z_\alpha)$ of $(\mathfrak{G}_\infty(k))^{\otimes n}$ has its coefficients in the prime field $\mathbf{F}_p$, we have, for $\mathrm{ht}(\alpha) \leqslant r$,

$$(Z_\alpha^{(r)})^p = Z_\alpha^{(r)} \qquad \text{if} \quad \alpha(i) = 0 \quad \text{for} \quad i > m,$$

and

$$(Z_\alpha^{(r)})^{p^l} = 0 \qquad \text{if} \quad \alpha(i) > 0 \quad \text{for some} \quad i > m.$$

Consider the elements $X_{r+1,1}^{(r)}$ for $1 \leqslant i \leqslant n$. As

$$V^{r+1}(X^{(r)}_{r+1,i}) = X^{(r)}_{0i}, \qquad \text{for} \quad 1 \leqslant i \leqslant m,$$

$$(X^{(r)}_{r+1,i})^p - X^{(r)}_{r+1,i} = Y_i \in \mathfrak{s}_r,$$

and for $i > m$,

$$(X^{(r)}_{r+1,i})^{p^l} = Y'_i \in \mathfrak{s}_r.$$

From the fact that $(Z^{(r)}_\alpha)$ is a privileged basis and that if $\alpha + \beta = p^{r+1}\varepsilon_i$, it follows that $\alpha$ and $\beta$ are multiples of $\varepsilon_i$, we see that

$$\Delta(Y_i) - 1 \otimes Y_i - Y_i \otimes 1 = 0 \quad \text{and} \quad \Delta(Y'_i) - 1 \otimes Y'_i - Y'_i \otimes 1 = 0,$$

in other words, both the $Y_i$ and $Y'_i$ are in $\mathfrak{g}_0$. By adding to each $x_l$ a linear combination of monomials of total degree $p^{r+1}$ as in II, §3, No. 1, Proposition 2, we get a new privileged basis $(Z'^{(r+1)}_\alpha)$ for $\mathfrak{G}$, which coincides with $(Z^{(r)}_\alpha)$ for $|\alpha| < p^{r+1}$, and is such that

$$X^{(r+1)}_{r+1,i} = X^{(r)}_{r+1,i} + \sum_{j=1}^{n} a_{ij} X^{(r)}_{0j},$$

where the $a_{ij}$ may be chosen *arbitrarily* in $k$. Let

$$Y_i = \sum_{j=1}^{n} b_{ij} X^{(r)}_{0j} \quad (1 \leqslant i \leqslant m), \qquad Y'_i = \sum_{j=1}^{n} c_{ij} X^{(r)}_{0j} \quad (m + 1 \leqslant i \leqslant n).$$

We then have

$$(16) \quad (X^{(r+1)}_{r+1,i})^p - X^{(r+1)}_{r+1,i} = \sum_{j=1}^{n} (a^p_{ij} - a_{ij} + b_{ij}) X^{(r)}_{0j} + \sum_{j=m+1}^{n} a^p_{ij} (X^{(r)}_{0j})^p - \sum_{j=m+1}^{n} (a_{ij} - b_{ij}) X^{(r)}_{0j} \qquad \text{for} \quad 1 \leqslant i \leqslant m$$

(17)

$$(X^{(r+1)}_{r+1,i})^{p^l} = \sum_{j=1}^{n} (a^p_{ij} + c_{ij}) X^{(r)}_{0j} + \sum_{j=m+1}^{n} c_{ij} X^{(r)}_{0j} \qquad \text{for} \quad m + 1 \leqslant i \leqslant n.$$

But from the Fitting theory it follows that we may assume the basis

$$(X_{0j}) = (X^{(r)}_{0j})_{m+1 \leqslant j \leqslant n}$$

of $\mathfrak{g}_0^{(1)}$ has been taken so that

$$(X_{0j}^{(r)})^p = \sum_{k=m+1}^{n} \lambda_{jk} X_{0k}^{(r)} \quad \text{with} \quad \lambda_{jk} = 0 \quad \text{for} \quad k \leqslant j.$$

It is then possible, first to determine the $a_{ij}$ with $1 \leqslant i \leqslant m$ so as to annihilate the right-hand side of (16), then the $a_{ij}$ with $i \geqslant m + 1$ and $1 \leqslant j \leqslant m$ so as to annihilate in the right-hand side of (17) the coefficients of the $X_{0j}^{(r)}$ for $1 \leqslant j \leqslant m$. It then follows that we have

$$(X_{r+1,i}^{(r+1)})^p = X_{r+1,i}^{(r+1)} \quad \text{for} \quad 1 \leqslant i \leqslant m,$$

and

$$(X_{r+1,i}^{(r+1)})^{p^{2l}} = 0 \quad \text{for} \quad m + 1 \leqslant i \leqslant n$$

Finally, by the process used in the proof of III, §2, No. 2, Theorem 2, a new modification of the privileged basis $(Z_\alpha'^{(r+1)})$ to $(Z_\alpha^{(r+1)})$, leaving invariant the $Z_\alpha'^{(r+1)}$ with $|\alpha| \leqslant p^{r+1}$, ensures the existence of a homomorphism $u_{r+1}$ of $(\mathfrak{G}_\infty(k))^{\otimes n}$ onto $\mathfrak{G}$ such that

$$u_{r+1}(Z_{\lambda(\alpha)}) = Z_\alpha^{(r+1)} \quad \text{for} \quad \operatorname{ht}(\alpha) \leqslant r + 1,$$

and the induction may proceed. It obviously yields a privileged basis $(Z_\alpha)$ of $\mathfrak{G}$ such that the $Z_\alpha$ with $\alpha(i) = 0$ for $i > m$ form a basis for the bigebra $\mathfrak{C}$ of a formal torus of dimension $m$, and the $Z_\alpha$ with $\alpha(i) > 0$ for some $i > m$ the basis of a biideal J supplementary to $\mathfrak{C}$; taking for $\mathfrak{R}$ the largest subbigebra contained in $k \cdot 1 + J$, we obviously satisfy the conditions of (i).

(ii) Now suppose that **G** is a formal torus, and let **H** be any formal subgroup of **G**. As the Lie algebra $\mathfrak{h}_0$ of H is a $p$-Lie subalgebra of $\mathfrak{g}_0$, the Fitting theory now shows the existence of a basis $(X_{0i})_{1 \leqslant i \leqslant n}$ of $\mathfrak{g}_0$ such that

$$X_{0i}^p = X_{0i} \quad \text{for} \quad 1 \leqslant i \leqslant n,$$

and the $X_{0i}$ for $1 \leqslant i \leqslant m$ form a basis for $\mathfrak{h}_0$. One then constructs inductively as in (i) a privileged basis $(Z_\alpha)$ of $\mathfrak{G}$, such that

$$X_{hi}^p = X_{hi} \quad \text{for every } h \quad \text{and} \quad 1 \leqslant i \leqslant n,$$

the $Z_\alpha$ for which $\alpha(i) = 0$ for $i > m$ form a privileged basis of $\mathfrak{H}$, and there is a homomorphism $u$ of $(\mathfrak{G}_\infty(k))^{\otimes n}$ onto $\mathfrak{G}$ such that the $Z_\alpha$ are the images by $u$ of the $Z_{\lambda(\alpha)}$ of the canonical basis of $(\mathfrak{G}_\infty(k))^{\otimes n}$; we leave the details to the reader. It is then clear that **H** is a formal torus and that the $Z_\alpha$ such that $\alpha(i) = 0$ for $1 \leqslant i \leqslant m$ form a privileged basis for the bigebra of a formal torus **H**′ which is supplementary to **H** in **G**.

We will say that the largest formal torus **C** in a commutative formal group **G** is the *core* of **G**, and that the largest subgroup **R** of **G** containing no formal torus $\neq$ **e** is the *p-radical* of **G**.

**Corollary 1.** *Let* **G** *be a commutative formal group,* $\mathfrak{G}$ *its bigebra. The following properties are equivalent:*

(a) **G** *is a formal torus.*
(b) *The Frobenius mapping* $Z \mapsto Z^p$ *of* $\mathfrak{G}$ *into itself is injective.*
(c) *The Frobenius mapping* $X \mapsto X^p$ *of the Lie algebra* $\mathfrak{g}_0$ *into itself is injective (or surjective).*

*In particular, a group isogenous to a formal torus is a formal torus.*

The last statement comes from the fact that the Frobenius mapping F of $\mathfrak{G}$ into itself is in fact *bijective* when **G** is a torus, and this immediately implies that if **G**′ is isogenous to **G**, the Frobenius mapping of the Lie algebra $\mathfrak{g}_0$ of **G**′ into itself is surjective (cf. II, §3, No. 6).

**Corollary 2.** *Let* **G**, **H** *be two formal commutative groups,* **u**: **G** → **H** *a formal homomorphism. Then the image by* **u** *of the core of* **G** *is contained in the core of* **G**′ *and the image by* **u** *of the p-radical of* **G** *is contained in the p-radical of* **G**′.

If we consider the corresponding homomorphism $u$: $\mathfrak{G} \to \mathfrak{H}$ of bigebras, the second assertion immediately follows from the relation

$$u(Z^{p^h}) = (u(Z))^{p^h}.$$

To prove the first, we may (by Corollary 1) assume that $\mathfrak{H} = \mathfrak{G}/\mathrm{J}$, where **G** is a torus and J is the biideal generated by $\mathfrak{N}^+$, $\mathfrak{N}$ is a *reduced* subbigebra of $\mathfrak{G}$ and $u$ is the natural homomorphism; but then we know that the group $\mathbf{G}/\mathbf{N}$ is a torus.

**Proposition 15.** *Let* $\mathbf{G}$ *be a formal Lie group,* $'\mathbf{Z}$ *its reduced center,* $\mathbf{N}$ *an invariant formal subgroup of* $\mathbf{G}$ *containing* $'\mathbf{Z}$. *If* $\mathbf{N}/'\mathbf{Z}$ *is a formal torus, then* $\mathbf{N} = {}'\mathbf{Z}$.

Let $m \leqslant q \leqslant n$ be the dimensions of $'\mathbf{Z}$, $\mathbf{N}$, and $\mathbf{G}$. We may assume that we have a privileged basis $(Z_\alpha)$ of $\mathfrak{G}$ such that the $Z_\alpha$ for which $\alpha(i) = 0$ for $i > m$ (resp. $i > q$) form a privileged basis of the bigebra $'\mathfrak{Z}$ (resp. $\mathfrak{N}$) of the subgroup $'\mathbf{Z}$ (resp. $\mathbf{N}$). Furthermore, we may assume that the $X_{0i}$ for $m + 1 \leqslant i \leqslant q$ are such that their classes mod $'\mathfrak{z}_0$ $(= \mathfrak{g}_0 \cap {}'\mathfrak{Z})$ satisfy the relations $\overline{X}{}_{0i}^p = \overline{X}_{0i}$, in other words

$$X_{0i}^p - X_{0i} \in {}'\mathfrak{z}_0$$

(which is contained in the center of the Lie algebra $\mathfrak{g}_0$). The proof is done in several steps.

1) We first prove that $\mathfrak{n}_0 = \mathfrak{g}_0 \cap \mathfrak{N}$ is in the center of $\mathfrak{g}_0$. Indeed, if $m + 1 \leqslant i \leqslant q$, we have

$$[X_{0i}^p - X_{0i}, Y] = 0 \qquad \text{for any} \quad Y \in \mathfrak{g}_0;$$

but by Jacobson's formula, this gives

$$(18) \qquad [X_{0i}, Y] = [X_{0i}^p, Y] = [X_{0i}, [X_{0i}, [\cdots [X_{0i}, Y] \cdots]]]$$

with $p$ brackets. As $\mathbf{N}/'\mathbf{Z}$ is commutative, we have $[U, V] \in {}'\mathfrak{z}_0$ for any two elements $U$, $V$ of $\mathfrak{n}_0$; as $\mathbf{N}$ is an invariant subgroup of $\mathbf{G}$, $[X_{0i}, Y] \in \mathfrak{n}_0$, and then, as $p \geqslant 2$, formula (18) first shows that $[X_{0i}, Y] \in {}'\mathfrak{z}_0$, and then that $[X_{0i}, Y] = 0$.

As a basis for an inductive argument, suppose now that we have proved that all elements $X_{hi}$ with $h < r$ and $m + 1 \leqslant i \leqslant q$ are in the center of the Frobenius subbigebra $\mathfrak{s}_{r-1}$ of $\mathfrak{G}$; the same is true of course for all the $X_\alpha$ of height $\mathrm{ht}(\alpha) < r$ such that $\alpha(i) = 0$ for $i > q$, and therefore also of the $Z_\alpha$ corresponding to the same indices (in other words, $\mathfrak{s}_{r-1} \cap \mathfrak{N}$ is in the center of $\mathfrak{s}_{r-1}$).

2) We next show that

$$[X_{ri}, Y] = [X_{ri}^p, Y] \qquad \textit{for} \quad m + 1 \leqslant i \leqslant q \quad \textit{and for any} \quad Y \in \mathfrak{s}_{r-1}.$$

By assumption, we have

$$X_{ri}^{p} - X_{ri} \in \mathfrak{g}_r \cap \mathfrak{N},$$

and as

$$\mathrm{V}^r(X_{ri}^{p} - X_{ri}) = X_{0i}^{p} - X_{0i} \in {}^r\mathfrak{z}_0,$$

we may write

$$X_{ri}^{p} - X_{ri} = S + T \quad \text{where} \quad S \in \mathfrak{g}_r \cap {}^r\mathfrak{Z} \quad \text{and} \quad T \in \mathfrak{s}_{r-1} \cap \mathfrak{N}.$$

But $[S, Y] = 0$ by definition, and $[T, Y] = 0$ by the inductive hypothesis.

3) Now we prove that

$$[X_{ri}, X_{0j}] = 0 \quad \textit{for} \quad m + 1 \leqslant i \leqslant q \quad \textit{and} \quad 1 \leqslant j \leqslant n.$$

From the fact that $(Z_\alpha)$ is a privileged basis, we deduce

$$\begin{aligned}\Delta([X_{ri}, X_{0j}]) &- 1 \otimes [X_{ri}, X_{0j}] - [X_{ri}, X_{0j}] \otimes 1 \\ &= \sum_{\alpha+\beta=p^r\varepsilon_i} Z_\alpha \otimes [X_{0j}, Z_\beta] + [Z_\alpha, X_{0j}] \otimes Z_\beta,\end{aligned}$$

where in the right-hand side $|\alpha| > 0$ and $|\beta| > 0$. But then the $Z_\alpha$ and $Z_\beta$ in that sum belong to $\mathfrak{s}_{r-1} \cap \mathfrak{N}$, and the inductive assumption shows that

$$[X_{ri}, X_{0j}] \in \mathfrak{g}_0;$$

as **N** is invariant in **G**, $[X_{ri}, X_{0j}]$ also belongs to the (right or left) ideal J generated by $\mathfrak{N}^+$ in $\mathfrak{G}$, hence to $\mathfrak{g}_0 \cap \mathrm{J} = \mathfrak{n}_0$. But then Jacobson's formula

$$[X_{ri}^{p}, X_{0j}] = [X_{ri}, [X_{ri}, [\cdots [X_{ri}, X_{0j}] \cdots]]] \tag{19}$$

shows (using the commutativity of **N**/$^r$**Z** and the relation $p \geqslant 2$) that $[X_{ri}^{p}, X_{0j}]$ belongs to the ideal I of $\mathfrak{G}$ generated by ${}^r\mathfrak{Z}^+$; on the other hand, we have seen in 2) that

$$[X_{ri}, X_{0j}] = [X_{ri}^{p}, X_{0j}],$$

hence

$$[X_{ri}, X_{0j}] \in \mathrm{I} \cap \mathfrak{g}_0 = {}^r\mathfrak{z}_0.$$

But then Jacobson's formula (19) gives $[X_{ri}^p, X_{0j}] = 0$ and finally $[X_{ri}, X_{0j}] = 0$.

4) We can now prove that

$$[X_{ri}, Y] = 0 \qquad \textit{for} \quad m+1 \leqslant i \leqslant q \quad \textit{and for any} \quad Y \in \mathfrak{s}_{r-1}.$$

It is enough to consider the case $Y = Z_\alpha$ with $\mathrm{ht}(\alpha) < r$; we use induction on $|\alpha|$, the result being proved in 3) for $|\alpha| = 1$. The inductive hypothesis first shows that for $\lambda \leqslant p^r\varepsilon_i$ and $\mu < \alpha$, $Z_\lambda$ and $Z_\mu$ commute, hence

$$\Delta([X_{ri}, Z_\alpha]) - 1 \otimes [X_{ri}, Z_\alpha] - [X_{ri}, Z_\alpha] \otimes 1 = 0$$

i.e., $[X_{ri}, Z] \in \mathfrak{g}_0$. But then, by 3) and Jacobson's formula

$$[X_{ri}^p, Z_\alpha] = [X_{ri}, [X_{ri}, [\cdots[X_{ri}, Z_\alpha]\cdots]]],$$

we conclude that $[X_{ri}^p, Z_\alpha] = 0$, and finally 2) shows that $[X_{ri}, Z_\alpha] = 0$.

5) The next step is to show that

$$[X_{0i}, X_{rj}] = 0 \qquad \textit{for} \quad m+1 \leqslant i \leqslant q \quad \textit{and} \quad 1 \leqslant j \leqslant n.$$

As we have $X_{0i}^p - X_{0i} \in {}^r\mathfrak{z}_0$, we may write

$$[X_{0i}, X_{rj}] = [X_{0i}^p, X_{rj}].$$

On the other hand, the inductive hypothesis shows as above that

$$[X_{0i}, X_{rj}] \in \mathfrak{g}_0;$$

we conclude again by Jacobson's formula.

6) We now prove by induction on $h < r$ that

$$[X_{hi}, X_{rj}] = 0 \qquad \textit{for} \quad m+1 \leqslant i \leqslant q \quad \textit{and} \quad 1 \leqslant j \leqslant n.$$

From the induction hypothesis it follows that $[Z_\alpha, X_{rj}] = 0$ for all $\alpha$ such that $\alpha(i) = 0$ for $i > q$ and $\mathrm{ht}(\alpha) < h$. We deduce from that by the usual argument that $[X_{hi}, X_{rj}] \in \mathfrak{g}_0$. But as in 2) we may write

$$X_{hi}^p - X_{hi} = S + T \qquad \text{where} \quad S \in \mathfrak{g}_h \cap {}^r\mathfrak{Z} \quad \text{and} \quad T \in \mathfrak{s}_{h-1} \cap \mathfrak{N};$$

by the inductive hypothesis $[T, X_{rj}] = 0$ hence

$$[X_{hi}, X_{rj}] = [X_{hi}^p, X_{rj}];$$

the usual method using Jacobson's formula and the relation $[X_{hi}, X_{rj}] \in \mathfrak{g}_0$ finally gives $[X_{hi}, X_{rj}] = 0$.

7) The final step in the proof consists in showing that

$$[X_{ri}, X_{rj}] = 0 \qquad \textit{for } m + 1 \leqslant i \leqslant q \quad \textit{and} \quad 1 \leqslant j \leqslant n.$$

From 3) and 6) it follows that $X_{ri}$ commutes with all elements of $\mathfrak{s}_{r-1}$, and $X_{rj}$ with all elements of $\mathfrak{s}_{r-1} \cap \mathfrak{N}$; the usual argument then shows that $[X_{ri}, X_{rj}] \in \mathfrak{g}_0$. On the other hand, we have

$$X_{ri}^p - X_{ri} = S + T \qquad \text{with} \quad S \in \mathfrak{g}_r \cap {}^r\mathfrak{Z} \quad \text{and} \quad T \in \mathfrak{s}_{r-1} \cap \mathfrak{N};$$

from 6) we deduce that $[T, X_{rj}] = 0$, hence

$$[X_{ri}, X_{rj}] = [X_{ri}^p, X_{rj}].$$

From 3) and Jacobson's formula, we finally conclude $[X_{ri}, X_{rj}] = 0$, thus completing the induction which proves that every element of $\mathfrak{N}$ is in the center of $\mathfrak{G}$. QED.

**Corollary 1.** *In a formal group* **G**, *a formal torus* **N** *which is invariant in* **G** *is contained in the reduced center* $^r$**Z** *of* **G**.

Indeed, (**N** ∨ $^r$**Z**)/$^r$**Z** is isogenous to **N**/(**N** ∧ $^r$**Z**), which is therefore a torus (Proposition 14 and Corollary 2 to Proposition 14); as **N** ∨ $^r$**Z** is an invariant subgroup of **G**, Proposition 15 is applicable.

**Corollary 2.** *Let* **G** *be a formal group,* $^r$**Z** *its reduced center,* **f**: **G** → **G**/$^r$**Z** *the natural formal homomorphism. For any formal torus* **T′** *in* **G**/$^r$**Z**, $^r$**f**$^{-1}$(**T′**) *is commutative and its p-radical is the p-radical of* $^r$**Z**.

As **f**($^r$**f**$^{-1}$(**T′**)) = **T′**, $^r$**f**$^{-1}$(**T′**)/$^r$**Z** is isogenous to **T′**, hence a formal torus, and Proposition 15 proves the commutativity of $^r$**f**$^{-1}$(**T′**). If **R** is the $p$-radical of $^r$**f**$^{-1}$(**T′**), **f**(**R**) is contained in the $p$-radical of **T′**, hence equal to

$\mathbf{e}$, which proves that $\mathbf{R} \subset {}^r\mathbf{Z}$, and as $\mathbf{R}$ contains the $p$-radical of ${}^r\mathbf{Z}$, it must coincide with it.

**11.** *Formal unipotent groups and maximal formal tori.* We will say that a formal group $\mathbf{G}$ is *unipotent* if there is a power $p^h$ of the characteristic such that $X^{p^h} = e$ for a generic point X of $\mathbf{G}$; this is equivalent to saying that, in each group $\mathbf{G}_A$, $z^{p^h} = e$ for every element $z \in \mathbf{G}_A$. That notion has evidently no counterpart over a field of characteristic 0. It is clear that subgroups, products, and quotients of unipotent formal groups are unipotent, and that a formal group isogenous to a unipotent formal group is unipotent. Furthermore, if $\mathbf{G}$ is a formal group, $\mathbf{N}$ an invariant (reduced!) formal subgroup of $\mathbf{G}$, then, if both $\mathbf{N}$ and $\mathbf{G}/\mathbf{N}$ are unipotent, so is $\mathbf{G}$: this follows from the fact that $(\mathbf{G}/\mathbf{N})_A = \mathbf{G}_A/\mathbf{N}_A$ (II, §3, Proposition 8) for any $A \in \boldsymbol{Alc}_k$ and from the preceding characterization of unipotent formal groups.

The definition of a unipotent formal group may also be expressed in the following way: if we define by induction the mapping $c^{(r)}: \mathfrak{G}^* \to (\mathfrak{G}^*)^{\hat{\otimes} r}$ as equal to $(c^{(r-1)} \hat{\otimes} 1_{\mathfrak{G}^*}) \circ c$, and the mapping (multiplication) $m^{(r)}: (\mathfrak{G}^*)^{\hat{\otimes} r} \to \mathfrak{G}^*$ as equal to $m \circ (m^{(r-1)} \hat{\otimes} 1_{\mathfrak{G}^*})$, the mapping $\mathfrak{G}^* \to \mathfrak{G}^*$ corresponding to $X^r$ is $m^{(r)} \circ c^{(r)}$; to say that $\mathbf{G}$ is unipotent means that, for some exponent $h$, $m^{(p^h)} \circ c^{(p^h)}$ is 0 in the augmentation ideal. If $c$ is identified with a system of $n$ power series

$$\begin{aligned}\varphi(X; Y) &= \varphi(x_1, \ldots, x_n; y_1, \ldots, y_n)\\ &= (\varphi_1(x_1, \ldots, x_n; y_1, \ldots, y_n), \ldots, \varphi_n(x_1, \ldots, x_n; y_1, \ldots, y_n))\end{aligned}$$

one defines by induction the systems of formal power series in $x_1, \ldots, x_n$,

$$\psi^{(1)}(X) = X, \quad \psi^{(2)}(X) = \varphi(X; X), \; \ldots, \quad \psi^{(r)}(X) = \varphi(X; \psi^{(r-1)}(X)), \; \ldots$$

and the condition expressing that $\mathbf{G}$ is unipotent is that the system $\psi^{(p^h)}(X)$ be reduced to 0 for some $h \geqslant 1$.

**Proposition 16.** (i) *If* G *is an affine unipotent group, then* $\Phi(G)$ *is a formal unipotent group.*

(ii) *If* G *is an affine group and* $\mathbf{H}$ *a formal unipotent subgroup of the formal group* $\mathbf{G} = \Phi(G)$, *then* $A(\mathbf{H})$ *is an affine unipotent subgroup of* G.

(i) The mapping $z \mapsto z^{p^h}$ is a morphism of the affine variety G into itself and if $v$ is the corresponding comorphism, and a regular system of parameters of the local ring of G at $e$ is identified to $n$ indeterminates in $\mathfrak{G}^*$, then $(\mathbf{v}(x_i))$ is the system of formal power series denoted above by $\psi^{(p^h)}(X)$. If G is unipotent, $z^{p^h} = e$ for an integer $h \geqslant 1$ and all $z \in G$, which means that the formal power series $\mathbf{v}(x_i)$ are all 0.

(ii) Suppose $G \subset k^N$, and let $P_i(t_1, \ldots, t_N)$ $(1 \leqslant i \leqslant N)$ be polynomials whose restrictions to G are identified with the functions associating to $z \in G$ the N coordinates of $z^{p^h}$. The closed set U of points $z \in G$ such that $z^{p^h} = e$ is then the set defined by the ideal generated by the $P_i$ (if $e$ is taken at the origin of $k^N$). But if **H** is unipotent, then, if ${}^t v_*$ is the natural homomorphism $k[[t_1, \ldots, t_N]] \to \mathfrak{H}^*$, corresponding to a system

$$\mathbf{v}(X) = (v_1(x_1, \ldots, x_n), \ldots, v_N(x_1, \ldots, x_n))$$

of N formal power series, the assumption $X^{p^h} = e$ for the generic point of **H** implies that

$$P_i(v_1(X), \ldots, v_N(X)) = 0 \qquad \text{for} \quad 1 \leqslant i \leqslant N;$$

hence $A(\mathbf{H}) \subset U$.

If **G** is a formal *commutative* group, to say that **G** is unipotent means that, for some integer $h \geqslant 1$, the endomorphism $[p^h]$ of the bigebra $\mathfrak{G}$ (III, §1, No. 3) is 0 in $\mathfrak{G}^+$. From the structure of commutative formal groups up to isogeny (III, §6, No. 1), it follows that there is in **G** a *largest* unipotent formal subgroup **U**, contained in the $p$-radical of **G**, and isogenous to a *direct product of truncated Witt groups* $\mathbf{G}_{r,\infty}(k)$. Furthermore, the $p$-radical **R** of **G** contains a largest *divisible* formal subgroup **D**, i.e., such that the restriction of the endomorphism $[p]$ to the bigebra **D** is *surjective*, and which is isogenous to a *direct product of simple groups* $\mathbf{G}_{m,n}(k)$ other than $\mathbf{G}_{1,0}(k) = \mathbf{G}_m(k)$; and **R** is isogenous to the direct product of **U** and **D**. The reduced subbigebra $\mathfrak{U}$ of the formal subgroup **U** may also be defined as the largest reduced subbigebra whose augmentation ideal is contained in the kernels of the endomorphisms $[p^r]$ for $r > 0$.

It should be observed that in general **R** is not the direct product of **U** and **D**. For instance, consider the formal group $\mathbf{G}_0$, which is the direct product of $\mathbf{G}_{1,\infty}$ and $\mathbf{G}_{1,h}$ with $h \geqslant 1$; there is a canonical privileged basis of the bigebra $\mathfrak{G}_0$ such that $X_{r1}^p = 0$ for all $r \geqslant 0$ and $X_{02}^p = 0$; it is then immediate to check that the subalgebra of $\mathfrak{s}_0$ generated by the element $Y_0 = X_{01} - X_{02}$ is a subbigebra $\mathfrak{N}_0$ of $\mathfrak{G}_0$ of height 0; if J is the biideal of

$\mathfrak{G}_0$ which is generated by $\mathfrak{N}_0^+$, then in $\mathfrak{G} = \mathfrak{G}_0/J$, the natural images of $\mathfrak{G}_{1,\infty}$ and $\mathfrak{G}_{1,h}$ are the bigebras of the largest unipotent formal subgroup and of the largest divisible formal subgroup, but these subbigebras have an intersection which is not reduced to $k \cdot 1$, since it contains the common image of $X_{01}$ and $X_{02}$. One may give similar examples to show **U** is only in general an *almost direct* product of truncated Witt groups and **D** an *almost direct* product of simple divisible subgroups (in contrast to Proposition 14).

**Proposition 17.** *Let* G *be an affine group; if* **T** *is a formal torus of* **G** $= \mathbf{\Phi}(\mathrm{G})$, *then* A(**T**) *is an affine torus in* G.

The affine group A(**T**) is commutative (Proposition 8) and connected, hence it is the direct product of a torus S and a unipotent group V. Let $v$ be the projection of A(**T**) onto V corresponding to that decomposition, having image V and kernel S; then the corresponding formal homomorphism

$$\mathbf{v} : \mathbf{\Phi}(\mathrm{A}(\mathbf{T})) \to \mathbf{\Phi}(\mathrm{V})$$

has a reduced kernel equal to $\mathbf{\Phi}(\mathrm{S})$ (Proposition 1); but $\mathbf{\Phi}(\mathrm{V})$ is a unipotent formal group (Proposition 16), hence $\mathbf{v}(\mathbf{T}) = \mathbf{e}$ (Corollary 2 to Proposition 14); this shows that $\mathbf{T} \subset \mathbf{\Phi}(\mathrm{S})$, hence $\mathrm{A}(\mathbf{T}) \subset \mathrm{S}$, and finally $\mathrm{A}(\mathbf{T}) = \mathrm{S}$.

**Proposition 18.** *Let* G *be an affine group,* **H** *a formal subgroup of* **G** $= \mathbf{\Phi}(\mathrm{G})$.

(i) *If* $\mathbf{T}_1$, $\mathbf{T}_2$ *are two maximal formal tori of* **H**, *there exists* $s \in \mathrm{A}(\mathbf{H})$ *such that* $\mathbf{Int}(s)(\mathbf{T}_1) = \mathbf{T}_2$.

(ii) *Any maximal formal torus of* **H** *can be written* $\mathbf{\Phi}(\mathrm{S}) \wedge \mathbf{H}$, *where* S *is a maximal torus in* A(**H**), *and conversely, any such reduced intersection is a maximal formal torus of* **H**.

(iii) *Any maximal formal torus in a maximal solvable formal subgroup of* **H** *is a maximal formal torus of* **H**.

(i) $\mathrm{A}(\mathbf{T}_1)$ and $\mathrm{A}(\mathbf{T}_2)$ are tori in A(**H**) by Proposition 17, hence, respectively, contained in maximal tori $\mathrm{S}_1$, $\mathrm{S}_2$. As there is an $s \in \mathrm{A}(\mathbf{H})$ such that $s\mathrm{S}_1s^{-1} = \mathrm{S}_2$, we may, by replacing $\mathbf{T}_1$ by $\mathbf{Int}(s)(\mathbf{T}_1)$, suppose that $\mathrm{S}_1 = \mathrm{S}_2$, and therefore we may suppose that $\mathbf{T}_1$ and $\mathbf{T}_2$ are both contained in the formal torus $\mathbf{\Phi}(\mathrm{S}_1)$. As both are maximal in **H**, if they were distinct, neither would be contained in the other, hence $\mathbf{T}_1 \vee \mathbf{T}_2$ would be distinct from $\mathbf{T}_1$ and $\mathbf{T}_2$; but as $\mathbf{T}_1 \vee \mathbf{T}_2$ is contained in the formal torus $\Phi(S_1)$, it is a formal torus, which contradicts the maximality of $\mathbf{T}_1$ and $\mathbf{T}_2$.

(ii) If $\mathbf{T}$ is a maximal formal torus of $\mathbf{H}$, it follows as in (i) that it is contained in $\Phi(S)$, where S is a maximal torus in $A(\mathbf{H})$; as $\Phi(S) \wedge \mathbf{H}$ is a formal torus, it must be $\mathbf{T}$. Conversely, let S be a maximal torus in $A(\mathbf{H})$; if $\mathbf{T} = \Phi(S) \wedge \mathbf{H}$ were not maximal, it would be properly contained in a maximal formal torus $\mathbf{T}_1$ of $\mathbf{H}$, which in turn would be of the form $\Phi(S_1) \wedge \mathbf{H}$ where $S_1$ is a maximal torus of $A(\mathbf{H})$. Then there would be $s \in A(\mathbf{H})$ such that $S = sS_1s^{-1}$, hence

$$\mathbf{Int}(s)(\mathbf{T}_1) \subset \Phi(S);$$

as $\dim(\mathbf{T}_1) > \dim(\mathbf{T})$, $\mathbf{T} \vee \mathbf{Int}(s)(\mathbf{T}_1)$, which is contained in $\Phi(S)$, would be a formal torus of $\mathbf{H}$ properly containing $\mathbf{T}$, contrary to the assumption that $\mathbf{T} = \Phi(S) \wedge \mathbf{H}$.

We shall see a little later that, for a maximal formal torus $\mathbf{T}$ of $\mathbf{H}$, in fact $A(\mathbf{T})$ is a *maximal torus of* $A(\mathbf{H})$ (No. 12, Corollary 1 to Proposition 23).

(iii) This follows at once from (ii) and from the corresponding statement for maximal tori and Borel subgroups in an affine group.

**Corollary 1.** *Let* $\mathbf{G}$ *be a formal group,* ${}^r\mathbf{Z}$ *its reduced center,* $\mathbf{f}: \mathbf{G} \to \mathbf{G}' = \mathbf{G}/{}^r\mathbf{Z}$ *the natural homomorphism. Then*:

(i) *For any maximal formal torus* $\mathbf{T}'$ *in* $\mathbf{G}'$, ${}^r\mathbf{f}^{-1}(\mathbf{T}')$ *is commutative and its core* $\mathbf{C}$ *is such that* $\mathbf{f}(\mathbf{C}) = \mathbf{T}'$.

(ii) *For any maximal formal torus* $\mathbf{T}$ *in* $\mathbf{G}$, $\mathbf{f}(\mathbf{T})$ *is a maximal formal torus in* $\mathbf{G}'$.

(iii) *All maximal formal tori of* $\mathbf{G}$ *are isomorphic.*

Assertion (i) follows from Corollary 2 to Proposition 15. If $\mathbf{T}$ is a maximal formal torus of $\mathbf{G}$, $\mathbf{f}(\mathbf{T})$ is a formal torus of $\mathbf{G}'$, and if it were properly contained in a formal torus $\mathbf{T}'$ of $\mathbf{G}'$, $\mathbf{T}$ would be properly contained in a formal torus of ${}^r\mathbf{f}^{-1}(\mathbf{T}')$. The last assertion follows from Proposition 18(i) and from the fact that two formal tori of same dimension are isomorphic.

**Corollary 2.** *In a formal group* $\mathbf{G}$, *the largest formal subgroup contained in all maximal formal tori of* $\mathbf{G}$ *is the core of the reduced center* ${}^r\mathbf{Z}$ *of* $\mathbf{G}$.

From Corollary 1 and from Proposition 18(ii), we see that this largest formal subgroup $\mathbf{C}$ is contained in $'\mathbf{f}^{-1}(\Phi(T_0) \wedge \mathbf{G}')$, where $\mathbf{G}'$ is a formal subgroup of some $\mathbf{GL}(N)$, $\mathbf{f}: \mathbf{G} \to \mathbf{G}'$ is surjective and has reduced kernel $'\mathbf{Z}$, and $T_0$ is the intersection of the maximal tori of the affine group $A(\mathbf{G}')$. But it is known that $T_0$ is a torus contained in the center of $A(\mathbf{G}')$, hence $\Phi(T_0) \wedge \mathbf{G}'$ is a formal torus contained in the center of $\mathbf{G}'$; by Proposition 15, $'\mathbf{f}^{-1}(\Phi(T_0) \wedge \mathbf{G}')$ is contained in $'\mathbf{Z}$. As any maximal formal torus of $\mathbf{G}$ contains the core of $'\mathbf{Z}$, this proves the corollary.

## 12. *Nilpotent formal groups.*

**Proposition 19** *Any unipotent formal group* $\mathbf{G}$ *is nilpotent.*

The quotient $\mathbf{G}' = \mathbf{G}/'\mathbf{Z}$ of $\mathbf{G}$ by its reduced center is unipotent, so we may assume that $\mathbf{G}$ is representable. But if $\mathbf{G}$ is a unipotent formal subgroup of $\Phi(L)$, where L is an affine group, $A(\mathbf{G})$ is a closed unipotent subgroup of L (Proposition 16), hence $A(\mathbf{G})$ is nilpotent, and so is $\mathbf{G}$ (Proposition 8).

**Proposition 20.** *In a representable nilpotent formal group* $\mathbf{G}$, *there is a largest formal torus* $\mathbf{C}$ *(the core of the reduced center* $'\mathbf{Z}$*), a largest unipotent formal subgroup* $\mathbf{G}_u$, *both of which are invariant subgroups of* $\mathbf{G}$, *and* $\mathbf{G}$ *is direct product of* $\mathbf{C}$ *and* $\mathbf{G}_u$.

We may assume that $\mathbf{G}$ is a formal subgroup of $\Phi(L)$, where L is an affine group. Then $A(\mathbf{G})$ is connected and nilpotent (Proposition 8), hence it is known that it is the direct product of its unique maximal torus T and its largest unipotent subgroup V, and it is clear that $\mathscr{D}(A(\mathbf{G})) = \mathscr{D}(V)$. We therefore have

$$\mathscr{D}(\mathbf{G}) = \Phi(\mathscr{D}(V)) = \mathscr{D}(\Phi(V))$$

by Theorem 2, and the commutative formal group $\mathbf{G}/\mathscr{D}(\mathbf{G})$ is contained in the group

$$\Phi(T) \times (\Phi(V)/\mathscr{D}(\Phi(V))).$$

Now $\mathbf{G}/\mathscr{D}(\mathbf{G})$ is the direct product of its core $\mathbf{C}'$ and its $p$-radical $\mathbf{R}'$

(Proposition 14), and as $\Phi(\mathrm{T})$ is a torus and $\Phi(\mathrm{V})/\mathscr{D}(\Phi(\mathrm{V}))$ a unipotent formal group, we have necessarily

$$\mathbf{C}' \subset \Phi(\mathrm{T}) \quad \text{and} \quad \mathbf{R}' \subset \Phi(\mathrm{V})/\mathscr{D}(\Phi(\mathrm{V})),$$

which already shows that $\mathbf{R}'$ is a unipotent formal group. The reduced inverse image $\mathbf{G}_u$ of $\mathbf{R}'$ in $\mathbf{G}$ contains

$$\mathscr{D}(\mathbf{G}) = \mathscr{D}(\Phi(\mathrm{V})),$$

and as $\mathbf{R}'$ and $\mathscr{D}(\Phi(\mathrm{V}))$ are unipotent, so is $\mathbf{G}_u$, hence $\mathrm{A}(\mathbf{G}_u) \subset \mathrm{V}$, and therefore

$$\mathbf{G}_u \subset \Phi(\mathrm{V}) \wedge \mathbf{G};$$

but if that inclusion were a proper one,

$$(\Phi(\mathrm{V}) \wedge \mathbf{G})/\mathscr{D}(\Phi(\mathrm{V}))$$

would be contained in the $p$-radical $\mathbf{R}'$ of $\mathbf{G}/\mathscr{D}(\mathbf{G})$ and have a dimension strictly greater than $\mathbf{R}'$, which is absurd; we therefore have

$$\mathbf{G}_u = \Phi(\mathrm{V}) \wedge \mathbf{G}.$$

If $f\colon \mathrm{A}(\mathbf{G}) \to \mathrm{V}$ is the natural projection, then we have

$$\mathscr{D}(\mathbf{G}) \subset \mathbf{f}(\mathbf{G}) \subset \Phi(\mathrm{V}),$$

and $\mathbf{f}(\mathbf{G}) \subset \mathbf{G}_u$; but as the dimension of $\mathbf{f}(\mathbf{G})/\mathscr{D}(\mathbf{G})$ is at most that of $\mathbf{G}/\mathscr{D}(\mathbf{G})$, and the latter is equal to $\mathbf{G}_u/\mathscr{D}(\mathbf{G})$, we must have $\mathbf{f}(\mathbf{G}) = \mathbf{G}_u$, hence $\mathbf{G}$ is contained in the direct product $\mathbf{G}_u \times \Phi(\mathrm{T})$. As it is reduced and contains $\mathbf{G}_u$, it is equal to $\mathbf{G}_u \times (\mathbf{G} \wedge \Phi(\mathrm{T}))$; but $\mathbf{G} \wedge \Phi(\mathrm{T})$ is a maximal formal torus of $\mathbf{G}$ (Proposition 18(ii)), and as $\mathbf{G}$ is invariant in $\Phi(\mathrm{A}(\mathbf{G}))$, it follows from Proposition 18(i) that $\mathbf{C} = \mathbf{G} \wedge \Phi(\mathrm{T})$ is the unique maximal formal torus of $\mathbf{G}$ and that $\mathbf{G} = \mathbf{C} \times \mathbf{G}_u$.

**Corollary 1.** *In a representable formal commutative group* $\mathbf{G}$, *the largest divisible subgroup of the p-radical is reduced to* $\mathbf{e}$.

This shows in particular that all *simple* formal commutative groups

$\mathbf{G}_{m,n}(k)$ other than the multiplicative group $\mathbf{G}_m(k) = \mathbf{G}_{1,0}(k)$ and the additive group $\mathbf{G}_a(k) = \mathbf{G}_{1,\infty}(k)$ are *not representable* (in contrast to Ado's theorem !). Of course the multiplicative and additive formal groups are representable; more generally, all truncated Witt groups $\mathbf{G}_{r,\infty}(k)$ are representable and algebraic, since this is true in general for all formal groups for which the comultiplication $c$ is given by formal power series which are *polynomials*: for such a comultiplication may be regarded as the continuous extension of itself, considered as a homomorphism of $\mathrm{A} = k[t_1, \ldots, t_{\mathrm{N}}]$ into $\mathrm{A} \otimes \mathrm{A}$ (hence defining an affine group).

**Corollary 2.** *If* L *is an affine group and* $\mathbf{G}$ *a formal subgroup of* $\mathbf{L} = \Phi(\mathrm{L})$, *then in the commutative formal groups* $\Phi(\mathrm{A}(\mathbf{G}))/\mathbf{G}$ *and* $\mathbf{G}/\mathscr{D}(\mathbf{G})$, *the largest divisible subgroup of the p-radical is reduced to* $\mathbf{e}$.

This follows from the fact that

$$\mathscr{D}(\mathbf{G}) = \Phi(\mathscr{D}(\mathrm{A}(\mathbf{G})))$$

(Theorem 2), hence $\Phi(\mathrm{A}(\mathbf{G}))/\mathscr{D}(\mathbf{G})$ is the formalization of the commutative affine group $\mathrm{A}(\mathbf{G})/\mathscr{D}(\mathrm{A}(\mathbf{G}))$, and is therefore representable; apply then Corollary 1 to the subgroup $\mathbf{G}/\mathscr{D}(\mathbf{G})$ and the quotient group $\Phi(\mathrm{A}(\mathbf{G}))/\mathbf{G}$ of $\Phi(\mathrm{A}(\mathbf{G}))/\mathscr{D}(\mathbf{G})$, together with the structure theorem of the commutative formal groups (III, §6, No. 1).

**Proposition 21.** *In order that a representable formal group* $\mathbf{G}$ *be unipotent, it is necessary and sufficient that it contain no formal torus other than* $\mathbf{e}$.

The necessity being obvious, we prove that the condition is sufficient, by showing that it implies that $\mathrm{A}(\mathbf{G})$ is unipotent. We first observe that

$$\mathscr{D}(\mathbf{G}) = \Phi(\mathscr{D}(\mathrm{A}(\mathbf{G})))$$

(Theorem 2) contains no formal torus $\neq \mathbf{e}$, hence the affine group $\mathscr{D}(\mathrm{A}(\mathbf{G}))$ does not contain any torus not reduced to $e$, and therefore it is known that $\mathscr{D}(\mathrm{A}(\mathbf{G}))$ is unipotent. It will be therefore enough to prove that the affine group

$$\mathrm{L} = \mathrm{A}(\mathbf{G})/\mathscr{D}(\mathrm{A}(\mathbf{G}))$$

is a commutative unipotent group. Suppose the contrary; as L is the direct product of its unique maximal torus S and its unique maximal unipotent subgroup V, this assumption means that $S \neq \{e\}$. Let $f$: $A(\mathbf{G}) \to L$ be the natural homomorphism; the image $\mathbf{f}(\mathbf{G})$ is a formal subgroup of $\Phi(L)$, and a direct product of its core $\mathbf{C} \subset \Phi(S)$ and its $p$-radical $\mathbf{R} \subset \Phi(V)$ (Corollary 2 to Proposition 14). If we had $\mathbf{C} = \mathbf{e}$, $\mathbf{f}(\mathbf{G})$ would be contained in $\Phi(V)$, hence $\mathbf{G}$ would be contained in the affine group $f^{-1}(V)$, which is properly contained in $A(\mathbf{G})$ (since $\dim(S) > 0$ by assumption), and this contradicts the definition of $A(\mathbf{G})$. The assumption $\dim(S) > 0$ therefore implies $\dim(\mathbf{C}) > 0$. Consider now the affine group $S_0 = f^{-1}(S)$ of $A(\mathbf{G})$ which is such that $\Phi(S_0)$ contains the invariant formal subgroup $\mathbf{C}_0 = {}^r\mathbf{f}^{-1}(\mathbf{C})$ of $\mathbf{G}$. If T is a maximal torus of $A(\mathbf{G})$, $f(T)$ is a maximal torus of $f(A(\mathbf{G})) = L$, hence equal to S, which proves that $T \subset S_0$. We are going to show that

$$\dim(\Phi(T) \wedge \mathbf{C}_0) > 0,$$

which will mean that $\mathbf{C}_0$ contains a formal torus of dimension $> 0$ and bring about the required contradiction. As we have $\mathbf{f}(\Phi(T) \vee \mathbf{C}_0) = \Phi(S) \vee \mathbf{C} = \Phi(S)$,

$$\dim(\Phi(S)) = \dim(\Phi(T) \vee \mathbf{C}_0) - \dim \mathscr{D}(\mathbf{G}) = \dim \Phi(T)$$

since by assumption $\mathscr{D}(\mathbf{G})$ contains no formal torus $\neq \mathbf{e}$, hence

$$\Phi(T) \wedge \mathscr{D}(\mathbf{G}) = \mathbf{e}.$$

Finally we get

$$\begin{aligned}\dim(\Phi(T) \wedge \mathbf{C}_0) &= \dim\Phi(T) + \dim \mathbf{C}_0 - \dim(\Phi(T) \vee \mathbf{C}_0)\\ &= \dim \mathbf{C}_0 - \dim \mathscr{D}(\mathbf{G}) = \dim \mathbf{C} > 0\end{aligned}$$

and this ends the proof.

**Proposition 22.** *In order that a formal group* $\mathbf{G}$ *be nilpotent, a necessary and sufficient condition is that it contain a unique maximal formal torus, which is then the core of the reduced center* ${}^r\mathbf{Z}$ *of* $\mathbf{G}$.

The necessity of the condition follows at once from Proposition 20 if

$\mathbf{G}$ is representable. In general, $\mathbf{G}' = \mathbf{G}/{}^r\mathbf{Z}$ is nilpotent, hence contains a unique maximal formal torus $\mathbf{T}'$ which is in its reduced center. As $\mathbf{T}'$ is invariant in $\mathbf{G}'$, its reduced inverse image $\mathbf{T}$ in $\mathbf{G}$ is an invariant subgroup of $\mathbf{G}$ containing ${}^r\mathbf{Z}$, and $\mathbf{T}/{}^r\mathbf{Z}$ is isogenous to $\mathbf{T}'$, hence a formal torus; it follows from Proposition 15 that $\mathbf{T} = {}^r\mathbf{Z}$, hence $\mathbf{T}' = \mathbf{e}$, and therefore (Corollary 2 to Proposition 14) the only formal tori $\neq \mathbf{e}$ in $\mathbf{G}$ are in ${}^r\mathbf{Z}$, hence contained in the core of ${}^r\mathbf{Z}$.

To prove the sufficiency of the condition, let us first assume that $\mathbf{G}$ is a formal subgroup of $\Phi(\mathrm{L})$, where L is an affine group. Let $\mathbf{T}$ be the unique maximal formal torus of $\mathbf{G}$. By Proposition 18(i) we have, for every $s \in \mathrm{A}(\mathbf{G})$, $\mathbf{Int}(s)(\mathbf{T}) = \mathbf{T}$, which shows that $\mathrm{A}(\mathbf{G}) \subset {}^a\mathscr{N}(\mathbf{T})$ (Proposition 9), hence $\mathbf{T}$ is an invariant formal subgroup of $\Phi(\mathrm{A}(\mathbf{G}))$ (and *a fortiori* of $\mathbf{G}$); this shows that $\mathbf{T}$ is in the reduced center of $\Phi(\mathrm{A}(\mathbf{G}))$ (Corollary 1 to Proposition 15), hence $\mathrm{A}(\mathbf{T})$ is in the center of $\mathrm{A}(\mathbf{G})$ (Proposition 8). Furthermore, from Proposition 18(ii) it follows that, for any maximal torus S of $\mathrm{A}(\mathbf{G})$,

$$\Phi(\mathrm{S}) \wedge \mathbf{G} = \mathbf{T},$$

and this implies that $\mathrm{S} \supset \mathrm{A}(\mathbf{T})$ by definition of the algebraic hull. Consider the affine group $\mathrm{L}' = \mathrm{A}(\mathbf{G})/\mathrm{A}(\mathbf{T})$, and the natural homomorphism $g \colon \mathrm{A}(\mathbf{G}) \to \mathrm{L}'$; the maximal tori of $\mathrm{L}'$ are the images $g(\mathrm{S})$ of the maximal tori of $\mathrm{A}(\mathbf{G})$. If $\mathbf{G}' = \mathbf{g}(\mathbf{G})$, $\mathrm{L}'$ is equal to $\mathrm{A}(\mathbf{G}')$, otherwise the inverse image of $\mathrm{A}(\mathbf{G}')$ by $g$ would contain $\mathbf{G}$ and be properly contained in $\mathrm{A}(\mathbf{G})$, which contradicts the definition of $\mathrm{A}(\mathbf{G})$. In addition we have

$$\mathbf{G}' \wedge \mathbf{g}(\Phi(\mathrm{S})) = \mathbf{e};$$

indeed,

$$\mathbf{G}' \vee \mathbf{g}(\Phi(\mathrm{S})) = \mathbf{g}(\mathbf{G} \vee \Phi(\mathrm{S}))$$

and as $\mathbf{G} \vee \Phi(\mathrm{S})$ contains the reduced kernel $\Phi(\mathrm{A}(\mathbf{T}))$ of $\mathbf{g}$, we have

$$\dim(\mathbf{G}' \vee \mathbf{g}(\Phi(\mathrm{S}))) = \dim(\mathbf{G} \vee \Phi(\mathrm{S})) - \dim(\Phi(\mathrm{A}(\mathbf{T}))).$$

But as $\mathbf{G} \wedge \Phi(\mathrm{S}) = \mathbf{T}$ and $\mathbf{G}$ is invariant in $\Phi(\mathrm{A}(\mathbf{G}))$, we have

$$\dim(\mathbf{G} \vee \Phi(\mathrm{S})) = \dim \mathbf{G} + \dim \Phi(\mathrm{S}) - \dim \mathbf{T},$$

hence

$$\dim(\mathbf{G}' \vee \mathbf{g}(\Phi(S))) = (\dim \mathbf{G} - \dim \mathbf{T}) + (\dim \Phi(S) - \dim(\Phi(A(\mathbf{T}))))$$
$$= \dim \mathbf{G}' + \dim \mathbf{g}(\Phi(S))$$

since the reduced kernel $\Phi(A(\mathbf{T}))$ of $\mathbf{g}$ is such that

$$\mathbf{T} \subset \Phi(A(\mathbf{T})) \subset \Phi(S) \qquad \text{and} \qquad \Phi(S) \wedge \mathbf{G} = \mathbf{T}.$$

As $\mathbf{G}'$ is invariant in $\Phi(L') = \Phi(A(\mathbf{G}'))$, we conclude from the preceding relation that

$$\dim(\mathbf{G}' \wedge \mathbf{g}(\Phi(S))) = 0.$$

As the maximal formal tori of $\mathbf{G}'$ have the form

$$\mathbf{G}' \wedge \mathbf{g}(\Phi(S)) = \mathbf{G}' \wedge \Phi(\mathbf{g}(S))$$

(Proposition 18(ii)), they are reduced to $\mathbf{e}$, and Proposition 21 proves that $\mathbf{G}'$ is *unipotent*, and as $\mathbf{G}'$ is isogenous to $\mathbf{G}/\mathbf{T}$, $\mathbf{G}$ is nilpotent ($\mathbf{T}$ being in its reduced center).

Now consider the general case of a formal group $\mathbf{G}$ which has a unique maximal formal torus $\mathbf{T}$; let ${}^r\mathbf{Z}$ be the reduced center of $\mathbf{G}$, and $\mathbf{u}$ the natural formal homomorphism $\mathbf{G} \to \mathbf{G}' = \mathbf{G}/{}^r\mathbf{Z}$; we have seen in Corollary 1 to Proposition 18 that $\mathbf{T}' = \mathbf{u}(\mathbf{T})$ is the unique maximal formal torus of $\mathbf{G}'$; as we have seen above that $\mathbf{G}'$ is then nilpotent, so is $\mathbf{G}$. In addition, ${}^r\mathbf{u}^{-1}(\mathbf{T}')$ is invariant in $\mathbf{G}$ and its quotient by ${}^r\mathbf{Z}$ is a torus; hence, by Proposition 15, we conclude that $\mathbf{T}' = \mathbf{e}$, $\mathbf{T} \subset {}^r\mathbf{Z}$, and in fact $\mathbf{G}/{}^r\mathbf{Z}$ is unipotent.

**Proposition 23.** *Let* L *be an affine group,* $\mathbf{G}$ *a solvable formal subgroup of* $\mathbf{L} = \Phi(L)$. *Then:*

(i) *There is a largest unipotent formal subgroup* $\mathbf{G}_u$ *of* $\mathbf{G}$, *which is invariant in* $\mathbf{G}$, *and for any maximal formal torus* $\mathbf{T}$ *of* $\mathbf{G}$,

$$\mathbf{G} = \mathbf{T} \vee \mathbf{G}_u \qquad \textit{and} \qquad \mathbf{T} \wedge \mathbf{G}_u = \mathbf{e}.$$

(ii) *In addition,* $A(\mathbf{G}_u)$ *is the largest connected unipotent subgroup of* $A(\mathbf{G})$, *and, for any maximal formal torus* $\mathbf{T}$ *in* $\mathbf{G}$, $A(\mathbf{T})$ *is a maximal torus in* $A(\mathbf{G})$. *Moreover, the formal group* $\Phi(A(\mathbf{G}))/\mathbf{G}_u$ *is commutative and is the direct product of the unipotent group* $\Phi(A(\mathbf{G}_u))/\mathbf{G}_u$ *and of the torus* $(\Phi(A(\mathbf{T})) \vee \mathbf{G}_u)/\mathbf{G}_u$.

Consider the derived group $\mathscr{D}(\mathbf{G})$, which is nilpotent (Proposition 11) and equal to $\Phi(\mathscr{D}(\mathrm{A}(\mathbf{G})))$ (Theorem 2). Let $f$ be the natural homomorphism of $\mathrm{A}(\mathbf{G})$ onto the affine commutative group

$$\mathrm{A}(\mathbf{G})/\mathscr{D}(\mathrm{A}(\mathbf{G})) = \mathrm{G}';$$

$\mathbf{f}(\mathbf{G})$ is then commutative, hence the direct product of its core $\mathbf{C}'$ and its $p$-radical $\mathbf{R}'$, which is unipotent by Proposition 20. Let

$$\mathbf{N} = {}^r\mathbf{f}^{-1}(\mathbf{R}') \supset \mathscr{D}(\mathbf{G}),$$

which is a solvable invariant formal subgroup of $\mathbf{G}$; furthermore, if $\mathbf{S}$ is a maximal formal torus of $\mathbf{N}$, we have $\mathbf{f}(\mathbf{S}) = \mathbf{e}$, hence $\mathbf{S} \subset \mathscr{D}(\mathbf{G})$. But as $\mathscr{D}(\mathbf{G})$ is nilpotent, it has a unique maximal formal torus (Proposition 22), and so therefore has $\mathbf{N}$, which proves that $\mathbf{N}$ is nilpotent (Proposition 22). Proposition 20 then shows that $\mathbf{N}$ is quasidirect product of its largest unipotent formal subgroup $\mathbf{G}_u$ and its unique maximal formal torus $\mathbf{S}$. Furthermore, if $\mathbf{U}$ is any unipotent formal subgroup of $\mathbf{G}$, and $\mathbf{g}: \mathbf{G} \to \mathbf{G}/\mathbf{N}$ the natural formal homomorphism, we have $\mathbf{g}(\mathbf{U}) \subset \mathbf{G}/\mathbf{N}$, and $\mathbf{G}/\mathbf{N}$ is isogenous to $\mathbf{f}(\mathbf{G})/\mathbf{R}'$, hence to $\mathbf{C}'$, and is therefore a torus; this implies $\mathbf{g}(\mathbf{U}) = \mathbf{e}$ (Corollary 1 to Proposition 14), which in turn implies $\mathbf{U} \subset \mathbf{N}$, and finally $\mathbf{U} \subset \mathbf{G}_u$. As $\mathbf{G}_u$ is thus the largest unipotent formal subgroup of $\mathbf{G}$ and $\mathbf{G}$ is invariant in $\Phi(\mathrm{A}(\mathbf{G}))$, we have

$$\mathbf{Int}(s)(\mathbf{G}_u) \subset \mathbf{G}_u \qquad \text{for any} \quad s \in \mathrm{A}(\mathbf{G}),$$

hence $\mathbf{G}_u$ is invariant in $\Phi(\mathrm{A}(\mathbf{G}))$, and *a fortiori* in $\mathbf{G}$.

Let now $\mathbf{T}$ be any maximal formal torus of $\mathbf{G}$; then $\mathbf{T} \wedge \mathbf{N}$ is a formal torus, hence contained in $\mathbf{S}$; $\mathbf{N}$ is invariant in $\Phi(\mathrm{A}(\mathbf{G}))$ (Proposition 6), and the same argument as above shows that

$$\mathbf{Int}(s)(\mathbf{S}) \subset \mathbf{S} \qquad \text{for any} \quad s \in \mathrm{A}(\mathbf{G}),$$

hence $\mathbf{S}$ is invariant in $\Phi(\mathrm{A}(\mathbf{G}))$, and *a fortiori* in $\mathbf{G}$. In the formal subgroup $\mathbf{T} \vee \mathbf{S}$, $\mathbf{S}$ is invariant and such that $(\mathbf{T} \vee \mathbf{S})/\mathbf{S}$, isogenous to $\mathbf{T}/(\mathbf{T} \wedge \mathbf{S})$, is a formal torus; it follows from Proposition 15 and its Corollary 1 that $\mathbf{S}$ is in the reduced center of $\mathbf{T} \vee \mathbf{S}$, and then that $\mathbf{T} \vee \mathbf{S}$ is a formal torus, hence $\mathbf{S} \subset \mathbf{T}$ by definition of $\mathbf{T}$; furthermore

$$\mathbf{T} \wedge \mathbf{N} = \mathbf{S} \qquad \text{and} \qquad \mathbf{T} \wedge \mathbf{G}_u = \mathbf{e}.$$

Observe that $\mathbf{N}/\mathbf{G}_u$ is isogenous to $\mathbf{S}$, hence a formal torus, and $\mathbf{G}/\mathbf{N}$ isogenous to $\mathbf{C}'$, hence also a formal torus; furthermore, $\mathbf{N}/\mathbf{G}_u$ is invariant in $\mathbf{G}/\mathbf{G}_u$, hence contained in the center of $\mathbf{G}/\mathbf{G}_u$ (Corollary 1 to Proposition 15), and finally $\mathbf{G}/\mathbf{G}_u$ is a formal torus by Proposition 15.

There is a maximal torus Q in A($\mathbf{G}$) such that $\mathbf{T} = \Phi(Q) \wedge \mathbf{G}$ (Proposition 18(ii)); we have $A(\mathbf{G}) = Q \cdot H_u$, where $H_u$ is the largest connected unipotent subgroup of A($\mathbf{G}$); as $A(\mathbf{G}_u)$ is unipotent and connected, it is contained in $H_u$; this implies that

$$\Phi(H_u) \wedge \mathbf{G} = \mathbf{G}_u,$$

for $\Phi(H_u) \wedge \mathbf{G}$ is a formal unipotent subgroup of $\mathbf{G}$ containing $\mathbf{G}_u$. Now the formal commutative group $\Phi(A(\mathbf{G}))/\mathbf{G}$ (Corollary to Theorem 2) contains $(\Phi(H_u) \vee \mathbf{G})/\mathbf{G}$, which is isogenous to $\Phi(H_u)/(\Phi(H_u) \wedge \mathbf{G})$, i.e., to $\Phi(H_u)/\mathbf{G}_u$; the latter is therefore commutative. On the other hand, $\mathbf{G}_u$ is invariant in $\Phi(A(\mathbf{G}))$ (Proposition 6). As $\mathbf{G}/\mathbf{G}_u$ is invariant in $\Phi(A(\mathbf{G}))$ and is a torus, it is in the reduced center of $\Phi(A(\mathbf{G}))/\mathbf{G}_u$ (Corollary 1 to Proposition 15); furthermore $(\Phi(A(\mathbf{G}))/\mathbf{G}_u)/(\mathbf{G}/\mathbf{G}_u)$ is isogenous to $\Phi(A(\mathbf{G}))/\mathbf{G}$, hence commutative; therefore $\Phi(A(\mathbf{G}))/\mathbf{G}_u$ is nilpotent. It is obviously the least upper bound of the invariant formal subgroup $\Phi(H_u)/\mathbf{G}_u$, which is unipotent, and of the subgroup $(\Phi(Q) \vee \mathbf{G}_u)/\mathbf{G}_u$, which is isogenous to

$$\Phi(Q)/(\Phi(Q) \wedge \mathbf{G}_u) = \Phi(Q),$$

i.e., a formal torus. We conclude (Proposition 20) that $\Phi(A(\mathbf{G}))/\mathbf{G}_u$ is the direct product of the formal torus $(\Phi(Q) \vee \mathbf{G}_u)/\mathbf{G}_u$ and of $\Phi(H_u)/\mathbf{G}_u$; but as the latter is commutative, $\Phi(A(\mathbf{G}))/\mathbf{G}_u$ is *commutative*, and $(\Phi(Q) \vee \mathbf{G}_u)/\mathbf{G}_u$ is its *core*. We therefore have

$$\mathbf{G}/\mathbf{G}_u \subset (\Phi(Q) \vee \mathbf{G}_u)/\mathbf{G}_u,$$

hence $\mathbf{G} \subset \Phi(Q) \vee \mathbf{G}_u$, i.e.,

$$\Phi(Q) \vee \mathbf{G}_u = \Phi(Q) \vee \mathbf{G}.$$

Now we have on one hand

$$\begin{aligned}\dim((\Phi(Q) \vee \mathbf{G})/\mathbf{G}_u) &= \dim((\Phi(Q) \vee \mathbf{G})/\mathbf{G}) + \dim(\mathbf{G}/\mathbf{G}_u)\\ &= \dim(\Phi(Q)/\mathbf{T}) + \dim(\mathbf{G}/\mathbf{G}_u)\end{aligned}$$

since $(\Phi(Q) \vee \mathbf{G})/\mathbf{G}$ is isogenous to

$$\Phi(Q)/(\Phi(Q) \wedge \mathbf{G}) = \Phi(Q)/\mathbf{T}.$$

On the other hand, as

$$\Phi(Q) \wedge \mathbf{G}_u = \mathbf{e},$$

$(\Phi(Q) \vee \mathbf{G}_u)/(\mathbf{T} \vee \mathbf{G}_u)$ is isogenous to $\Phi(Q)/\mathbf{T}$, hence we have

$$\begin{aligned}\dim((\Phi(Q) \vee \mathbf{G})/\mathbf{G}_u) &= \dim((\Phi(Q) \vee \mathbf{G}_u)/\mathbf{G}_u)\\ &= \dim(\Phi(Q)/\mathbf{T}) + \dim((\mathbf{T} \vee \mathbf{G}_u)/\mathbf{G}_u)\end{aligned}$$

and the comparison of these relations proves that

$$\dim((\mathbf{T} \vee \mathbf{G}_u)/\mathbf{G}_u) = \dim(\mathbf{G}/\mathbf{G}_u)$$

hence $\mathbf{T} \vee \mathbf{G}_u = \mathbf{G}$.

Finally, we deduce from the last relation that

$$A(\mathbf{G}) = A(\mathbf{T}) \cdot A(\mathbf{G}_u)$$

(Corollary 1 to Proposition 5); as

$$A(\mathbf{T}) \subset Q, \quad A(\mathbf{G}_u) \subset H_u, \quad A(\mathbf{G}) = Q \cdot H_u, \quad Q \cap H_u = \{e\},$$

this is only possible, for reasons of dimension, if $A(\mathbf{T}) = Q$ and $A(\mathbf{G}_u) = H_u$. QED.

**Corollary 1.** *Let* L *be an affine group,* $\mathbf{G}$ *a formal subgroup of* $\mathbf{L} = \Phi(L)$. *Then, for any maximal formal torus* $\mathbf{T}$ *of* $\mathbf{G}$, $A(\mathbf{T})$ *is the only maximal torus* S *of* $A(\mathbf{G})$ *such that* $\mathbf{T} \subset \Phi(S)$.

It follows from Proposition 18 that one may assume that $\mathbf{T}$ is contained in a maximal formal solvable subgroup $\mathbf{H}$ of $\mathbf{G}$; then $A(\mathbf{H})$ is a Borel subgroup of $A(\mathbf{G})$ (Proposition 13), and it follows from Proposition 23 that $A(\mathbf{T})$ is a maximal torus of $A(\mathbf{H})$, hence also a maximal torus of $A(\mathbf{G})$. If S is another maximal torus of $A(\mathbf{G})$ such that $\mathbf{T} \subset \Phi(S)$, then $A(\mathbf{T}) \subset A(\Phi(S)) = S$, hence $A(\mathbf{T}) = S$.

**Corollary 2.** *The notations being as in Corollary 1, for any maximal solvable formal subgroup* $\mathbf{H}$ *of* $\mathbf{G}$, *the largest unipotent formal subgroup of* $\mathbf{H}$ *is also a maximal unipotent formal subgroup of* $\mathbf{G}$. *The largest unipotent formal subgroup of* $\mathbf{Rad}(\mathbf{G})$ *is the largest unipotent invariant formal subgroup* $\mathbf{Rad}_u(\mathbf{G})$, *called the unipotent radical of* $\mathbf{G}$.

If $\mathbf{V}$ is a unipotent formal subgroup of $\mathbf{G}$ containing the largest unipotent subgroup $\mathbf{H}_u$ of $\mathbf{H}$, it is contained in a maximal solvable formal subgroup of $\mathbf{G}$, hence there is an $s \in \mathrm{A}(\mathbf{G})$ such that

$$\mathbf{Int}(s)(\mathbf{V}) \subset \mathbf{H}$$

(Proposition 13); but then $\mathbf{Int}(s)(\mathbf{V}) \subset \mathbf{H}_u$ by definition, hence

$$\dim(\mathbf{V}) \leqslant \dim(\mathbf{H}_u),$$

and as by assumption $\mathbf{V} \supset \mathbf{H}_u$, we must have $\mathbf{V} = \mathbf{H}_u$. The second assertion is trivial.

The structure of nonrepresentable solvable formal groups is still unclear. I conjecture that, for *any* formal group $\mathbf{G}$, the direct product $\mathbf{C} \times \mathbf{D}$ of the core $\mathbf{C}$ of the reduced center ${}^r\mathbf{Z}$ and the largest divisible subgroup $\mathbf{D}$ of the $p$-radical of ${}^r\mathbf{Z}$ is an *almost direct factor* of $\mathbf{G}$. Then the reduced center of $\mathbf{G}/(\mathbf{C} \times \mathbf{D})$ would be unipotent by Proposition 15 (although $\mathbf{G}/(\mathbf{C} \times \mathbf{D})$ needs not be representable), and it is easily seen that Proposition 23(i) would still be valid for that group.

# Bibliography


[1] I. BARSOTTI, Moduli canonici i gruppi analitici commutativi, *Ann. Scuola Norm. Sup. Pisa*, **13** (1959), 303–372.

[2] I. BARSOTTI, Metodi analitici per varieta abeliane in caratteristica positiva, *Ann. Scuola Norm. Sup. Pisa*, **18** (1964), 1–25; **19** (1965), 277–330 and 481–512; **20** (1966), 101–137 and 331–365.

[3] S. BOCHNER, Formal Lie groups, *Ann. of Math.*, **47** (1946), 192–201.

[4] A. BOREL, *Linear Algebraic Groups*, Benjamin, New York, 1969.

[5] N. BOURBAKI, *Eléments de mathématique: Algèbre*, Chaps. 1–3, Hermann, Paris, 1970.

[6] N. BOURBAKI, *Eléments de mathématique: Algèbre*, Chap. VIII: Modules et anneaux semi-simples, Hermann, Paris, 1958.

[7] N. BOURBAKI, *Eléments de mathématique: Algèbre commutative*, Chap. III: Graduations, filtrations et topologies; Chap. IV: Idéaux premiers et décomposition primaire, Hermann, Paris, 1961.

[8] E. CARTAN, *Oeuvres complètes*, Vol. $I_1$, Gauthier-Villars, Paris, 1952.

[9] P. CARTIER, Groupes algébriques et groupes formels. Dualité, *Coll. sur la théorie des groupes algébriques*, Bruxelles, 1962.

[10] P. CARTIER, Groupes formels associés aux anneaux de Witt généralisés, *C. R. Acad. Sci. Paris*, **265** (1967), 50–52.

[11] P. CARTIER, Modules associés à un groupe formel commutatif. Courbes typiques, *C. R. Acad. Sci. Paris*, **265** (1967), 129–132.

[12] P. CARTIER, Relèvements des groupes formels commutatifs, *Sém. Bourbaki* 1968/69, exposé 359, *Lecture Notes in Mathematics*, No. 179, Springer, Berlin, 1971.

[13] P. CARTIER, Groupes formels, fonctions automorphes et fonctions zêta des courbes elliptiques, *Actes du Congrés Intern. des Math.*, 1970, Vol. 2, pp. 291–299, Gauthier-Villars, Paris, 1971.

[14] C. CHEVALLEY, *Théorie des groupes de Lie, tome II: Groupes algébriques*, Hermann, Paris, 1951.

[15] C. CHEVALLEY, *Classification des groupes de Lie algébriques*, Séminaire C. Chevalley 1956–58, Secrétariat math., 11, Rue P. Curie, Paris, 1958.

[16] M. DEMAZURE and P. GABRIEL, *Groupes algébriques*, Vol. I, Masson, Paris, 1970.

[17] J. DIEUDONNÉ, Groupes de Lie et hyperalgèbres de Lie sur un corps de caractéristique $p > 0$ (V), *Bull. Soc. Math. France*, **84** (1956), 207–239.

[18] J. DIEUDONNÉ, Lie groups and Lie hyperalgebras over a field of characteristic $p > 0$ (VIII), *Amer. Journ. of Math.*, **80** (1958), 740–772.

[19] J. DIEUDONNÉ, Witt groups and hyperexponential groups, *Mathematika*, **2** (1955), 21–31.

[20] J. DIEUDONNÉ, On the Artin-Hasse exponential series, *Proc. Amer. Math. Soc.*, **8** (1957), 210–214.

[21] J. DIEUDONNÉ, Sur la réduction canonique des couples de matrices, *Bull. Soc. Math. France*, **74** (1946), 59–68.

[22] J. DIEUDONNÉ, Linearly compact spaces and double vector spaces over sfields, *Amer. Journ. of Math.*, **73** (1951), 13–19.

[23] B. DITTERS, Sur une série exponentielle non commutative définie sur les corps de caractéristique *p*, *C. R. Acad. Sci. Paris*, **268** (1969) 580–582.

[26] A. FRÖHLICH, *Formal Groups, Lecture Notes in Mathematics*, No. 74, Springer, Berlin, 1968.

[27] P. GABRIEL, Exposé VII/A: Etude infinitésimale des schémas en groupe et groupes formels; Exposé VII/B: Groupes formels, in *Schémas en groupes I: Propriétés générales des Schémas en groupes, Séminaire de Géométrie Algébrique du Bois-Marie* 1962/64 (SGA 3) *Lecture Notes in mathematics*, No. 151, Springer, Berlin, 1970.

[28] A. GROTHENDIECK and J. DIEUDONNÉ, *Eléments de Géométrie algébrique* I, Springer, Berlin, 1971.

[29] A. GROTHENDIECK, Groupes de Barsotti-Tate et cristaux, *Actes du Congrès Intern. des math.*, 1970, Vol. 1, pp. 431–436, Gauthier-Villars, Paris, 1971.

[30] T. HONDA, Isogeny classes of abelian varieties over finite fields, *Journ. Math. Soc. Japan*, **20** (1968), 83–95.

[31] N. JACOBSON, *The theory of rings*, Math. Surveys No. II, Amer. Math. Soc., New York, 1943.

[32] N. JACOBSON, *Lie Algebras*, Wiley-Interscience, New York, 1962.

[33] M. KAROUBI, Cobordisme et groupes formels (d après D. QUILLEN et T. tom DIECK), *Sém. Bourbaki*, 1971/72, exposé 408.

[34] N. KATZ, Travaux de Dwork, *Sém. Bourbaki* 1971/72, exposé 409.

[35] M. LAZARD, La non-existence des groupes de Lie formels non abéliens à un paramètre, *C. R. Acad. Sci. Paris*, **239** (1954), 942–945.

[36] S. LEFSCHETZ, *Algebraic Topology*, Amer. Math. Soc. Colloquium Publ., Vol. XXVII, New York, 1942.

[37] J. LUBIN, One-parameter formal Lie groups and *p*-adic integer rings, *Ann. of Math.*, **80** (1964), 464–484.

[38] J. LUBIN and J. TATE, Formal complex multiplication in local fields, *Ann. of Math.*, **81** (1965), 380–387.

[39] Yu. I. MANIN, The theory of commutative formal groups over fields of finite characteristic, *Russian Math. Surveys*, **18** (1963), No. 6, 1–83.

[40] D. MUMFORD, *Abelian varieties*, Oxford Univ. Press, London and New York, 1970.

[41] D. MUMFORD, Bi-extensions of formal groups, *Algebraic Geometry* (Bombay Colloq., 1968), Oxford Univ. Press, London and New York, 1969.

[42] T. ODA, The first de Rham cohomology group and Dieudonné modules, *Ann. Ecole Norm. Sup.*, (4), **2** (1969), 63–125.

[43] O. ORE, Theory of non commutative polynomials, *Ann. of Math.*, **34** (1933), 480–508.

[44] J. P. SERRE, Commutativité des groupes formels de dimension 1, *Bull. Soc. Math. France*, **91** (1967), 113–115.

[45] J. P. SERRE, Groupes $p$-divisibles (d après J. Tate), *Sém. Bourbaki*, 1966/67, exposé 318.

[46] J. P. SERRE, Sur les groupes de Galois attachés aux groupes $p$-divisibles, *Nuffic Summer School at Driebergen*, Springer, Berlin, 1967.

[47] J. TATE, $p$-divisible groups, *Nuffic Summer School at Driebergen*, Springer, Berlin, 1967.

[48] J. TATE, Endomorphisms of abelian varieties over finite fields, *Invent. Math.*, **2** (1966), 134–144.

[49] J. TATE, Classes d'isogénie des variétés abéliennes sur un corps fini, *Sém. Bourbaki*, 1968/69, exposé 352, *Lecture Notes in Mathematics*, No. 179, Springer, Berlin, 1971.

[50] S. TÔGÔ, Note on formal Lie groups, *Amer. Journ. of Math.*, **81** (1959), 632–638.

[51] S. TÔGÔ, Note on formal Lie groups II, *Journ. Sci. Hiroshima Univ*. Ser. A–1, **25** (1961), 353–356.

[52] M. DEURING, *Algebren*, Ergeb. der Math., Bd. IV, Heft 1, Springer, Berlin, 1937.

# Index

In the following Index, Roman numerals indicate the chapter, and arabic numerals the section and subsection within the chapter.